PETER ARNELL

PORTFOLIO

1980–2020

HATJE
CANTZ

Editor
Nadine Barth

Managing Editor
Dan Zuzunaga

Project Management
Sonja Altmeppen
Claire Cichy, Hatje Cantz

Text
Tristan Thom

Copyediting
Aaron Bogart

Design Director
Liyuan Tong

Designers
Joanne Min, Yuko Yamazaki

Typeface
Adobe Garamond Pro
and Helvetica Neue

Production
Stefanie Kruszyk, Hatje Cantz

Image Preparation
Vincenzo Piscitelli

Reproductions
LVD Gesellschaft für
Datenverarbeitung mbH, Berlin

Paper
GardaMatt Ultra, 150 g/m²

Printing and Binding
Printer Trento S.r.l

Published by
Hatje Cantz Verlag GmbH
Mommsenstraße 27
10629 Berlin
www.hatjecantz.com
A Ganske Publishing Group Company

ISBN 978-3-7757-4667-0
Printed in Italy

Frontispiece
Peter Arnell, photographed in his first
studio at 100 Grand Street, NY, 1980

Dedication Page Illustration by Anita Arnell

PETER ARNELL

PORTFOLIO

1980–2020

VOLUME I

Proceeds from the sales of this publication will be donated to Special Olympics in honor of its founder Eunice Kennedy Shriver.

For Anita, Lily, Lucy, and Thomas

In a life filled with extraordinary blessings and good fortune, being your father has been the greatest privilege of all.

AUTHOR'S NOTE

Peter Arnell

I've always been resistant to labels, hesitant to take up the mantles sometimes bestowed upon me, disinclined to categorize myself or my work as any one thing. In short, I've never wanted to be restricted, either in the ways that others perceive me, or in the way that I perceive myself. The result of that unwillingness to commit singularly to any one field or discipline has been a body of work that defies easy categorization, a career that has exceeded my hopes and dreams, and a life that has been rewarding beyond measure.

This book is a celebration of imagination, creativity, and action. But if there has been any one through line to my journey over these past forty years, it would be this: following the currents of inspiration and intuition wherever they may lead, often far beyond the bounds of convention, predictability, common sense, and accepted wisdom.

Each project you'll encounter in this book, whether it be a simple hand sketch or a mass-produced automobile; the earliest germ of a concept, or a fully realized commercial project—are the result of the same essential act: nurturing imaginative impulse, and then pushing tirelessly to build it, give it form, definition, and clarity. In short, to bring an idea to life.

I tend to work quickly because I've learned that the best ideas and solutions are almost always radically simple. The hard work that follows in building and expanding ideas is of course anything but simple or fast, but the core impulse driving even the most complex project should always be reducible to something that can be grasped in an instant. If something needs to be explained in complex, overwrought terms—frankly, if it needs to be explained at all, it has less of a chance to succeed.

Assembling this book has been a monumental undertaking—hence the reason for waiting forty years—and one that has raised many thoughts and questions on how to pay proper tribute to the countless individuals who have worked alongside me, pouring their energy, passion, vision, grit, and determination into everything contained within these pages.

As I take the time now to revisit the work that I have been privileged enough to be a part of, assembled here in one place for the first time, I feel a profound sense of pride. But it's not pride in myself; rather, it's pride of having led and been part of a team, and the satisfaction of being able to learn from and share in the collective talent of so many remarkable individuals.

Looking at the breadth of work compiled here, what I feel most of all is gratitude: to have lived such an extraordinarily fortunate life full of incredible opportunities; to have been given so much trust, belief, and good faith.

Contained within these pages is the work of countless individuals from every background and every discipline: designers, engineers, architects, graphic artists, illustrators, photographers, printers, re-touchers, filmmakers, writers, thinkers, strategists, entrepreneurs, leaders, legends, kindred spirits, and so many others who simply cannot be categorized, whom I have been blessed to know, and whose awe-inspiring talents have come together so brilliantly in the work that you see here. I am endlessly inspired by their willingness to pour everything into what they do, and so deeply in awe of their capacity to create, inspire, and imagine. While I can do my best to pay proper tribute to the work and our clients, I know that I will never be able to fully express what it has meant to find myself living, working, and creating alongside such remarkable partners.

AT THE SPEED OF IMAGINATION

Michel Marriott

From nothing. Everything.

The indisputable truth is that the act of creation, and by extension, its earthly manifestations of conception, invention, and reinvention, have always occupied a revered place in the hierarchy of human heritage.

From the Cradle of Humankind nestled in the southern tip of Africa to astronauts' boot prints on the dusty plains of the moon, people have, for eons, been on an endless journey of imagination and innovation. And we have enshrined the human experience by exulting and preserving the evidence of invention wherever we find it.

What are our millions of years of relics and artifacts—cave drawings and Picasso, stone hammers and laser drills, ancient lodestone compasses and Global Positioning Systems in our pockets—if not Exhibit A in a case for the unique wonder of the animal that first stood on the sun-baked savanna to free its hands to grasp the world with its mind?

Perhaps as a consequence of this great lasso of genetic predisposition and learned allure, we are uniquely drawn to the fruits of invention: the more elegant, the more soulfully reassuring. And as for their inventors? We often imbue them with an aura of awe cast in the reflected light of the celestial, the divine, even the magical. And, yet, their gifts are often grounded and appreciated in the practical as well.

For instance, a walker can traverse twenty to thirty miles in a single day. But, assisted by the material magic of a bicycle, a cyclist becomes a human hybrid of grit and gears, capable of traveling many orders of magnitude miles in the same day. This simple, common piece of machinery is a triumph of imagination that no other animal on the planet can match.

"Imagination is more important than knowledge," Albert Einstein once said. "For knowledge is limited, whereas imagination embraces the entire world, stimulating progress, giving birth to evolution."

Little wonder, then, that a nineteenth-century America striving to distinguish itself from its European roots embraced its greatest inventor of that age, Thomas Alva Edison, and anointed him the "Wizard of Menlo Park."

Little wonder, then, that generations upon generations of us flock to the world's museums and galleries, monumental and modest, to be enthralled and inspired by displays of invention, transcendent and eternal.

And little wonder, now, that the readers of this book—*Peter Arnell Portfolio 1980–2020*—will be inescapably drawn to the works of the lifelong, perpetually creative career of Peter, a conjurer of form, fashion, language, and the aesthetic of attraction. It is a beckoning to invention made tangible in the lives of us all, where we live, far and wide.

It is also a beckoning into the inventor humble enough to understand that learning never ends. It is an open window to the works of a man who realized early that he always had to do his homework, listen well, and regularly consult master creatives, both those who are living and those who have been immortalized in books—like the thousands that line the walls of his New York City office and nearby art-filled home on the Hudson River.

"All these objects—scale, patina, color, material, history, knowledge, all the stuff that I have—affects, dramatically, my work," Peter said, surrounded by his home's collection of books, art, and what he calls "artifacts of aesthetic engineering." There is, he said, "a mirroring, a dialogue that I have with every one of these pieces."

For forty years, Peter, a Brooklyn-born grandson of a fishmonger who raised him to know that if you sell bad fish customers won't come back, has reshaped the common world with the uncommon. Over the decades, he has enlivened everything he has touched with a zestful, often eye-winking, and whimsical originality. It is a vision that, quite plainly, looks beyond conventional boundaries of genre, design, and style; and it is an approach, which he's confessed, is no approach at all, at least in the conventional—which is to say limiting—sense of the term.

Some of the most stunning examples of this spill across the pages of this book's double volumes like rare stones on a jeweler's velvet display. Among the first is the iconic, black-and-white photograph of New York's Brooklyn Bridge receding into the morning mist. It quickly became a stellar symbol and projection of his vision for an emerging new world that then materialized in the clothing lines of Donna Karan in the 1980s.

Sandwiched between this retrospective of carefully selected works are the objects, devices, images, and ideas. Whether a New York manhole cover bearing the famously ubiquitous Con Edison logo of his design, or a deliciously sleek lines of automobiles, or ranges of cookware, footwear, wristwatches, or household appliances, or even plans for a dazzling sports arena he conceived of along with Frank Gehry, his collaborator on the project, each one of these projects calls on us to see and feel in surprising, satisfying, and most importantly new ways.

Photography. Architecture. Graphic Design. Advertising. Marketing. Brand Identity. Branded Entertainment and Product Creation. They are all, more or less, billboards along Peter's global highway of invention and reinvention.

Given the depth and breadth of Peter's work collected here, we might be forgiven for asking what is the elemental particle that permits him to span so many plains of expression? In other words, what is the unified field theory of Peter's imaginative universe?

In his own words, it may well be his simple search for truth and beauty.

"There is this never-ending battle to find the truth in things," Peter told me as we sat in his New York conference room beneath a full-sized model of the Soviet Sputnik 1 satellite suspended like a space-age chandelier. "Beauty somehow becomes the ultimate, of course, moment for the entire world. Right? That's where the world pays attention. Beauty is a very powerful tool. It seduces, but it also soothes."

In much the same way, the intersection of truth and beauty is where Peter and his work are most vividly alive. Getting there, in a sense, is what he lives for: certainly not the mundane destination of fame, nor profit. Answering his own higher calling is what animates Peter's work at its most artful. The result is what dominates these volumes.

Yet, he maintains, "None of this was planned. I continue on in my daily life in an unplanned, unstructured way, while building in a world of planning and structure. It's a conundrum." It is also part of Peter's process, a sort of unseen stretched canvass of an internal trampoline for his ideas.

For instance, during the 1990s, before Samsung became the South Korean consumer electronics juggernaut it is today, its leadership turned to Peter, among others, to plant and grow the company's brand in the lucratively fertile market of American consumers. At the time, according to Peter, its best-tech-forward candidate was a boxy, reasonably designed microwave oven when millions of people weren't entirely sold on whether microwaving ("nuking") food was entirely healthy.

One evening, while taking a taxicab home from a meeting gone sour, Peter noticed the driver at the wheel. He invited the man to his home studio for a photoshoot, sensing in the slim, sinewy Russian immigrant a sleek solution to an aching problem.

Peter took a series of black-and-white photographs of the shirtless driver holding Samsung's microwave over his hip. Man and machine, both side-by-side fit, sexy, and exuding health, became the establishing shot of a remarkably mind-changing print and billboard campaign.

In short, Peter enabled Samsung—long seen as a massive, yet soulless, manufacturer of reliable, yet unremarkable, products—to speak to consumers through a brand language of young, hip, and cool.

This was a Peter Arnell upgrade to the old design adage of form follows function: form follows function *and feelings.*

"It was a clear standout," Thomas Rhee, a former senior executive at Samsung notes, recalling the arc of Arnell inspiration in Geoffrey Cain's 2020 book, *Samsung Rising: The Inside Story of the South Korean Giant That Set Out to Beat Apple and Conquer Tech.* "We were sensing a new thing at the time."

Similarly, at a later dinner with Rhee, Peter was swept up in a brainstorm for a freshly compelling Samsung slogan for the newly proliferating digital age. Scribbling variations of the word "digital" on the backs of napkins at their table, Peter hit upon giving the word a double "L," making it "DigitAll."

His compulsive creativity led, in that case, to a yearslong Samsung campaign: "DigitAll: Everyone Is Invited."

Not surprisingly, Peter is the man drawing dots, and then turning those dots into dioramas of inventive landscapes. Peter moves through this world—the streets, parks, cafés, museums, libraries, restaurants—thinking, keeping journals, sketching ideas like the old masters but digitally into his iPhone, which he points, like the new masters, photographing a lusty swell of a contemporary building's façade at twilight, or the light play over a friend's face sharing an elevator ride with him to dinner.

His early fascination with cameras became much more when he soon realized that, "Wow, this machine can actually let people see what I see," he recalled. That initiating conversation with lens and light deepened in a spell of serendipity upon meeting iconic fashion editor Carrie Donovan.

While he was the self-described kid editing architecture books in the 1980s for the likes of Michael Graves, Robert Venturi, and Frank Gehry, Donovan asked him to pick up a camera and shoot cover stories for her at *The New York Times* and to go to Paris for a month to shoot the first major couture house to open in twenty years.

"She must have seen in me a fresh eye outside of fashion looking in," Peter said. Yet, he recently confessed that while he knew quite a bit about photography, he didn't, in first stirrings of his career, know much about being a photographer.

"But I never feared going somewhere I didn't know anything about. I picked up a camera and shot a bridge. I didn't know how to use a camera, but I knew what I wanted. The camera was a tool," he said.

Peter once told a reporter, "All my thinking comes out in my photographs." More specifically, he recently elaborated to me, "Black-and-white photography is the way I see the world."

"The lines of form are better seen in black and white," he later added. "The form is the truth."

Ironically, his world, particularly its convulsing convergence of creativity and commerce, is anything but black and white. This is part of his central conundrum of being a change agent.

"Finding the way in," he explained "means to produce beyond expectation, in a way that works, and is accessible to the people you are producing for."

But this, he added, also produces a discordant tension between the creator and client in a service business like his, a business in which the deliverable is change and disruption rooted in his instincts and intuition—often not shared, at least in the beginning, by the clients.

"The very notion of managing disruption in a relationship is different than managing produced disruption for a company and its image, its product, or whatever," Peter said.

"So, how do your transfer instincts, outpace expectation, and in the end find a client, a partner, a whatever, who actually becomes an advocate of the journey and the mission to get it through the company?"

That is the challenge, Peter concluded, and added, "If they knew what they wanted, knew how to achieve it, well, I wouldn't have a business."

The process is, as he is fond of saying, "heavy." And he has been unabashed in evoking everything from Hindu numerical harmony to Cartesian coordinate systems to make his point, and help make real his visions. But there is something more to his eliciting ancient wisdoms and systems. "I want to build enduring ideas," he explained, noting that he has never been interested in trends, but rather in leaving the world with original designs, perceptions, and tastes that endure.

This drive is fundamental to the DNA of Peter's invention, lending him the ability to read the genetic code of nascent concepts, services and products. It is, no doubt, a vital part to his capacity to craft embracive products, brands, and marketing plans.

Examples abound at the expectation-shattering speed of imagination.

As Chrysler's chief innovation officer, he single-handedly conceived of the endearing, twenty-first-century form factor and digital fluency of his Peapod electric car now wound tightly into the design DNA of practically all electric cars.

Similarly, he looked beneath the mundane shapes and skins of conventional home safety and security devices, like fire extinguishers, and in 2007 reinvented the category into a super brand for Home Depot. From its DNA up, Peter developed a stylish line of ergonomically-forward fire extinguishers and smoke and carbon monoxide detectors—

and an integrated marking plan—under the rubric of HomeHero.

Along the way, Peter even physically reinvented himself. For much of his adult life, he weighed more than 400 pounds on a frame, he recalled, built to accommodate a 150-pound man. Through a strict diet of his own design—which started, as he said with "oranges, many oranges," his irrepressible New York determination transformed him. He lost 256 pounds.

For years now, Peter has maintained a fashionably svelte 150-pound profile, putting to practice a quote from Socrates: "The secret of change is to focus all your energy, not on fighting the old, but on building the new."

The result was something that former President Bill Clinton could not help but notice and comment upon during a speech he gave in New York's Harlem in 2006. He talked passionately about the need for companies to produce healthier food for young people, and for people to eat better, smarter.

At one point in his remarks, he urged Peter, who was seated on the front row, to stand. When he did, Clinton said, "When I first met Peter a few years ago, he weighed 356 pounds. Now he is 150 pounds—and he doesn't even have any stretch marks."

Apparently, restless creativity—even when it comes to the body of a creator—abhors a vacuum as much as nature does. Call the phenomenon, for lack of a better term, synaptic refraction, a mind that splits creative insight into many, related expressions of that insight. Other restless creators throughout history have either ignored or defied strict categories to astounding effect. It is difficult to imagine anyone forcing a Leonardo da Vinci, for instance, to choose a lane and stay in it. Either paint on canvas or paint murals on plaster walls. Either study the anatomical wonders beneath the skin or build fifteenth-century war armaments, or design futuristic flying machines. In a similar vein, who would today demand that Elon Musk restrict his innovations to high-performance electric cars or high-soaring spacecraft?

This book showcases the products, literally and figuratively, of his partnerships with scores of major manufacturers, technology companies, philanthropic organizations, municipalities, and national governments. There are too, of course, the immediately recognizable brands, among them: Gucci, Fendi, Chanel, Donna Karan, Disney, Samsung, Sony, Chrysler, Reebok, McDonald's, Pizza Hut, Playboy, Pepsi, KFC, Nestlé, Tropicana, Levi's and AT&T. Each sought Peter's elemental particle to define, redefine, invent or reinvent, in a conservation of discovery captured in these pages, all the while surmounting the corporate tendency to segregate and silo its assets. Peter, the troublemaking maverick, attacked the problem of hard boundaries by being interdisciplinary before it was commonly defined, even before the word "brand" came to mean what it would by the midpoint of his career. Even before Apple.

An architect by education and orientation (his grandfather was also a Freemason), Peter, early on, had a propensity to build environments, spaces, and journeys that both encompass and beckon consumers to a brand. For instance, he has said that he instantly understood why the word "country" was added to one of Philip Morris's brands of cigarettes. "It's where you want to be. You have to know how to create an invitation to a destination where the mythical meets the authentic."

And throughout this process there, too, are the trace elements of the many personal and professional relationships that Peter enthusiastically nourishes and engenders in his work. A world-class conversationalist and collaborator, Peter is seldom, if ever, a stranger in any room, at any table, in any situation, or with anyone, including, over the years, Dr. Jonas Salk, Bill Clinton, Michael Jackson, Steven Spielberg, Muhammad Ali, Tom Brady, John Carl Warnecke, Frank Gehry, Walter Cronkite, Eugene Shriver, Eunice Shriver, Timothy Shriver, Maria Shriver, Quincy Jones, Carrie Donovan, Stevie Wonder, Céline Dion, Miky Lee, B.B. King, Sylvester Stallone, Rudolph Giuliani, Jay-Z, Buzz Aldrin, Donna Karan, André Leon Talley, Willie Nelson, Tommy Hilfiger, Gwyneth Paltrow, Martha Stewart, Oprah Winfrey, Yao Ming, Ray Lewis, Christy Turlington, Dawn Mello, Aretha Franklin, and Tina Turner.

It was his friendship, for example, with Tina Turner which exploded into a highly successful, twenty-million-dollar cross-marketing campaign when Peter introduced the legendary rocker—and her glamorous, trademark legs—to Hanes hosiery in the mid-1990s. His resulting commercials and print advertising, with tie-ins to the singer's national tour and related new album, resulted in a fully integrated alchemy of a rather traditional Southern American company with a globally celebrated icon of soul and rock. That synthesis still reverberates for both the entertainer and Hanes.

For the last forty years, as this book attests, Peter Arnell's proven powers of rarefied imagination and invention have, much like his first images of DKNY's Brooklyn Bridge, spanned from known shores into the mystifying, mesmerizing, and tantalizing space where magic—brilliant, bold, and timeless—is born.

WHAT DESIGN AND FOOTBALL HAVE IN COMMON

Tom Brady

From the moment I met Peter in his office, surrounded by his books, drawings, and collectibles, I've had a deep appreciation for his unmatched work ethic and creative mind. Peter and I spoke for hours (which seemed like minutes) that day and I felt we still had so much more to discuss. Our conversation wasn't centered around me the football player; rather, it was about Peter and his life's commitment to "getting it right." So you may ask: What do design and football have in common? My answer would most likely be: almost everything.

There are endless commonalities in our careers that have each spanned several decades. Study the opponent, come up with a plan, practice the plan, reshape the strategy, compete and execute, learn and grow, adjust in the field, and always try to get it right. Both of our ultimate end goals: to win.

Peter is a great one. He has accomplished so much and continues to work so hard, yet always believes his best is ahead of him. He wants everyone around him to be successful. He pushes others to be their best and isn't afraid of telling it like it is. He gets results and finds joy in the simple things. He is a great dad and has assembled a great team. He has lifelong friends and endless stories. Success comes in so many areas for Peter, and all of these elements play a special role in his life and success.

Peter has been winning for decades, but similarly to what I've learned on the football field, with wins also come losses. However, each of the losses in our careers have made us stronger. They have made us work harder and become more disciplined. They've required us to become more conscious of the mistakes in our processes and dig deeper to get better each and every day. The joy for us is being in the arena. Preparing, practicing, executing, and living our lives to the fullest. Oftentimes our work is scrutinized by so many, but neither of us shares a fear of not getting it right, because in the end, we both always give it our best. Our satisfaction comes in the work and the journey alongside our team.

What will happen next? In Peter's case, I imagine the best completions and boldest moves are yet to come. That's what I believe about my friend Peter. And I will be there to watch and learn every step of the way.

A SHADOW OF BRILLIANCE

Peter Lindbergh

The very second I had agreed to write a personal essay for Peter's monograph, I started feeling ambivalent about it. I was obviously quite honored to be considered, yet frightened by the challenge. In fact, the talent I like to believe I have behind a camera does not translate before a blank sheet of paper.

I first thought I'd make up a lie to step out of this self-inflicted situation. However, that would imply not acknowledging one of Peter's most upsetting qualities to me: he always has the strongest arguments and subsequently walks away with the last word—almost always. Which leads me to one of the lessons I've learned by working with him over the years: do not go with Peter to places where you have to "stretch the truth," for whatever lie you will struggle to put up, he'll eventually and inexorably find out. The reason is because he is always light-years ahead and would know what I was thinking anyway.

As a result, the same way Peter is moved by an unstoppable and inexplicable amount of energy, there was no way for me to back up smartly and get away with it. I owed it to him, out of friendship, out of respect, and … I love him!

Right from the start of our relationship, there was a sense of explosive creativity and intense power present, and a search for authenticity. Peter developed and created Donna Karan's campaigns from a radically different point of view, finding a totally new way to define a woman rather than the details of a clothing collection. It was an absolutely brilliant move. It's probably why our collaborations were seen as trailblazing—and I believe they are still quite relevant today. Indeed, timelessness and the pursuit of truth are defining aspects of both our works, apart and together.

Out of fairness, I must say that Peter started Donna's campaigns with Denis Piel, a photographer who was able to translate Peter's concept into beautiful, outstanding images. I came along a little later for the 1992 Presidential campaign and also when DKNY was launched—for which Peter also came up with a legendary logo that you'll surely find in these pages. It is then that we very successfully "kidnapped" New York City and made it Donna's very own. I am bringing this up to show that Arnell was the supersonic brain behind these great campaigns communicating quite a lot of extraordinary situations, establishing my friend Arnell as a true visionary with a 360-degree mind. It was always about transcending the trends and the clothes in order to convey a point of view, without any way of adjusting ourselves, rather uncompromisingly. But let's not forget the more personal and human qualities of Peter. Like many of us, Arnell needs to feel loved, especially from everybody working with him—don't try to fact check this with Peter, it is something he would never admit. One of his many ways to find that out was to organize meetings at two or three in the morning, to discuss supposedly urgent changes in current projects. The few times I actually took part in these nocturnal work sessions, their atmosphere reminded me of a sports team's pep talk, for instance when soccer players stick their heads together before a decisive penalty shot.

One probably much more precise way would be to define these meetings as voodoo ceremonies. There was undeniable passion passing through the air. It was his way to get to the bottom of things, but, just as importantly, the confirmation that everyone present would do anything for him.

What I particularly value and admire is that from the very moment Arnell knew the right solution for a problem, there was no room left for either hesitation or compromise. He willfully took whole responsibility, always going to the moon and back in order to bring his vision to life. It's when he is searching for ideas or solutions that he owns the world. There is an invariable will to experiment and anyone who wishes to bring even the unthinkable to the table is more than welcome.

With Peter Arnell, as long as there is a shadow of brilliance not too far away, he connects, and things are moving forward—and incredibly fast!

Yes Peter, that's you and that's how it was.

With love from a friend.

WILL AND EMOTION

André Leon Talley

Powerful. Exceptional. Propelling. A photographic visionary. A man as big as his towering concepts for media imaging. I am drawn to his work as a humanist.

I am proud to know Peter.

We first met, in 1986, when he was only 27 years old. He was just about to make a name for himself in advertising.

Tina Brown, then the editor of *Vanity Fair* sent me packing to his downtown Manhattan office to collaborate, on the first ever *Vanity Fair* ad campaign. The concept was "all things to some people." The idea: a gathering of great talents over lunch in a Manhattan watering hole, or in other words, fashionable eatery.

From that moment on, Peter and I became great friends. His other great friend is his photography. For his work for special and important clients, Arnell directs and moves like Orson Welles, across the sands of a dream achieved.

His dreams in pictures are built from the miracle moments of time and history. When Arnell received the CFDA award for his special vision for Donna Karan's hosiery advertising, few people who saw the sensual, erotic legs and feet of the model would ever know that the image was inspired by the feet of the Pietà, the magic of Michelangelo, in the Vatican.

The human spirit and drive of great minds drives Peter to approach his subjects with a sensitivity and passion, which unfurls, like a beautiful flower in nature, into a glory, a triumph of will and emotion. Peter Eric Arnell, made in Brooklyn, has a tough mind and a tender heart, to quote the words of the late Dr. Martin Luther King, Jr. It is the tender heart, his compassion for the noble spirit that he seeks to record in his voyages.

On the giant arena stage, his vision is to capture the nobility of Tina Turner. It is the nobility in a face that expresses many rivers crossed, yet many rivers of life ahead.

One does not have to speak of what he photographs; it is the thought and deep inner human quality that is the foundation of all he captures.

I think beauty is an ethical aesthetic for Peter. Therefore, Bernini Putti or the silent and eternal humility of clasped hands on a memorial tomb can trigger for him, a sense of timeless rapture.

As I feel in all great visionaries, and I think I know a few great living ones, the important thing is to approach creativity with the freshness and innocence of a child. This Peter does. He gets off on himself. He can excite himself with the eye of the camera and give back to the world, the same joy he gets from it.

AN AMAZING ODYSSEY

Frank Gehry

Dear World,

I've been asked to write a foreword to a book about Peter Arnell. I'm doing it under some duress and with some difficulty. The duress for me is that I don't really know how to talk about it. I've just experienced it. You have to experience it to know it. The difficulty is that it's so hard to write about somebody who is so much into everything and who is such a voracious creator.

For a guy who is difficult to understand sometimes, he's an irascible genius. He's relentless in his creativity. The length and breadth of his accomplishments are extraordinary.

He's lived and worked all over the world. His work encompasses branding, corporate identity architecture, wayfinding, product design, and marketing. He's built brands from the ground up for Donna Karan, for Samsung, for Chrysler, for Gwyneth Paltrow. He's advised Pepsi, and Unilever, and Nespresso, and Gucci, and Chanel, and Tommy Hilfiger, and the Special Olympics. He's worked with almost every brand you've ever heard of. He's worked with the All-Time Greats: artists, creators, and icons including Muhammad Ali, Tina Turner, Steven Spielberg, Michael Jackson.

He's an author. He's an architect. He's a photographer. He's an inventor.

All of his photographs are beyond-belief beautiful. His photographs say everything.

It's hard to put to words what this person has done and what he's meant to all of us with his work, with his persona, and with his generosity.

Open the book and prepare to go for a ride on an amazing odyssey.

Make it different
Make it simple
Make it matter

FEDERICO FELLINI
VIA MARGUTTA, 110
ROMA

Rome, June 4, 84

Dear Mr. Arnell,

thank you for the lovely volumes, which I looked over with great interest. They proved very useful for an idea for the décor for a sequence in my next picture.

I hope to meet you some day, either in Italy or in America. I thank you again for your kind thought, and send you all my best wishes,

VOLUME I

Selected Works

Rizzoli
Alessi
Vanity Fair
David Hockney's Cameraworks
EAB
Swid/Powell
Bergdorf Goodman
Donna Karan
Donna Karan Hosiery
Donna Karan Intimates
Donna Karan Bath & Body
Donna Karan Parfum
Anne Klein
Christian Lacroix
Ellesse
Fendi
DKNY
DKNY Coverings
DKNY Jeans
DKNY ASAP
DKNY Footwear
Christian Lacroix Parfum
Byblos Vis à Vis
O, Wonderful
Chanel No. 5
Chanel Allure
Banana Republic
Ray-Ban
Hanes
Samsung
Nespresso
Apple & Paul Rand
IS**C by Isaac Mizrahi
Ungaro
Esprit
Loewe
Henri Bendel
Misty Harbor
Fila
Fila Women
Ray-Ban *Men in Black*
Casio G-Shock
Casio Baby-G
Tommy Hilfiger
Iceberg
Rena Lange
Guggenheim Museum
The Art of the Motorcycle
Lifetime Television
Progressive Insurance
Mikimoto
Marui
FarmClub
Sony Media Experience
Kid Robot
Indulge
Thalia
Concord
Parmalat
José Cuervo
1800 Tequila
Playboy Intimates
Virgin Playboy Resort
ConEdison
NYPD
FDNY
Martha Stewart Everyday
Michael Jackson's *Invincible*
Jeep *Band of Brothers*
Tribute 9/11
Fashion for America
Dodge Aerosmith
Gateway
Aiwa
Bank of America
Wynn Resorts
Davidoff
Rockport
Rbk
Rbk S. Carter Collection
Rbk G-Unit
Reebok Yao Ming
Reebok IPUMP
Reebok Vector
Reebok's Terry Tate Office Linebacker
Reebok Olympians
Chrysler
Dodge Car Show
Jeep *Tomb Raider*
McDonald's
Kmart
M&M's
M&M's Minis
3M Post-it
Mr. Post-it
Seeds of Change
Visine Gentle Mist
Pfizer Wobball
OMO Splats
Lipton Teaspoon
Sudafed Sudacare
Elements
7 WTC
Liberty National Gold Course
Pix & Trix Xplorer
Whiskas
Disney Mouseum
Mickey Mouse Watches
Disney Pets
G.O.A.T

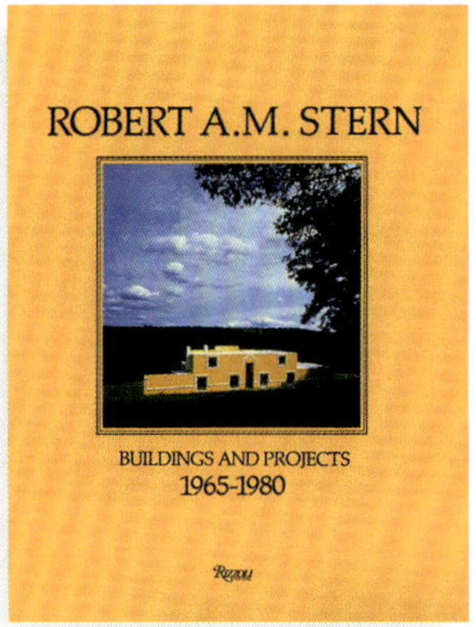
ROBERT A.M. STERN
BUILDINGS AND PROJECTS
1965-1980

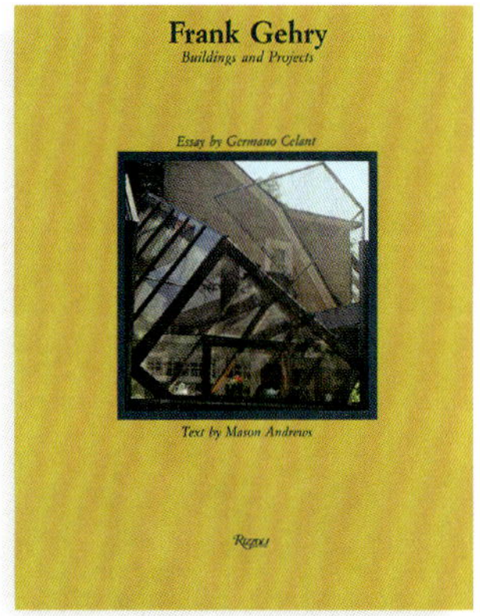
Frank Gehry
Buildings and Projects
Essay by Germano Celant
Text by Mason Andrews

Charles Gwathmey and Robert Siegel
Buildings and Projects 1964–1984

Aldo Rossi
Buildings and Projects
Essays by Vincent Scully and Rafael Moneo

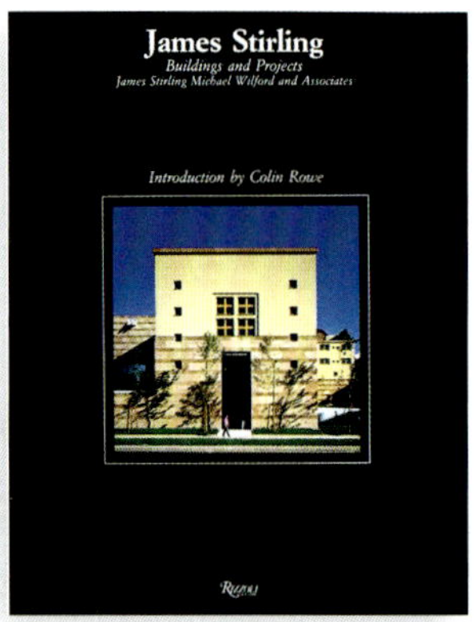
James Stirling
Buildings and Projects
James Stirling Michael Wilford and Associates
Introduction by Colin Rowe

A Tower for Louisville
The Humana Competition

A Center for the Visual Arts
The Ohio State University Competition

Mississauga City Hall
A Canadian Competition

Southwest Center
The Houston Competition

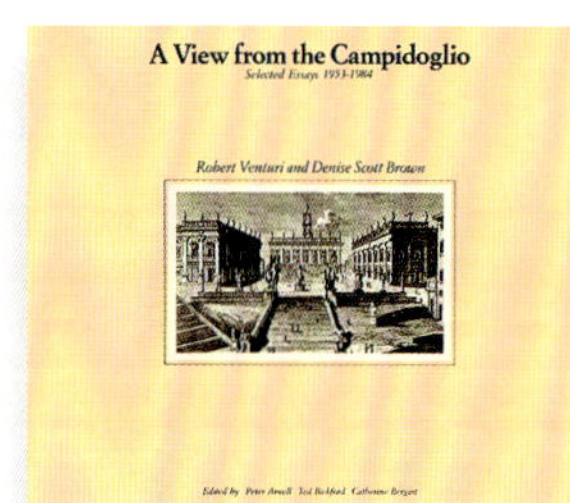
A View from the Campidoglio
Selected Essays 1953-1984
Robert Venturi and Denise Scott Brown
Edited by Peter Arnell Ted Bickford Catherine Bergart

RIZZOLI

1983

For Rizzoli, the famed Italian publishing house, Peter and Ted Bickford, the cofounder of Arnell-Bickford, authored, edited, and designed a series of books dedicated to the work of the most influential architects of the then-emerging postmodern movement in architecture. Over a period of roughly two years, Peter and Ted traveled extensively across America and Europe to work intimately with some of the most prolific architects of the day, gaining access to their studios, archives, and workshops; making pilgrimages to their built projects; touring in-progress construction; photographing, documenting, interviewing, and otherwise compiling vast troves of material and unparalleled insights into their work and process.

The resulting series of books was an unrivaled record of the then-emerging postmodern style in architecture, providing a detailed overview of the built work and scholarship of figures such as Frank Gehry, Michael Graves, Robert A.M. Stern, James Stirling & Michael Wilford, Aldo Rossi, Charles Gwathmey & Robert Siegel, and Robert Venturi & Denise Scott Brown. These *oeuvres complètes* came to the attention of renowned architectural patron and preservationist Phyllis Lambert, and she provided Peter and Ted with a grant to produce compendiums of competition entries for landmark projects like the Mississauga City Hall in Ontario, Canada, the Southwest Center in Houston, Texas, the Humana Tower in Louisville, Kentucky, and the Center for Visual Arts at Ohio State University.

As the first comprehensive scholarly overview of the work of this emerging generation of architects, this series proved to be an indispensable resource to scholars and students who revered these figures, but had scant access to their body of work in the pre-Internet age. By providing a detailed, intimate, and exhaustive look at work that had previously been inaccessible, these books played a considerable role in expanding both architectural pedagogy and practice throughout the 1980s and beyond.

Today, these books stand as some of the earliest and most complete contributions to the scholarship around the most significant late twentieth-century architects, many of whom continue to shape the form and character of the built environment to this day.

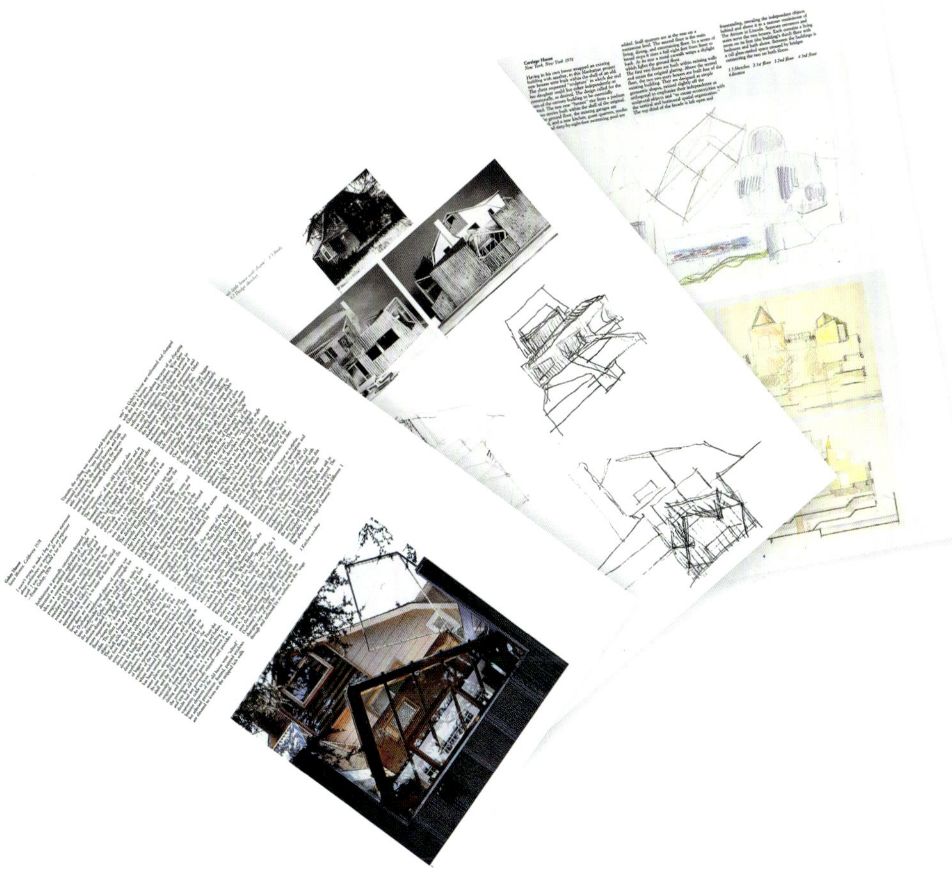

OFFICINA
ALESSI
EPB ITALY

ALESSI

1983

For the launch of the Officina Alessi collection in 1983, a seminal design of Austrian architect Josef Hoffman was reimagined with the aid of early 3-D computer imaging. Known as the Rosenschale bowl, the piece was based on a soap dish designed by Hoffman and fellow Austrian Koloman Moser—one of the foremost artists of the Vienna Secession movement, and a cofounder, alongside Hoffman, of the Wiener Werkstätte. The rare, one-of-a-kind, handcrafted piece had originally been produced for the Palais Stoclet in Brussels, a private residence that Hoffman designed for Belgian financier Adolphe Stoclet. No copies of the original were known to exist, nor were any manufacturing details or engineering specifications available to help recreate it. Partnering with pioneering special effects studio Digital Effects, a photograph from the archive of the Museum of Applied Arts in Brussels was analyzed by a Harris 500 computer, and a simplified 3-D model was "lifted" from the archival image. The computer software was able to deduce the contours of the bowl and create a set of working drawings that became the starting point for its eventual industrial production.

This approach, though previously used on other types of historical artifacts, was the first time an industrially produced product was brought to market in this manner. A critical reconstruction rather than a faithful replica, the reimagined Rosenschale bowl closely adhered to the original design, while evolving its form factor and scale to make it more functional for daily use. The techniques used to recreate the bowl offered a new level of legitimacy and authority to Alessi's manufacturing prowess, and led to relationships with the estates and archives of numerous other twentieth-century designers, including many prominent figures of the Bauhaus. In the years to come, many other classic "lost" designs were reconstructed through similar processes: a canny use of modern production technology that both honored the past and looked confidently toward the future.

VANITY FAIR

1983

Tina Brown, the legendary editor of *Vanity Fair,* engaged Peter to create the campaign that would be accompanying the magazine's much anticipated relaunch in 1983. This inaugural campaign captured the attention of New York's social elite with the cheeky and irreverent line, "All Things for Some People."

The final campaign featured a who's who crowd of celebrities, socialites, journalists, dignitaries, and other social luminaries like John Richardson, Dominick Dunne, André Leon Talley, Liza Minelli, Bianca Jagger, and of course, Brown herself, all mingling together: at once pulling back the curtain on this rarefied social strata, and mythologizing the larger-than-life personalities who populated it.

The campaign was so successful with readers that the format immediately grew into its own monthly feature—V.F. Camera—and became an integral part of the magazine's coverage of contemporary celebrity and style by giving an unfiltered behind-the-scenes look at the figures shaping our culture.

Gregory Swimming
Los Angeles
1982

CAMERAWORKS

1984

David Hockney's *Cameraworks,* a collection of photo collages made by the artist throughout the 1970s and early 1980s, celebrated the artistic potential of mainstream consumer photography tools, including the then-ubiquitous square Polaroids as well as four-by-six prints shot on inexpensive point-and-shoot cameras and developed at neighborhood one-hour photo labs. Hockney's photographic "joiners" challenged the conventional perspectives of the static medium of photography, and attempted to represent the experience of moving through the world, seeing things in the round from multiple viewpoints over spans of time.

Working alongside Hockney, and in close collaboration with Knopf editor Raymond Foye, art critic Lawrence Weschler, and printer Ken Tyler, Peter designed and coedited this now-iconic publication. Spending countless hours in Hockney's studio arranging and rearranging hundreds of images that had been shot from the same vantage point, the concept for the book design was to create a canvas for the collages that would allow readers to viscerally feel the enveloping and immersive quality that these compositions had in reality, even at a greatly diminished scale. The meticulous works presented scenes and events unfolding over time from simultaneous perspectives: in essence, a collapsing of both time and space into a coherent single image that challenged viewers to consider new ways of seeing the world around them as both observer and observed.

Today, particularly in the context of nearly instantaneous social media image-making practices, Hockney's modes of visualization and his ways of seeing and representing the ephemera of life have only come to seem all the more powerful and prescient.

EAB
The bank for Long Islanders
Long Island Swordsmen
East Williston, Long Island

EAB
The bank for Long Island
Nassau Model Railroad Club
Williston Park, Long Island

EAB
The bank for Long Islanders

EAB

1984

European American Bank (now a subsidiary of CitiGroup) had long been the dominant regional player in personal banking for much of Long Island. Facing increasing competition from national and international financial institutions, a strategy was developed to help EAB distinguish itself by reinforcing its core point of differentiation: that it was a bank of and for Long Islanders.

The resulting campaign announced EAB as "The Bank for Long Islanders" and presented a series of unusual and arresting portraits of actual Long Islanders, albeit ones without ordinary jobs, hobbies and interests. The campaign was shot by the award-winning photographer and film director Neal Slavin, who had gained notoriety through his exquisitely staged group portraiture, and in particular through his landmark publication *Britons*—an extraordinary collection of photographs of every conceivable group or association in the UK, executed over an eight year period. Seeking to capture the inner workings of these groups of people, Slavin's group photographs drew out natural hierarchies, group dynamics, and other idiosyncrasies that otherwise would have remained invisible to the outside world.

Building upon this signature gesture, Slavin captured beautifully composed shots of groups of Long Islanders for EAB's campaign: firefighters, fencers, baseball players, synchronized swimmers, and even model train enthusiasts. The arresting images honored the real Long Islanders who EAB served, while cleverly encouraging a dual double-take: first at the image itself, and then at the friends and neighbors all around whose private worlds might be anything but expected.

EAB
The bank for Long Islanders
Empire Hose Company #3
Merrick, Long Island

A Salute to Long Island
EAB
The bank for Long Islanders

10:00

10:05

10:11

10:16

10:20

10:26

26 MINUTES IN THE LIFE OF TODAY'S MOST PRODUCTIVE PC.

The machine is AT&T's PC 6300 PLUS. At its heart, a bit of technological lightning that permits the computer to do several tasks at once.

This AT&T breakthrough is called Simul-Task. It mightly expands the usefulness the PC experts already place in the first rand of today's more powerful "80286" machines.

One mangager likes Simul-Task to having an assistant handle one chore while you get on with another. With Simul-Task aboard, the AT&T PC 6300 can combe through a database for the figures you need while you update a memo... or print out a report while you set up a spreadsheet.

SIMUL-TASK: THE SOUL OF A BETTER PC

There exist few machines that can match the power of AT&T, PC 6300 PLUS. The speed is intoxicating, the graphics razor sharp. Look hard enough and you may find a similar machine. But you will find none that apply its power to one of your existing PC programs - and to other jobs at the same time.

Catnat emptor. You can pay plenty for software that let you look at several programs at once. It's the difference between window shopping - and getting things done.

The secret: Simul-Task unbocks the full potential of the "80286" processor chip. So you have all the benefits of today's PC software, with out the limitations of today's PCs.

Simul-Task gives anyone who sits down at an AT&T PC 6300 PLUS the ability to do more - and do it faster - than users of other PCs. But there is more to the Simul-Task story.

Item: Simul-Task opens the door to better date communications. Two ketstrokes can tap you in to a mainframe 1,500 miles away, or the PC down the hall.

Item: Even as you run one, two, or three business programs, the PC 6300 PLUS is open to receive and store electronic mail or date someone else is sending to you. No matter what programs you are running, your " mail" is just three keystrokes away.

Item: You put your computer on "automatic pilot," running time-consuming jobs like database sorts after hours. Rather like setting a VCR to tape tomorrow night's movie.

COMPUTERS WITH THE FUTURE BUILT IN

Most of today's 19 million PCs are islands unto themselves. But tomorrow is coming up fast. For some companies, the next step is tapping a remote database. For others, it is networking one PC to another. Or uniting all their computers in a larger constellation.

Here Simul-Task shines. For many companies, AT&T"s PC 6300 Plus with Simul-Task can be the bridge between today's personal computer applications... and tomorrow's compangwide applications.

When the time comes to take the next step, you'll be ready. AT&T makes the pieces fit.

Hour much will all thin cost? The exact figure depends on how you equip your system, but most companies have been pleasantly surprised. And right now (for a limited time) you can get even more for your money.

To learn more, and to arrange for a Simu-Task demonstration, please contact your AT&T Account Executive, stop by any authorized AT&T supplier or telephone 1 800 247 1212.

AT&T
The right choice.

AT&T

1984

For an early ad promoting the integration capabilities of AT&T's Information Systems, a series of national business-to-business communications was created in collaboration with AT&T's creative agency Ogilvy & Mather Partners, the O.&M. boutique subsidiary that specialized in promoting business services and premium products. Peter was approached by Joe Bracken, then the president and CEO of Ogilvy & Mather Partners, who invited him to lead the creative development of a new campaign for AT&T.

Following their collaboration on *Cameraworks*, David Hockney encouraged Peter to adapt his signature photo collage technique to create a series of images that would convey the speed, dynamism, and intelligence that AT&T's enterprise solutions offered. Each collage was produced by taking hundreds of photographs over the course of a typical workday at AT&T headquarters, then reassembling these images into a single composition that pulled back the curtain and showcased AT&T's advanced technologies at work.

BERGDORF GOODMAN

1984

Since its founding in 1899, Bergdorf Goodman has consistently set the standards of fashion and taste for the stylish set both in New York and around the world. In the early 1980s, the iconic department store embarked upon a complete renovation of its Fifth Avenue flagship store located on the site of the historic Cornelius Vanderbilt II House. To capitalize on this fresh start and help to reinvigorate the conservative reputation of the store, Bergdorf Goodman fashion director Dawn Mello started bringing in young creatives from a wide range of non-fashion backgrounds.

Peter had come to Mello's attention through his involvement in David Hockney's *Cameraworks* and she imagined that he might be a good candidate to design a book celebrating the history of Bergdorf's. But in the course of their first meeting, it quickly became clear that the ideas being discussed were much larger than a single commemorative book, and Mello hired Peter to develop an ad concept to mark the upcoming reopening of the store.

At the time, a controversial issue with the renovation was the decision to install an escalator between the main and second floors where an elevator bank had once been. To a certain set of Bergdorf's customers (the so-called Carriage Trade) the presence of an escalator was an unwelcome intervention that diminished the rarefied white-glove service that Bergdorf's had always stood for. Seeking to simultaneously appease that small but important group of legacy customers, while also inviting in a younger and more diverse new customer base, a simple but impactful composition was devised where an image of the new escalator was applied to the golden doors of the old elevator. Inspired by the famous painting *The Son of Man* by surrealist artist René Magritte, the result was a witty, striking, slightly surreal superimposition of tradition and modernity that welcomed in the new guard without alienating Bergdorf's longtime customers.

Subsequent print ads were developed that took the form of film strips: a way of introducing movement, dynamism, and vitality to a static medium, that, in a sense, brought the product to life. Developed in collaboration with legendary photographers Denis Piel and Roberto Schezen, along with stylist Jacques Denois, these communications introduced a clever, esoteric, youthful spirit that helped to cement Bergdorf's as New York's singular fashion destination for a new generation.

BERGDORF GOODMAN
"It's my favorite
store in New York.
First and foremost in my
mind, Bergdorf Goodman
represents quality."
"The boutique
is a beautiful reflection
of the shop in Paris.
The ideas are all there."
Maud Frizon
February '84
Maud Frizon
at Bergdorf Goodman
Third Floor

MAGIE NOIRE
BERGDORF GOODMAN
The Yves Saint Laurent
Reptile Collection
at Delman
Ring lizard pump
in natural, amber,
rose $194
Creme kidskin sling
with natural ring lizard
trim $182
Delman Shoe Salon
on Four

BERGDORF GOODMAN
"I've known Bergdorf's
for as long as I've
been in New York,
about eighteen years.
One of the things I really like
about it is that I always know
where to go when I want
something specifically.
Lingerie, something for
my children, a really
wonderful evening dress.
There's a pampered feeling,
which is nice when you
don't have a lot of time."
"When they asked me
what I wanted for the new shop,
I knew I didn't want anything
associated with
traditional jewelry stores.
I thought it should look
somewhere between the inside
of a Rolls Royce and
a Zen garden."
"On a more abstract level,
I think of Bergdorf's as being
something like
pastel sapphires.
That's just very personal.
My house is light
and when I associate
anything with
something positive,
I never think of dark and shiny.
I suppose too that I make
that association because
of my collection."
Angela Cummings
February '84
Our Angela Cummings Boutique
for fine jewelry
Opening Valentine's Day,
Tuesday February 14,
with a completely new collection
designed specially
for Bergdorf Goodman

DONNA KARAN

1984

As Donna Karan prepared to launch her namesake label in 1984, her singular ambition was to create a new brand that spoke to a generation of young, ambitious, professional women by offering them chic and versatile clothing that blurred the boundary between their professional and personal lives. Patti Cohen, who was Donna's head of communications and public relations—and employee number one—had been introduced to Peter's work at Bergdorf Goodman by Dawn Mello, and decided to bring him in for an interview. By the end of that first interview—which spanned nearly twelve hours—the core identity of the Donna Karan brand had been set: namely, to hold up a mirror to this new generation of working women.

Armed with a point-and-shoot Contax camera, Peter roamed around his (and Donna's) native city, capturing images that embodied the power and strength of Donna's vision for a new kind of woman … and he found it in the city itself: in the concrete, steel, and asphalt; in the rush of cars, and the surge of pedestrians. Returning to one of the most memorable sites of his own childhood—a spot at the base of the Brooklyn Bridge where his grandfather had worked—he captured what became one of the defining images of the Donna Karan brand: the Brooklyn Bridge, piercing the mid-winter mist rising off of the East River, stretching boldly toward the city just beyond, and all the hope and promise that it represented. Quoting both Edward Steichen's haunting, largely unpeopled images of New York's built environment, and Berenice Abbott's life-filled photos that appeared in her seminal 1939 publication *Changing New York,* the reportage-style images that Peter (and subsequently, Denis Piel) captured of the city evoked drama and romance, but also grit, determination, and ambition: an unforgettable invitation to the brand that didn't even show a single garment or model.

Subsequent campaigns, brought to life with the help of numerous legendary photographers like Arthur Elgort, Wayne Maser, and Torkil Gudnason, expanded and interated upon Donna Karan's central mandate of women's empowerment. Perhaps most poignant of all was the 1992 series *In Women We Trust* shot by Peter Lindbergh, where the journey of a female politician to the highest office in the country is shown quite matter-of-factly: first, on the campaign trail; next, winning her election; and ultimately, taking the oath of office as president of the United States.

DONNAKARAN
NEWYORK

DONNAKARAN
NEWYORK

leave me alone! I'm having a crisis!

DONNAKARAN
NEWYORK

DONNAKARAN
NEWYORK

DONNAKARAN NEW YORK
The President and Mrs. Reagan
request the pleasure
Ms. Karan
on Tuesday,
PURE WOOL
DONNAKARAN NEW YORK
DONNAKARAN

DONNA KARAN BACKSTAGE

1987

The Donna Karan brand quickly rose to the highest reaches of the fashion world, attracting the top talents of the day both behind the camera and on the runway.

Recognizing that what Donna and her team were creating was a once-in-a-generation phenomenon, Peter feverishly documented every aspect of the heart and soul of the company: a discreet but ubiquitous presence with his camera trained on every moment, and every little detail. During these early years at Donna Karan, everything was an asset; everything had value to be communicated; every ounce of creativity needed to be captured. The spontaneity, the energy, the exquisite chaos: that was the essence of Donna Karan.

Pictured here, a few of the day's top supermodels can be seen in an unguarded moment backstage at a Donna Karan runway presentation in the late 1980s.

DONNA KARAN HOSIERY

1987

Donna Karan Hosiery was launched in 1987 with a series of communications that proposed an entirely new way of looking at a product that had acquired a reputation as something staid and uncomfortable.

The images featured in the campaign showed both bare and hosiery-covered legs, shot in black and white, under dramatic, high-contrast lighting conditions. That the bare legs were scarcely distinguishable from the hosiery-covered ones in many images was precisely the point: the product was thin, smooth, sheer, and as close as possible to wearing nothing at all. Yet it was also strong—as attested to by the image of a woman's fingers slipping beneath her stockings, the fabric stretched to nearly sheer. Rather than something oppressive, it suddenly became something intimate, even sensual: a sort of invisible armor for the young professional woman.

Donna Karan Hosiery also helped to bring the Donna Karan brand to a far wider audience, offering every woman an entry point into the widely coveted brand with an affordable, ubiquitous wardrobe staple.

The campaign went on to be honored at the 1987 Council of Fashion Designers of America Awards, with a special award created to acknowledge the tremendous impact of these landmark communications—the first time the CFDA gave an award honoring design excellence in marketing and branding.

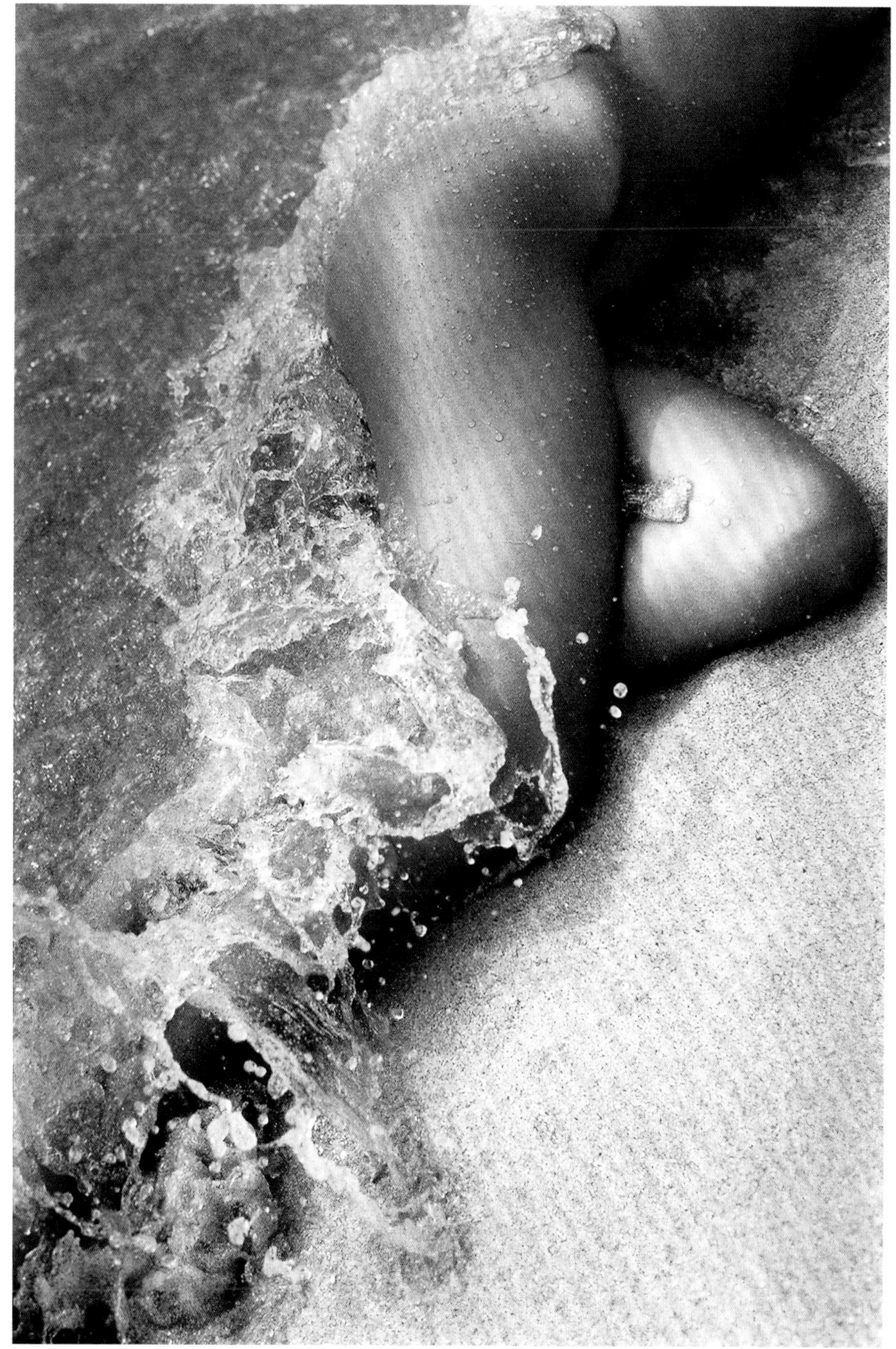

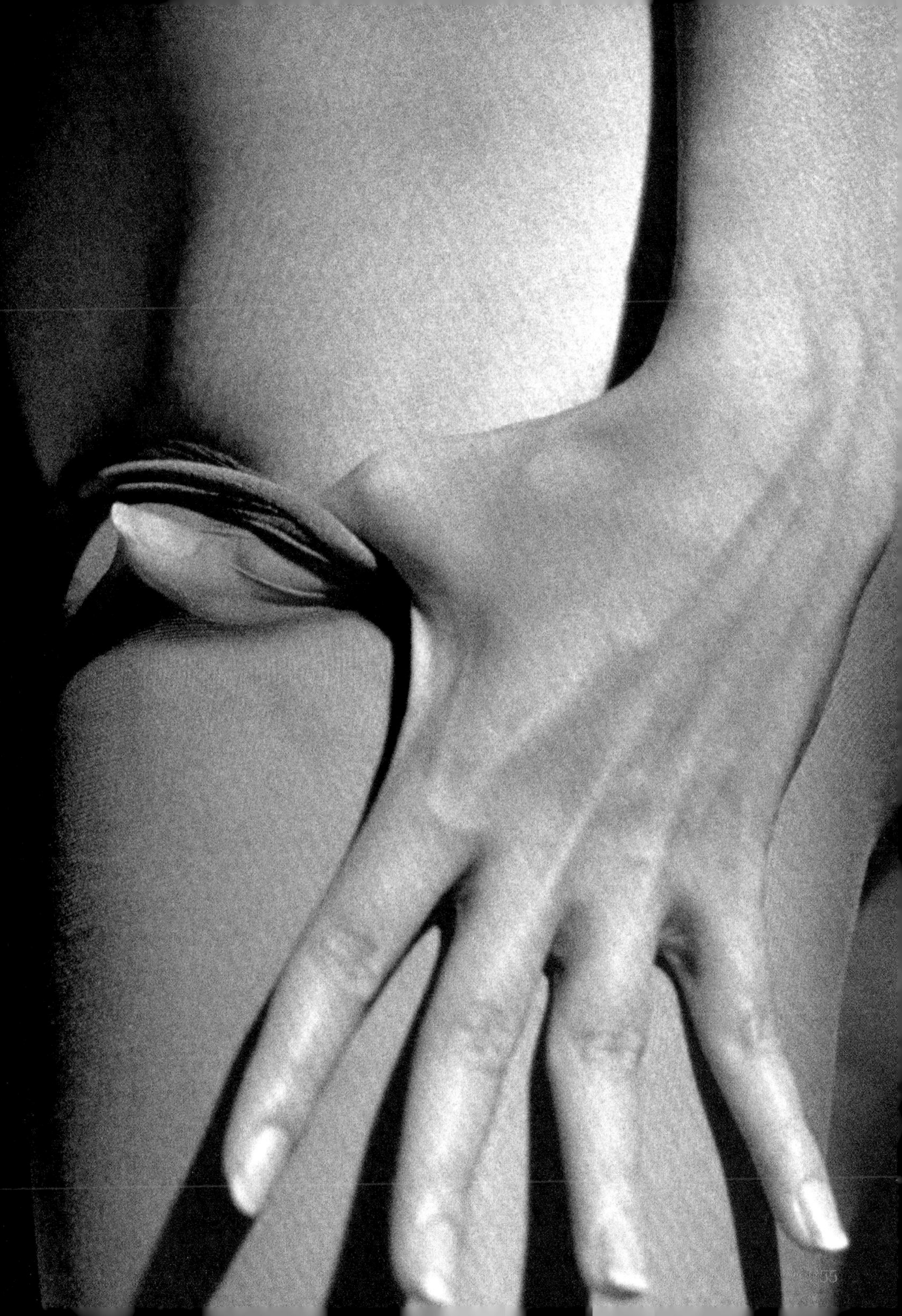

DONNA KARAN INTIMATES

1987

For the launch of Donna Karan Intimates, a campaign was developed that captured and built upon that spirit of empowerment that the Donna Karan brand had become renowned for. In the context of the late 1980s, an era that saw the rapid and widespread participation of women into the upper echelons of the corporate world, Donna Karan's clothes had immediately become something of a uniform for a generation of ambitious, high-achieving, professional women who refused to cow to sexist stereotypes that often treated femininity as an impediment to being taken seriously. And just as the main line had challenged prevailing ideas about what a professional woman should look like or how she should comport herself, the launch of Intimates was designed to shift the perceptions around intimate apparel from something that a woman wore for the benefit of others to something that they wore to foster their own sense of self.

In a radical departure from the contrived, posed, hyper-feminine style that was prevalent in the marketing of intimate apparel at the time, a series of launch communications was developed that showed women comfortable, relaxed, and natural rather than the expected preening seductiveness that had become a cliché of this market segment. The elegant informality of the models, coupled with the stark, naturalistic style of the photographs, eschewed the expected soft, gauzy, airbrushed look, contrasting with nearly everything else on the market. On-set, the models were given minimal direction, and instead encouraged to behave as though they were at home, alone, enjoying a private moment to themselves. The intimate quality of these images drew out a different sort of beauty, one that was unfiltered, unstudied, and entirely without artifice.

By depicting scenes that were, in essence, intimate snapshots of a woman's private life, a new and different sort of aspirational appeal was created around a product that had historically been marketed by encouraging women to see themselves through the eyes of another. In these images, the women wearing the lingerie are shown to be relaxed, self-assured, and completely at ease, eschewing performative sex appeal in favor of being truly self-assured, confident, and, above all else, comfortable in one's own skin.

DONNA KARAN BATH & BODY

1987

Donna Karan Bath & Body was launched as both a complement and counterpoint to the sense of dynamism, ambition, and empowerment that the main line had quickly become synonymous with. This new range was positioned as an invitation to the woman who does it all to take the time to relax and recharge. In the midst of a hectic life these products offered an entry point into a new and necessary indulgence: the ritual of relaxation.

Building on the strategy pioneered by the launch communications for the Donna Karan New York ready-to-wear line, which focused on the emotion of the brand rather than its image, communications for the Bath & Body line similarly drew upon on the emotional resonance of the new products, rather than their practical characteristics. Beautiful black-and-white photography captured women in private, unguarded moments, indulging in necessary self-care that will allow them to return to their work, their family, and their life restored and rejuvenated.

In what ultimately became the singular shot of the campaign, a woman rests her head on the edge of the bathtub, her face obscured by her soap-covered arms: a moment of pure relaxation. This iconic image perfectly captured the thesis of the brand: without any product being shown, or even a single word of copy, it's instantly clear that a moment of quiet reflection and escape from the demands of daily life are the greatest luxuries of all.

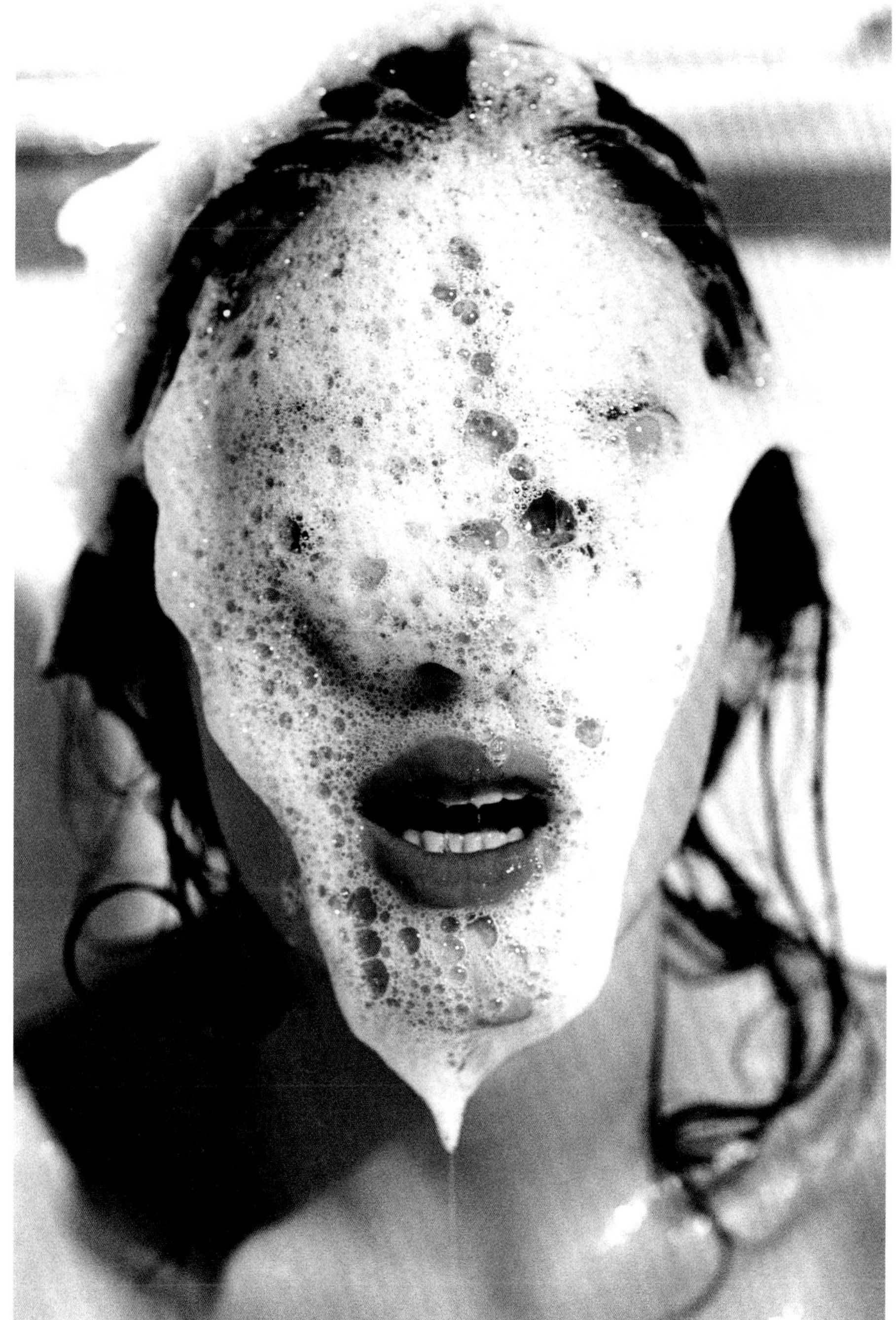

DONNA KARAN PARFUM

1987

The Donna Karan Parfum marketing strategy was built around the idea of drawing out and enhancing the natural sensuality of every woman. The launch communications captured a series of personal, private moments that depicted a striking intimacy and vulnerability.

At the heart of this campaign was a simple but radical gesture: showing women enjoying the fragrance themselves. So many perfume advertisements focused on building appeal, allure, and attraction for the benefit of others, or on promising to transport the wearer to some glamorous locale or fantastical realm. But by featuring simple but striking everyday moments—a close-up view of a gentle embrace, or a woman, wrist raised to her nose, inhaling the fragrance for her own pleasure—the campaign removed the typical tropes of seduction and grand fantasy that dominated the sector, and focused instead on capturing ordinary scenes of extraordinary intimacy that women everywhere could viscerally feel and envision for themselves.

These communications subtly but powerfully upended the expectations of the segment by stripping away all the usual trappings of fragrance marketing and placing the emphasis where it rightly belonged: on the woman wearing it.

ANNE KLEIN

ANNE KLEIN

1986

The communications campaign for Anne Klein featured supermodels Amber Valetta and Christy Turlington in timeless, classic, yet modern clothing appropriate for the professional woman. The no-nonsense campaign showcased beautifully tasteful looks, both aspirational and accessible, suited to the fast growing market for working women.

ANNE KLEIN

CHRISTIAN LACROIX

1987

In the spring of 1987, Carrie Donovan, the style editor at *The New York Times Magazine,* invited Peter to go to Paris to document the behind-the-scenes lead up to Christian Lacroix's inaugural haute couture collection for a special feature that was slated to coincide with the collection's debut in the summer of that year. Donovan, keenly aware of the historical nature of Lacroix's presentation, was eager to document it, but wanted to eschew all of the polished, highly managed "exclusives" that every media outlet was busily trying to secure. She was interested in capturing the reality of the moment, and she wanted someone who understood the world of fashion but wasn't a creature of it; someone who grasped the enormity of the moment but who wouldn't be deferent in their recording of it; someone who was, in the best ways, a bit of an outsider.

What began as a straightforward assignment soon blossomed into a genuine friendship—with Lacroix later remarking of their initial meeting that Peter had "parachuted into my life." With complete and unrestricted access to Lacroix, the photos that Peter shot captured a side of the designer that both confirmed and confounded the larger-than-life persona that he had constructed: in the fleeting, unguarded, vulnerable moments the full humanity of this creative genius was, for the first time, on full display. Whether in the atelier, or at Lacroix's country home in Arles, or holding court in the bar of Le Meurice, Peter's lens stripped back the veneer of Lacroix's carefully managed public image to show the actual man who existed beneath the splendor, and drama, and excess: an immense artistic genius who was, all the same, a fundamentally simple and humble man.

Though originally planned as a cover story for *The New York Times Magazine,* the story was bumped in favor of a feature about then-president Ronald Reagan. Donovan felt that this extraordinary collection of photographs deserved to be seen in its entirety, so she reached out to John Fairchild, the publisher and editor in chief of *Women's Wear Daily,* and told him that she was sitting on this trove of extraordinary images that the world needed to see. Fairchild immediately agreed to publish the piece as a cover feature, and offered Donovan complete editorial oversight over the final product. Ultimately published in *Women's Wear Daily* in November, 1987, *The Private Side of Lacroix* offered a bracingly intimate portrait of a man who many knew, but few truly understood.

Haute Couture
hiver

THE PRIVATE SIDE OF LACROIX

PARIS — Dinner with Christian Lacroix isn't quite what one expects. As he sits unnoticed at a quiet corner table in a Right Bank Russian restaurant that became his canteen during the Patou years, the year's most celebrated couturier affects no airs.

Continued on page 30

THE PRIVATE SIDE OF LACROIX

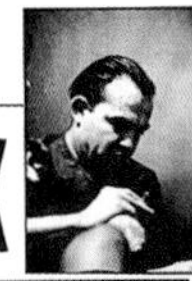

Continued from page 26

For once, ubiquitous right-hand man Jean-Jacques Picart is absent. When the waiter indicates they've run out of his favorite pimento vodka, Lacroix makes no fuss. Instead, he graciously conceals his disappointment and then whispers, "I should have brought a bottle from home."

Success, it seems, has not spoiled shy, soft-spoken Lacroix. Between bites of blinis and beluga caviar, he chats with candor and humility on everything from his wife and work, to farming and his passion for the polka. And he rejects the notion that fashion is an art. "I am not a *createur,*" says the down-to-earth designer. "I revive what I feel like and what I think women want. Every one of my dresses possesses a detail that can be connected to something historic, something from a past culture. We don't invent anything."

Those awaiting something dramatically different in his next collection are in for a surprise. Provence, the land of lavender and olive groves he so successfully transplanted to the runway this past July, will remain the theme. "The region is rich with many facets," explains the native Arlesian, "and by translating them differently every season, we can keep it as the base."

Lacroix's other great inspiration is Francoise Rosenthiel, whom Lacroix introduces as his wife although they have never married. He recounts, in detail down to the shoes she wore, their first encounter 14 years ago.

"It was love at first sight. I had been in Paris for only 10 days. I saw this girl with a short, square haircut, very red lips, a black fur coat, T-shirt, black and white pearls, flowered skirt and python shoes. I found it all very pretty — the colors, the complexion and the voice." Gifted with an uncanny memory for images, he describes her outfit on the second date with equal precision, then waxes sentimental. "I've never really known Paris without her. Francoise is the ultimate muse. She moves me," says Lacroix. Very much his complement, she is the embodiment of what he appreciates in women. "I like people with personality, people who are out of the ordinary, naughty people. I don't like lukewarm water."

Chic, daring and opinionated, fiery Rosenthiel fits the bill beautifully. While Lacroix invariably seeks the good side of people, Francoise is "very, very, very, very, very, very tough — worse than anyone." Lacroix claims she's capable of sustaining any thesis — even if she doesn't believe it — for the mere pleasure of hearing her own voice. Very verbal and thus often the life of the party, she is at times *"tres fatiguante,"* he proclaims with a contagious chuckle. "I prefer to listen than talk," he adds.

Neither one wants children. Lacroix confesses he belongs to that school of thought that children should be seen and not heard. He admires the British tradition of child rearing — from a distance, kept with the dogs somewhere.

Their "minuscule" sixth arrondissement apartment, according to Lacroix, contains not a single object of art, not a stick of furniture, not one painting of value — only books, sketches and documents piled up on two walls.

Lacroix has no qualms about being married to an older woman, perhaps because he appears older than his 36 years. He does, however, admit that she was *gentillement furieuse* when the press recently revealed she's six years his senior. Rosenthiel, once a style coordinator for Hermes, has no official function at the new couture house. Lacroix has other plans for her.

"I hope she will write one day. She is very intellectual and reads an enormous amount." With more of his characteristic modesty, he hastens to add, "I am not an intellectual. I have books on cinema and books of photos and images. I don't read. I look at the pictures." Mischievous schoolboy that he is, Lacroix sketches throughout the twice-weekly English lessons he doesn't really need. He also sketches in front of the TV set. His wife labels him "a teleholic." He switches the set on at 8 a.m. and off at bedtime.

"I am very much the voyeur," confesses Lacroix. "In the south of France in the evening, people leave their windows open, and I like to watch them in their undershirts eating their soup." Impassioned by the insides of people's homes, "and not just luxurious ones," Lacroix loves to see how others live.

To pursue his voyeuristic tendencies, he takes to Paris flea markets, where he has gathered a collection of old family photo albums. "I collect albums of people I don't know. At first, I'd buy them out of pity. I found it sad to see a family empty out the house and sell everything when the grandparents died. But then I started buying only those with a sentimental or esthetic charm." He mentions one with a girl in a chalet dancing like Isadora Duncan and another with a rather handsome boy in a Tudor castle.

While he reads nary a novel or philosophical work, the designer devours intimate diaries and biographies. "I am fascinated by the daily intimacies and ordinary, everyday things in people's lives." Intrigued by the most sentimental and routine aspects of the lives of people of genius, he is currently entrenched in Jean Cocteau's letters to Jean Marais. "It is a beautiful love story, even if there were infidélities. Theirs was a more respectful love — respect for each other's freedom."

How does Lacroix react to fame? With an air of introspection, he wrinkles his heavy brow and quotes Madame de Stael: "Success is the glorious funeral for happiness." His full, rounded face growing familiar in Paris and other fashion-conscious capitals, Lacroix finds recognition intimidating. Baffled by autograph seekers, he considers signing pretentious, but refusing to sign even more so. "When you ask a movie star for an autograph you're getting something of the star. The interest that I hold for others is not my being but my work," he adds, calling the showbiz and star-system side of fashion "a pity."

Partner Picart is nonetheless a mixture of Midas and Machiavelli in marketing. They both appear all too well aware of the ephemeral nature of fashion fame, but perhaps hope to outsmart the system. "It's become almost a reflex to do an article on Lacroix," observes the designer, saying that they are contemplating curtailing this excessive press coverage. "I am of the mind to refuse interviews. I'm afraid we'll tire people out, that too much has been said about us already and that it will start to affect our work. It's like anesthesia," he concludes, "and everyones anesthetized."

One inevitable result of renown in fashion is being copied the world over. Lacroix, perhaps already the most imitated designer, appears unperturbed. "It's the day they stop copying that it's time to start worrying," he reasons. "I'm not looking to be uncopiable. Fashion is in the air. There are

PHOTOS AND LAYOUT BY PETER ARNELL

those that get it more quickly than others, but we're all a bit on the same wavelength."

Lacroix has not escaped that other great sacrifice brought by fame — time. Denying reports he and Picart each earn close to a quarter of a million dollars, he insists "that's a lot of money, and, in any case, I wouldn't have the time to spend it." But a faintly disguised smile is detected when the prospect of his soon becoming a millionaire is mentioned.

At the time of this interview, he had not begun to design his deluxe ready-to-wear. The deadline, which seemed to render him slightly insecure, forced Lacroix to cancel his mid-September trip home. But he removed all doubts that his couture looks could be turned into salable rtw when his show in Paris had retailers salivating Oct. 16. It may seem odd he'd think of leaving Paris at the height of rtw season to be initiated into Arles's "brotherhood of friends of rice growers." But to Lacroix it's a "great honor," as his late father had supplied farm equipment to the rice-growing region. He also had to forgo the local bullfight, a ritual he describes: "One feels very strongly we're the descendants of those who watched the gladiators killing each other. We like the blood. There is a sexual aspect to it. Bacchanal."

Self-proclaimedly more Provencal than Parisian, Lacroix traces his love of life's contrasts back to his roots. "It's a region where you have two extremes: very straight protestant and intolerant on one hand, while the other extreme, to which my family and I tended, reflects what our ancestors the Greeks and Romans thought — 'everything is natural.' One doesn't make such a distinction between man and woman, rich and poor. This attitude was much healthier."

Two words recur throughout Lacroix's relaxed conversation on his work and life — honesty and sincerity. While only the ultimate test of time will tell, it momentarily seems he has somehow managed to keep these values intact.

"I must go on with my original formula — to work with honesty and sincerity. If I am always honest in my work, and I succeed, great. But if I fall flat on my face, I'll say that's the way it is. It couldn't have been otherwise. I did what I had to do."

— DENNIS THIM

RED
Black
Cult"

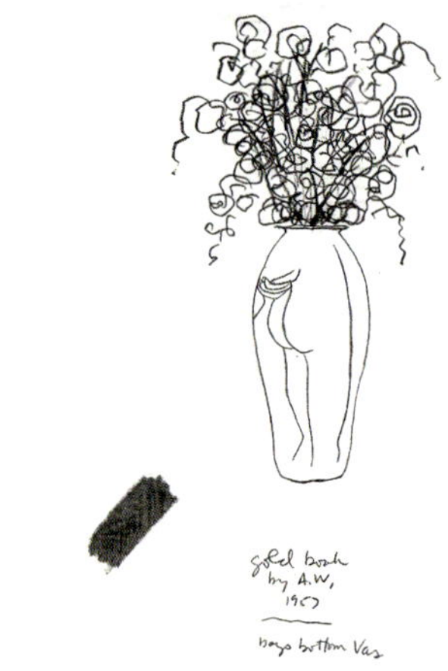
by A.W,
bottom Vas

Series of Hand +
Flower Canisters
in Porcelain

Stacking Coffee
Cups
as in
Flowers

SWID/POWELL

1988

In the late 1980s Swid/Powell, the New York-based home goods brand, decided to approach the recently established Warhol Foundation with a concept for translating the work of Andy Warhol into a range of home goods. The ubiquity of Warhol in popular culture coupled with the artist's lifelong willingness to blur the boundaries between fine art and commercial interests presented an opportunity to craft a series of products that existed between art and functional object.

As part of a proposal for a comprehensive partnership, conceptual designs were developed for a collection of tableware and decorative objects that captured Warhol's singular style, sense of humor, and eye for uncommon beauty. The final range of proposed products included bowls, plates, vases, clocks, and hand mirrors that all paid homage to the most iconic artworks, images, and moments of Warhol's storied career. Designs were even developed for a cookie jar whose top was a perfect porcelain replica of Warhol's infamous wig (designed by legendary hair stylist Christiaan Houtenbos)—a wry nod to Warhol's proclivity for deliberately wearing his wig askew.

Mass-produced, yet high-minded; functional yet impractical; beautiful yet uncanny, the proposed product range straddled the line between art and commerce with the artist's signature deftness, wit, and splash of absurdity. Though the collection never entered the marketplace, the designs served as a reminder of Warhol's unrivaled ability to transform the mundane ephemera of modern life into something transcendent and even revelatory.

ELLESSE

1989

Throughout the 1970s and 80s, Italian sportswear brand Ellesse had built a reputation for combining athletic performance with elevated, fashionable styling. They were one of the first major brands to understand the appeal of athletic wear as part of an everyday wardrobe, and to anticipate the way in which it would become elevated into a holistic lifestyle expression.

To help build the brand's North American presence, an immersive retail environment was conceived that would bring the unique culture, style, and message of Ellesse to life for customers worldwide. A new kind of shopping experience was envisioned to elevate the brand and evolve the notion of what sportswear could be. These new boutiques would not only provide customers with a new way to shop Ellesse, but they would also provide them with a new way to think about Ellesse.

The centerpiece of these retail environments was the Ellesse Tower: a multistory, fully inhabitable structure that showcased Ellesse's range of clothing, shoes, and accessories in a dynamic, immersive, and kinetic presentation. A nod to the towers that preside over the main square of almost every city and town in Italy, the Ellesse Tower explicitly connected the brand with the iconic imagery and symbolism of its country of origin, while also serving as a clear metaphor for its bold, confident, soaring aspirations.

The retail program also comprised a flexible system of dynamic in-store fixtures inspired by the form of Ellesse's Semi-Palla, or "half-ball," logo. By manipulating and adapting that iconic brand mark across scales and mediums, these fixtures—including a shoe wall, a shoe tower, hanging units, shelving units, customized shoe stands, proprietary hangers, flooring tiles, sneaker displays, pedestals, stools, tables, packaging, and modular T-shirt display units—proved out the idea that a brand's legacy and heritage can always be elevated and exploited to reveal new expressions of its longstanding DNA. Each of these elements, many of which were designed to include operable moving components that could be manipulated by customers, captured the spirit of motion and activity that was at the heart of the Ellesse brand, while also celebrating the unrivaled beauty of Italian art, design, and culture.

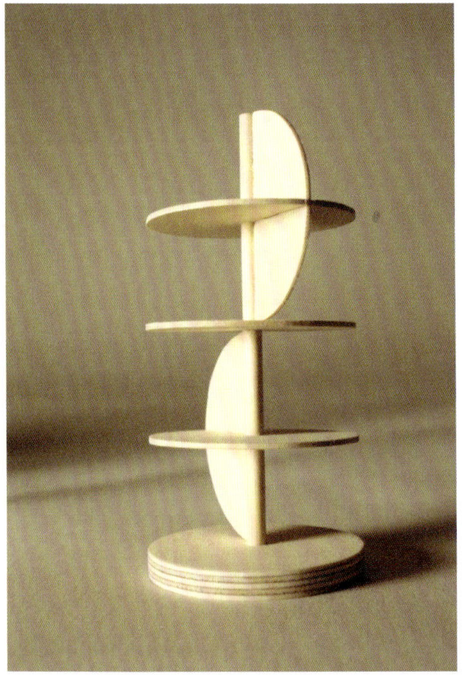

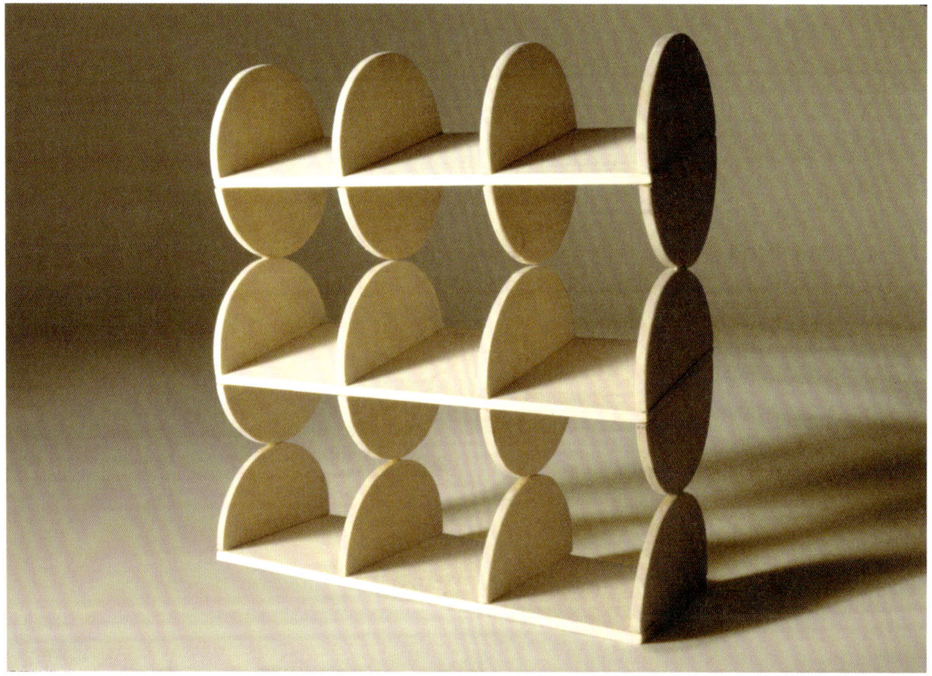

ellesse

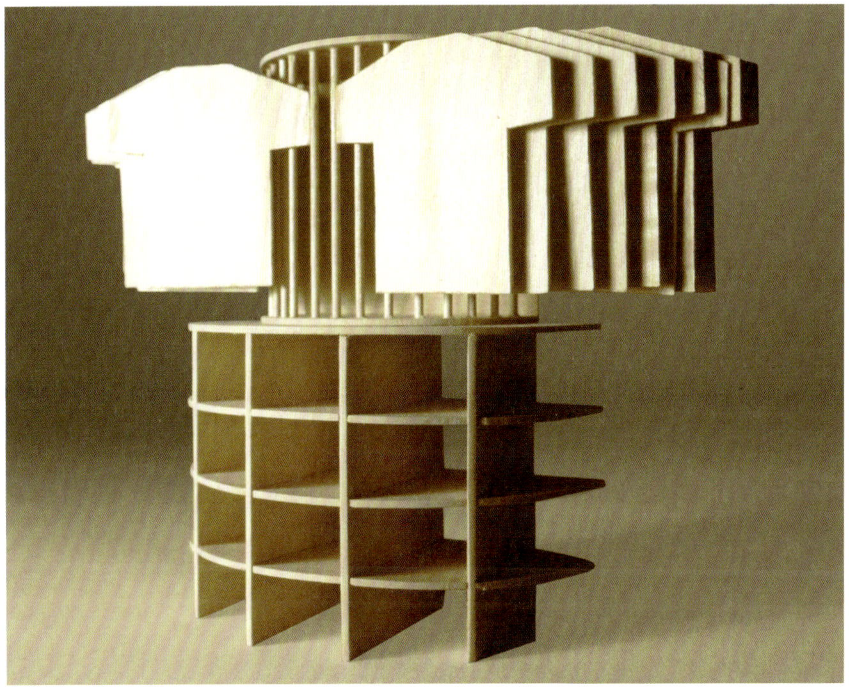

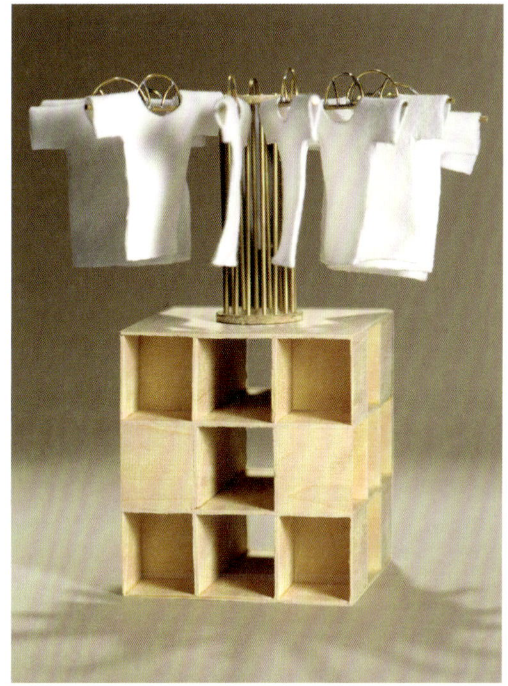

ellesse
100% cotton

ellesse
JUST FOR FEET
'92
"ELLESSE TOWER"

ellesse

CARLA

FENDI

1989

When Fendi began preparing plans to open their first flagship store in New York, the president of the iconic Italian luxury brand, Carla Fendi, set out to find a partner who could engineer a fresh and unexpected image for her family's design house. She charged her longtime deputy, Franco Saborelli, with conducting the initial search for a New York creative agency who could deliver on her vision. Saborelli vetted dozens of agencies before selecting a final few candidates, one of whom was Peter's firm. When Carla Fendi came to New York to conduct the final round of interviews, after she had met with all the candidates, she pulled Saborelli aside to ask for his preference, which he refused to give, claiming that it ought to be her decision and didn't want to sway her in any way. However, he did agree to write down his first choice on a slip of paper while insisting that she only look at it after she had made her decision. As it would turn out, the ploy was unnecessary: for both, their first choice was Peter.

For the Madison Avenue flagship store, a series of window displays were created that paid tribute to the brand's Roman roots in an unexpected and arresting way. Inspired by the courtyard of the Capitoline Museum in Rome that is filled with fragments of monumental Roman statuary—heads, hands, feet, and torsos, all many times larger-than-life—a series of 1:1 scale plaster casts of similar fragments were installed in the boutique's Fifth Avenue windows, along with the tagline "All Roads Lead to New York."

After this initial collaboration, the partnership continued for many years, ultimately expanding into retail design, product design, merchandising, brand strategy, and print communications, among many other projects and initiatives.

FENDI

FENDI MANNEQUINS

1989

For Fendi's New York City flagship location on Fifth Avenue, a new in-store display system was developed. The result was a range of mannequins based on the form of a traditional wooden clothespin. A clever and slightly surreal manipulation of scale drew an unexpected link between the elegant simplicity of this ubiquitous utilitarian object and the refinement of Fendi's exquisitely crafted luxury goods.

Inspired by the anthropomorphic qualities of a classic wooden clothespin and the corresponding childlike impulse to play with them as though they are actors on the stage of some great imaginative drama, the mannequins had a somewhat uncanny quality to them that was nonetheless perfectly in keeping with the brand's reputation for playful, irreverent beauty.

The dramatic scaling up of a ubiquitous utilitarian object—a gesture inspired by the sculptures of Dadaist artists like Marcel Duchamp and Man Ray, as well as the Pop Art sensibilities of Claes Oldenburg's supersized replicas of the mundane, everyday—was a provocative but elegant expression of a brand that had always stood for a more vital and exuberant sort of elegance.

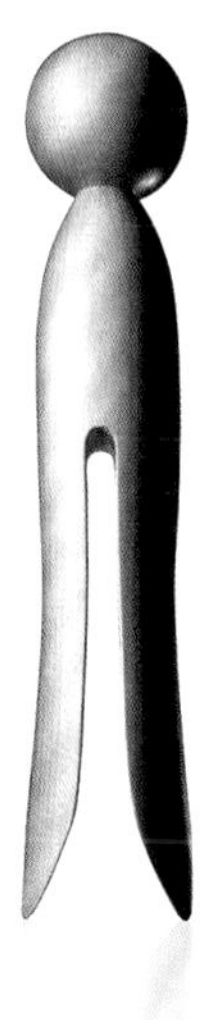

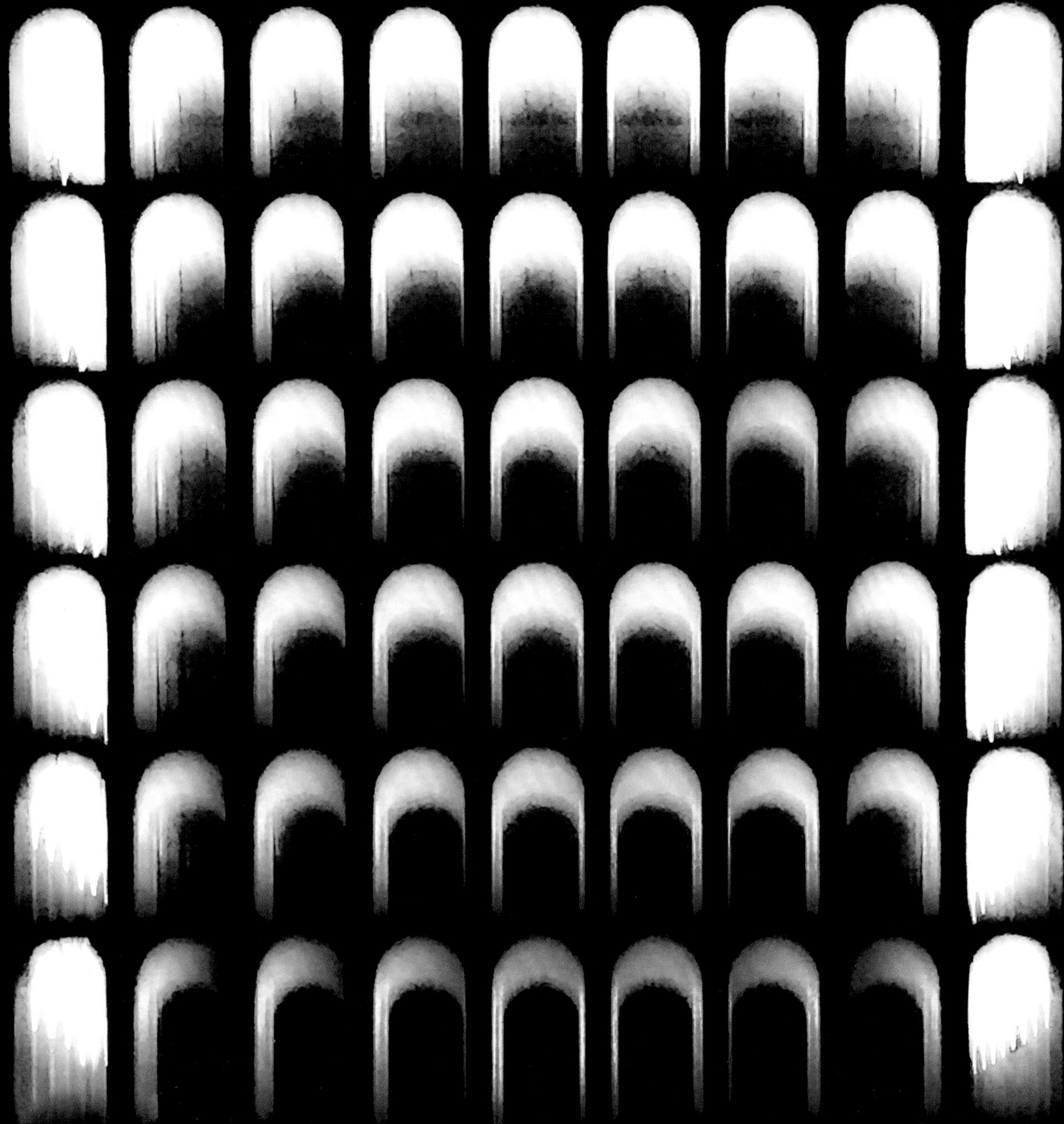

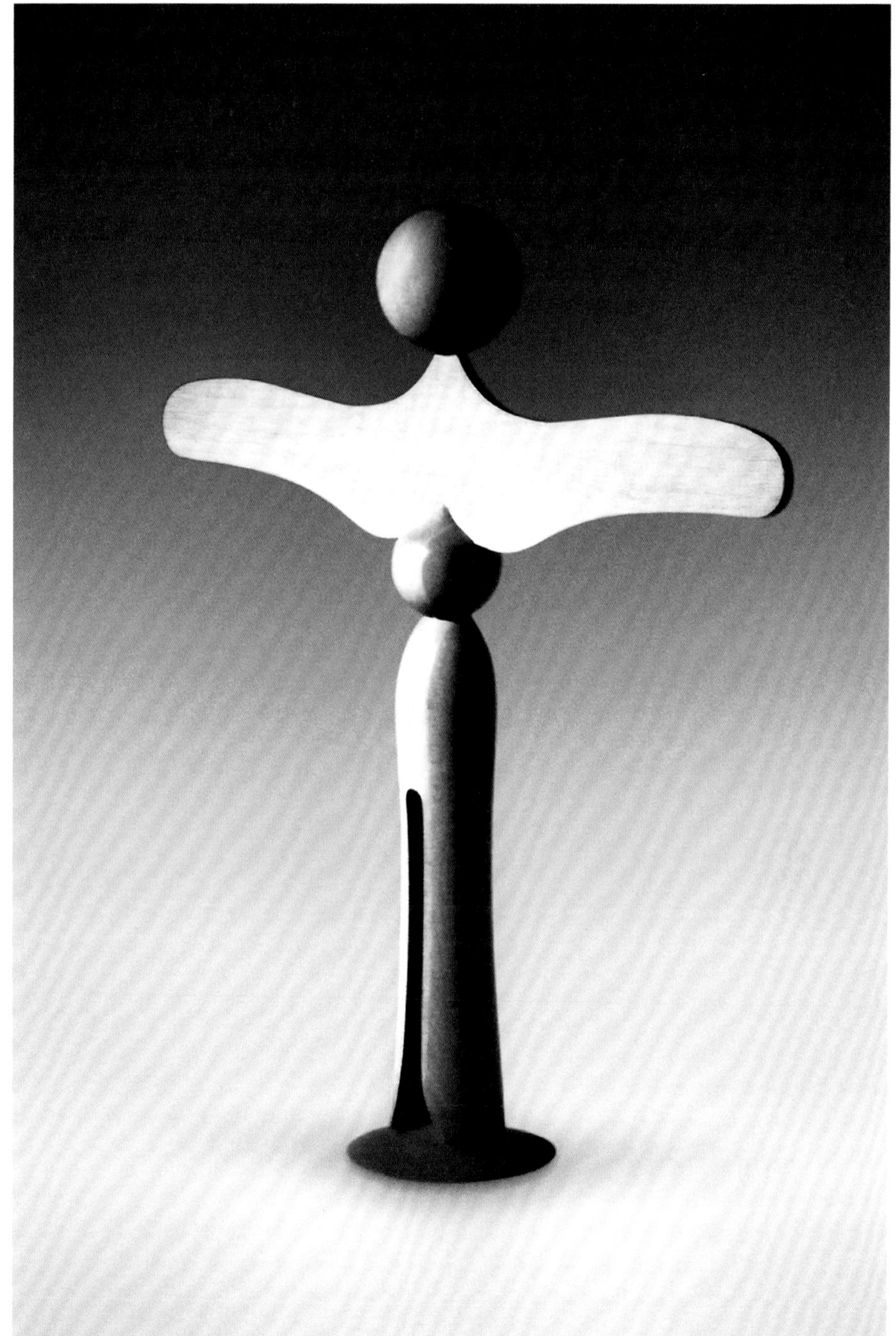

38 years; Anonymous, 41 years; Anonymous, 33 years; Anonymous, 24 years; Anonymous, 2 years; Anonymous, 38 years; Anonymous, 42 years; Anonymous, 28 years; Anonymous, 33 years; Anonymous, 4
Anonymous, 43 years; Anonymous, 33 years; Anonymous, 50 years; Anonymous, 63 years; Anonymous, 37 years; Anonymous, 32 years; Anonymous, 54 years; Anonymous, 92 years; Anon
9 years; Anonymous, 50 years; Anonymous, 38 years; Anonymous, 27 years; Anonymous, 29 years; Anonymous, 38 years; Anonymous, 30 years; Anonymous, 20 years; Anonymous, 48 years; Anonymous, 3
Anonymous, 91 years; Anonymous, 41 years; Anonymous, 77 years; Anonymous, 51 years; Anonymous, 52 years; Anonymous, 13 years; Anonymous, 33 years; Anonymous, 67 years; Anon
36 years; Anonymous, 33 years; Anonymous, 62 years; Anonymous, 55 years; Anonymous, 54 years; Anonymous, 38 years; Anonymous, 69 years; Anonymous, 39 years; Anonymous, 28 years; Anonymous, 7
Anonymous, 73 years; Anonymous, 51 years; Anonymous, 54 years; Anonymous, 39 years; Anonymous, 32 years; Anonymous, 13 years; Anonymous, 36 years; Anonymous, 49 years; Anon
28 years; Anonymous, 69 years; Anonymous, 30 years; Anonymous, 25 years; Anonymous, 49 years; Anonymous, 20 years; Anonymous, 87 years; Anonymous, 82 years; Anonymous, 37 years; Anonymous, 6
Anonymous, 36 years; Anonymous, 35 years; Anonymous, 34 years; Anonymous, 36 years; Anonymous, 18 years; Anonymous, 30 years; Anonymous, 34 years; Anonymous, 40 years; Anon
18 years; Anonymous, 33 years; Anonymous, 31 years; Anonymous, 34 years; Anonymous, 7 years; Anonymous, 56 years; Anonymous, 55 years; Anonymous, 35 years; Anonymous, 19 years; Anonymous, 2
Anonymous, 36 years; Anonymous, 34 years; Anonymous, 30 years; Anonymous, 35 years; Anonymous, 40 years; Anonymous, 42 years; Anonymous, 37 years; Anonymous, 51 years; Anon
73 years; Anonymous, 32 years; Anonymous, 18 years; Anonymous, 30 years; Anonymous, 44 years; Anonymous, 25 years; Anonymous, 34 years; Anonymous, 74 years; Anonymous, 39 years; Anonymous, 4
Anonymous, 42 years; Anonymous, 54 years; Anonymous, 38 years; Anonymous, 35 years; Anonymous, 62 years; Anonymous, 47 years; Anonymous, 35 years; Anonymous, 77 years; Anon
32 years; Anonymous, 35 years; Anonymous, 27 years; Anonymous, 16 years; Anonymous, 31 years; Anonymous, 15 years; Anonymous, 56 years; Anonymous, 16 years; Anonymous, 93 years; Anonymous, 4
Anonymous, 23 years; Anonymous, 19 years; Anonymous, 12 years; Anonymous, 38 years; Anonymous, 57 years; Anonymous, 39 years; Anonymous, 67 years; Anonymous, 40 years; Anon
47 years; Anonymous, 88 years; Anonymous, 3 years; Anonymous, 16 years; Anonymous, 43 years; Anonymous, 41 years; Anonymous, 33 years; Anonymous, 40 years; Anonymous, 30 years; Anonymous, 7
Anonymous, 66 years; Anonymous, 40 years; Anonymous, 38 years; Anonymous, 84 years; Anonymous, 34 years; Anonymous, 41 years; Anonymous, 79 years; Anonymous, 63 years; Anon
33 years; Anonymous, 28 years; Anonymous, 31 years; Anonymous, 3 years; Anonymous, 50 years; Anonymous, 36 years; Anonymous, 32 years; Anonymous, 38 years; Anonymous, 32 years; Anonymous, 3
Anonymous, 35 years; Anonymous, 3 years; Anonymous, 49 years; Anonymous, 33 years; Anonymous, 7 years; Anonymous, 31 years; Anonymous, 18 years; Anonymous, 17 years; Anonymous, 73 years; Ano
65 years; Anonymous, 30 years; Anonymous, 19 years; Anonymous, 40 years; Anonymous, 35 years; Anonymous, 2 years; Anonymous, 49 years; Anonymous, 63 years; Anonymous, 11 years; Anonymous, 3
Anonymous, 52 years; Anonymous, 19 years; Anonymous, 39 years; Anonymous, 67 years; Anonymous, 70 years; Anonymous, 84 years; Anonymous, 60 years; Anonymous, 2 years; Anon
37 years; Anonymous, 9 years; Anonymous, 34 years; Anonymous, 39 years; Anonymous, 30 years; Anonymous, 63 years; Anonymous, 30 years; Anonymous, 56 years; Anonymous, 37 years; Anonymous,
Anonymous, 95 years; Anonymous, 38 years; Anonymous, 70 years; Anonymous, 38 years; Anonymous, 66 years; Anonymous, 60 years; Anonymous, 84 years; Anonymous, 13 years; Anon
35 years; Anonymous, 30 years; Anonymous, 63 years; Anonymous, 36 years; Anonymous, 69 years; Anonymous, 67 years; Anonymous, 33 years; Anonymous, 12 years; Anonymous, 27 years; Anonymous
Anonymous, 32 years; Anonymous, 5 years; Anonymous, 56 years; Anonymous, 40 years; Anonymous, 60 years; Anonymous, 32 years; Anonymous, 72 years; Anonymous, 37 years; Anon
42 years; Anonymous, 38 years; Anonymous, 11 years; Anonymous, 31 years; Anonymous, 48 years; Anonymous, 13 years; Anonymous, 19 years; Anonymous, 36 years; Anonymous, 89 years; Anonymous,
Anonymous, 43 years; Anonymous, 83 years; Anonymous, 22 years; Anonymous, 33 years; Anonymous, 34 years; Anonymous, 51 years; Anonymous, 22 years; Anonymous, 32 years; Anon
71 years; Anonymous, 7 years; Anonymous, 31 years; Anonymous, 64 years; Anonymous, 36 years; Anonymous, 69 years; Anonymous, 66 years; Anonymous, 24 years; Anonymous, 67 years; Anonymous,
Anonymous, 75 years; Anonymous, 37 years; Anonymous, 24 years; Anonymous, 33 years; Anonymous, 51 years; Anonymous, 32 years; Anonymous, 84 years; Anonymous, 32 years; Anon
33 years; Anonymous, 26 years; Anonymous, 56 years; Anonymous, 34 years; Anonymous, 91 years; Anonymous, 17 years; Anonymous, 49 years; Anonymous, 33 years; Anonymous, 24 years; Anonymous, 4
Anonymous, 62 years; Anonymous, 37 years; Anonymous, 35 years; Anonymous, 68 years; Anonymous, 34 years; Anonymous, 72 years; Anonymous, 61 years; Anonymous, 37 years; Anon
2 years; Anonymous, 27 years; Anonymous, 39 years; Anonymous, 39 years; Anonymous, 28 years; Anonymous, 46 years; Anonymous, 30 years; Anonymous, 58 years; Anonymous, 19 years; Anonymous, 2
Anonymous, 49 years; Anonymous, 37 years; Anonymous, 37 years; Anonymous, 71 years; Anonymous, 72 years; Anonymous, 32 years; Anonymous, 32 years; Anonymous, 44 years; Anon
35 years; Anonymous, 91 years; Anonymous, 53 years; Anonymous, 35 years; Anonymous, 76 years; Anonymous, 95 years; Anonymous, 34 years; Anonymous, 41 years; Anonymous, 38 years; Anonymous, 5
Anonymous, 29 years; Anonymous, 64 years; Anonymous, 9 years; Anonymous, 31 years; Anonymous, 46 years; Anonymous, 39 years; Anonymous, 62 years; Anonymous, 31 years; Anon
56 years; Anonymous, 31 years; Anonymous, 37 years; Anonymous, 60 years; Anonymous, 66 years; Anonymous, 11 years; Anonymous, 56 years; Anonymous, 27 years; Anonymous, 94 years; Anonymous, 8
Anonymous, 71 years; Anonymous, 38 years; Anonymous, 63 years; Anonymous, 35 years; Anonymous, 10 years; Anonymous, 14 years; Anonymous, 87 years; Anonymous, 33 years; Anon
6 years; Anonymous, 14 years; Anonymous, 25 years; Anonymous, 37 years; Anonymous, 37 years; Anonymous, 39 years; Anonymous, 15 years; Anonymous, 19 years; Anonymous, 36 years; Anonymous, 3
Anonymous, 28 years; Anonymous, 67 years; Anonymous, 28 years; Anonymous, 88 years; Anonymous, 5 years; Anonymous, 34 years; Anonymous, 6 years; Anonymous, 29 years; Anonymous, 38 years; Ano
20 years; Anonymous, 24 years; Anonymous, 32 years; Anonymous, 39 years; Anonymous, 24 years; Anonymous, 49 years; Anonymous, 87 years; Anonymous, 36 years; Anonymous, 43 years; Anonymous,
Anonymous, 43 years; Anonymous, 64 years; Anonymous, 39 years; Anonymous, 73 years; Anonymous, 29 years; Anonymous, 3 years; Anonymous, 31 years; Anonymous, 49 years; Anon
2 years; Anonymous, 42 years; Anonymous, 30 years; Anonymous, 69 years; Anonymous, 49 years; Anonymous, 69 years; Anonymous, 63 years; Anonymous, 68 years; Anonymous, 58 years; Anonymous, 3
Anonymous, 61 years; Anonymous, 68 years; Anonymous, 59 years; Anonymous, 53 years; Anonymous, 59 years; Anonymous, 68 years; Anonymous, 32 years; Anonymous, 12 years; Anon

DIFFA

1989

By the late 1980s, the AIDS crisis had become impossible to ignore. But a lack of federal funding coupled with still-widespread misinformation about the virus meant that much of the essential work in terms of both research and care was being done by private individuals and organizations who refused to perpetuate these stigmas and insisted upon doing the right thing.

The Design Industries Foundation for AIDS was founded in 1984 by a group of New York designers to drive awareness around the growing HIV crisis in New York and beyond, and to raise funds to provide treatment, direct care services, preventive education programs, and advocacy for individuals impacted by HIV/AIDS.

In February 1989, the editor in chief of *Metropolitan Home* magazine, Dorothy Kalins, partnered with DIFFA to shine a spotlight on this crisis by devoting an entire issue of the design-focused magazine to raising money and awareness in the fight against AIDS.

This ad, which appeared in the special issue of *Metropolitan Home,* exposed the horrifying toll of AIDS by recording each of the thousands of AIDS-related deaths that had occurred in America since 1981 as "anonymous." This full-page ad filled with a seemingly endless roll call of "anonymous" victims and their ages—done principally out of respect for the victims and their families—also underscored the fact that AIDS did not discriminate, and very pointedly called out the shameful lack of action on part of local, state, and federal authorities who were allowing this plague to continue to ravage communities across the country.

This fundraising event showcased the works of many top designers, artists, and creative luminaries including David Hockney, Michael Graves, Ettore Sottsass, and Wolfgang Puck, and ultimately raised more than $3 million to benefit direct patient services, education, housing, and legal and financial aid.

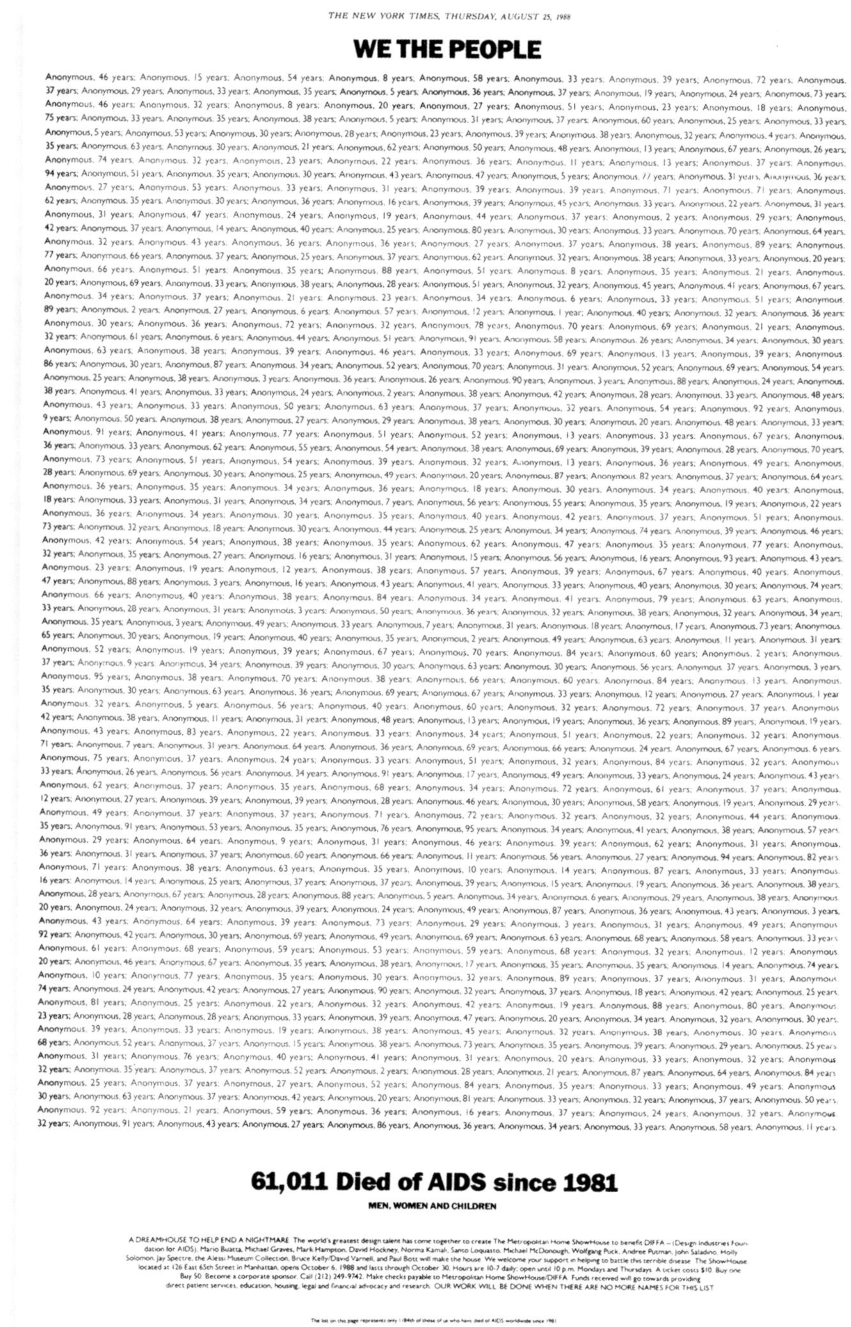

DKNY

DKNY

1989

By the late 1980s, the runaway success of Donna Karan's eponymous label had transformed the fledgling fashion house into one of the most recognizable brands in the world. It was clear that there was a tremendous opportunity to launch a sister line that would elevate and expand upon the tremendous cachet of the Donna Karan brand, while being more accessible to a broader and more diverse customer base. Unlike other secondary lines which, at the time, tended to be lower-quality imitations of the original line, DKNY was conceived of as an extension of the main line that while different—markedly more casual and affordable—was decidedly not lesser. As Donna herself remarked: "I like caviar, but I also like pizza. Why can't fashion give women both, all in one brand?" After all, the same woman who was able to afford luxury fashions also had a need for simple, casual pieces as part of her daily wardrobe. In essence, DKNY was the pizza to Donna Karan's caviar. Two sides of the same woman.

To introduce the new brand to the world, a seven-story mural was installed on the side of a building on Houston Street and Broadway, at the entrance to New York's SoHo shopping district. The mural, painstakingly hand-painted, depicted a bird's-eye view of all of the city, with a supersized Statue of Liberty triumphantly foregrounding the scene—all contained within the frame of the brand's four letters: D-K-N-Y. Though the composition represented an impossible view, it also somehow captured the image and, more importantly, the idea of New York City that so many people already held in their mind's eye: while the Statue of Liberty is physically dwarfed by the New York skyline, symbolically and emotionally, it looms large over the city as a beacon of hope to the entire world. And just as Lady Liberty has beckoned out to the "huddled masses yearning to breathe free" (as immortalized by Emma Lazarus's poem "The New Colossus," which appears on the statue's pedestal), so too was DKNY conceived of as something of a lodestar for a new generation of women: a symbol of their hopes, dreams, and aspirations, and an invitation into a better and more equal world.

The mural immediately became a New York City landmark in its own right, and immediately elevated the new DKNY logo into one of the most recognizable brand identities in the city, the country, and, arguably, the world. The mural stayed in place for over twenty years, ultimately becoming as much a part of New York as the famous cityscape that it depicted.

DKNY
DOWN TOWN
UP TOWN

DKN

DKNY
SHOW A HIT

DKNY

ESSEX

October 27, 1988

Dear Peter,

How can I ever thank you enough for making DKNY bigger than the sky? Peter, you constantly amaze me, and continue to inspire me to strive to new dimensions.

All my love,

Donna Karan

DK:bw

Mr. Peter Arnell
Arnell Bickford Associates
100 Grand Street
New York, New York 10013

DKNY COVERINGS

1990

Following the highly successful launch of the DKNY ready-to-wear line, a wide range of opportunities to extend the brand's reach and bolster its cachet were immediately explored. As part of this rapid expansion effort, a schematic diagram of a woman was created with annotations noting the different areas of her body that represented a potential new market for the brand: among them accessories, footwear, eyewear, and, of course, hosiery.

Having previously partnered with Sara Lee Hosiery to create Donna Karan Hosiery, DKNY Coverings was similarly positioned as an accessible entry point to the DKNY brand for every woman. The ubiquity of hosiery and the prime real estate that most department stores afforded it made DKNY Coverings displays, in effect, brand billboards that could capture the eye of every shopper coming into the store, reaching a vast, untapped customer base who might otherwise not have come into contact with the brand.

Like the Donna Karan Hosiery line before it, provocative and arresting black-and-white photographs of both bare and hosiery-covered legs made the product all the more desirable by minimizing its presence. The casual, unstudied style of the communications and the informal, conversational language on the packaging represented a dramatic contrast to the contrived, formal way that the hosiery was typically marketed, and helped to further cement DKNY's status as a young, fresh, and innovative brand.

DKNY
COVERINGS

DKNY JEANS

1990

The exploding popularity of designer denim in the early 1990s represented the perfect opportunity for DKNY to expand the brand's appeal to new market segments and younger generations, while still offering elegant, high-quality casual items that would be embraced by DKNY's existing clientele.

DKNY Jeans was launched with a campaign that captured the relaxed, but still elegant look that had already made DKNY so appealing to so many different women. In an iconic shot, supermodel Christy Turlington is shown wearing a pair of jeans while relaxing in a bathtub: a nod to the common practice of breaking in jeans that implicitly promised a perfect fit. The jeans themselves held tight to the brand's New York-centric ethos with simple, chic designs—and even sported buttons detailed to appear like miniature New York City manhole covers.

DKNY
ASAP
DKNY
ASAP

DKNY ASAP

1990

DKNY ASAP was a pilot program designed to bring working women in Manhattan critical wardrobe pieces in under two hours. The thousands of bike messengers who already blanketed the streets of the city during working hours provided a ready-made distribution network: but instead of the usual packages, letters, and documents, they would carry wardrobe essentials for the woman who didn't have time to get to the store during her busy work week.

Decades before the advent of online shopping and same-day or even one-hour delivery, DKNY ASAP anticipated the on-demand world that was to come and understood the value of a streamlined shopping and distribution model that delivered simple, high-quality, perfectly-curated essentials to customers everywhere.

DKNY
WALK

DKNY FOOTWEAR

1990

For the launch of DKNY Footwear, "DKNY" replaced the "DON'T" in the familiar "DON'T WALK" electronic crosswalk signs at strategic locations throughout New York City.

The resulting "DKNY WALK" that flashed up on a red light was a clever, whimsical, even slightly disorienting visual pun that captured the attention of thousands of pedestrians waiting at some of the city's busiest intersections.

This activation was yet another instance of the DKNY brand being embedded into the fabric of the city itself—as much a part of New York as any towering building, iconic landmark, or humble and ubiquitous piece of infrastructure.

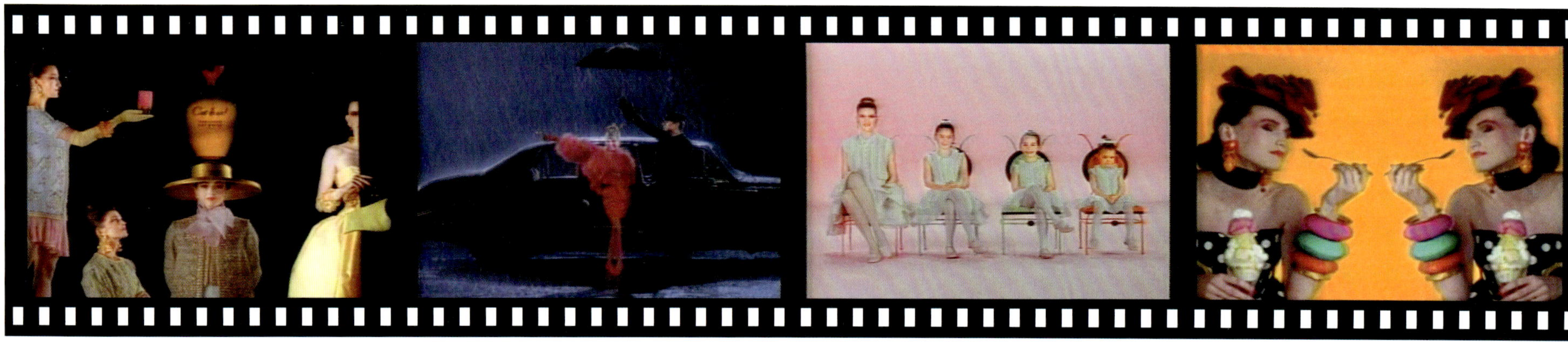

C'est la vie!
le parfum de
CHRISTIAN LACROIX

CHRISTIAN LACROIX PARFUM

1990

When Christian Lacroix decided to launch his first range of fragrances—Christian Lacroix Parfum—he approached Peter to help develop every aspect of his new fragrance C'est La Vie!

Since their first meeting in 1987, the two had remained close friends and collaborators, and there were few other individuals to whom Lacroix would have entrusted this all-important extension of his tightly controlled brand. Jean-Jacques Picart, the legendary fashion executive who had brokered the creation of Lacroix's haute couture maison with LVMH, enthusiastically championed Peter's involvement in the development of the new fragrance's brand identity and marketing, having been particularly impressed with his vision for both Donna Karan and DKNY.

From the fragrance's identity—an elegant handwritten representation of Lacroix's initials—to the sumptuous and whimsical launch communications, to the exuberant TV spot inspired by the classic 1957 Audrey Hepburn film *Funny Face* and its inedible musical number "Think Pink!," the launch of C'est La Vie! successfully brought to life the distinctive sense of joy and exuberance that the Lacroix brand had come to represent and made it accessible to women everywhere.

VIS à VIS

BYBLOS

1992

Founded in 1973 by Sergio and Donatella Girombelli, by the late 1980s Italian clothing brand Byblos had built a devoted following who were drawn to its chic, youthful styles, bold use of color, pattern, and print, and luxurious embroidery and detailing. As the launching pad for a generation of highly influential Italian designers, including Gianni Versace, and later Alan Cleaver and Keith Varty, Byblos was deeply linked with the style of its home country: a modern, expressive, and seductive style rooted in a palette of rich cultural influences.

To prime the brand for international expansion, a comprehensive new brand identity system was developed to perfectly capture this spirit of passion and vitality: a bold new logo mark, proprietary patterns inspired by the art, architecture, and nature of Italy, and vivid, joyful, evocative communications. And in a clever merging of brand identity and brand voice, the Byblos "B' logo mark also served as an irresistible imperative to customers: B young; B happy; B sexy; B Byblos.

And when Byblos began exploring the launch of a more accessible secondary line, it needed a name that captured all of the passion, vitality, and beauty that Byblos was known for. The essence of Byblos was all about enjoying life and being present and engaged with those around you: living life fully, unapologetically, and face-to-face. And with that, Byblos Vis à Vis was born.

O, WONDERFUL!

1992

To celebrate the 1992 Summer Olympics in Barcelona, global sponsor Bausch + Lomb commissioned Peter to document Olympic hopefuls at their training facilities around the world in the months and weeks leading up to the games. As a Worldwide Olympic Partner, Bausch + Lomb wanted to create a commemorative program that offered something more enduring and emotional than the usual battery of merchandise and commercial tie-ins, something that actually captured the true Olympic spirit and honored the higher ideals of the Games.

The series of photographs that resulted vividly captured these extraordinary young athletes pushing themselves to the very limit of human achievement—the endless hours of practice, the incredible individual effort, the concentration, the drive, and the indomitable spirit of these young men and women who would stop at nothing to achieve greatness. Traditionally, Olympic photographic reportage had focused on a singular moment of glory or the achievement of perfection, showing athletes as almost superhuman entities. These images chose instead to show the agonizing hard work required of every single Olympian, and the irrepressible sense of hope that propelled them.

At once larger than life and deeply humanizing, this collection of striking images was featured in a traveling exhibition, *The Magic of the Olympics,* as well as in an accompanying exhibition catalogue that was produced by the legendary fine art printer Sidney Rapoport, renowned for his peerlessly meticulous work for iconic photographers like Robert Frank and Richard Avedon. Titled *O Wonderful, wonderful and most wonderful, wonderful! And yet again wonderful ...*—a line made famous by Shakespeare's *As You Like It*—the publication offered an unflinchingly intimate portrait of these incredible athletes, and captured the enormous labor and tireless commitment that brought them to their moment of glory.

WHAT DO I WEAR IN BED?
WHY, CHANEL NO.5, OF COURSE.

CHANEL NO. 5

1994

By the early 1990s, the prestige of Chanel No. 5 had begun to wane from over-distribution and inconsistent marketing efforts. Alain Wertheimer, the owner of Chanel, and Arie Kopelman, the president and COO, sought out a fresh set of eyes to look at the iconic fragrance in a new way. In order to bring back the elegance and exclusivity of the nearly century-old luxury eau de parfum, a bold new campaign was devised that leveraged the brand's heritage and pop culture iconography to revamp the product for a new generation of women.

Generations of women had made Chanel No. 5 their signature scent, but perhaps none so memorably as Marilyn Monroe, who, when asked what she wore to bed, famously quipped "Why Chanel No. 5, of course!" Building off of that indelible association between the two icons, a technologically groundbreaking TV campaign was devised that revived Monroe as an endorser for her preferred scent. Working in partnership with the legendary filmmaker and photographer Jean-Paul Goude, the spot featured the model Carol Bouquet (a mainstay in Chanel fragrance campaigns since the 1980s) "morphing" into Monroe via CGI technology. The spot opens in black and white as Bouquet, eating popcorn, watches Monroe singing on screen in a film. Suddenly, through the morphing technique, she finds herself transformed into Monroe. The spot then turns to color, and her popcorn box becomes a giant Chanel No. 5 bottle. A voice that sounds like Monroe's breathily whispers "No. 5" and the words reverberate throughout the theater. A painstaking process that involved poring over thousands of hours of footage of Monroe to extract single frames, one by one, the overall effect was shockingly true to life and transcended what could have become a gimmick: Bouquet *was* Marilyn.

A subsequent print campaign shot by renowned photographer Ellen von Unwerth followed up the success of the Marilyn spot with a series of images that featured Bouquet, dressed in pearls and an elegant little black dress, relaxing in Coco Chanel's apartment at 31 Rue Cambon. Chanel's apartment—pristine, elegant, and refined—pointedly contrasted with Bouquet's more casual, informal, and relaxed bearing, yet taken together, the tension pointed to a new direction for the Chanel brand: respectful, and even reverent, of the past, but thoroughly modern in outlook.

3
2
1

N°5
CHANEL
PARFUM
N°5

N°5
PARIS

CHANEL

CHANEL ALLURE

1995

In the mid-1990s, as Chanel began to develop a new fragrance offering—its first in nearly eighty years—their ambition was to create a scent that would capture a more relaxed type of elegance to appeal to chic and modern young women everywhere. The resulting scent was crafted as a sexy, sophisticated, youthful fragrance, with a lighter, fresher character that blended florals, citron, and vanilla.

Working in close collaboration with Arie Kopelman, president and COO of Chanel, and Jean Zimmerman, executive vice president, the name "Allure" was ultimately chosen for its obvious connotations of desire and sensuality, and for its suggestion of a mysterious quality that transcended physical beauty. A sleek new bottle was also designed, taking many of its cues from the iconic Chanel No. 5 bottle, which it would be appearing next to in perfume counters and boutiques worldwide, while still being distinctive enough to be instantly identifiable as something new and distinct.

A striking campaign was also crafted in collaboration with the legendary American photographer Herb Ritts. Featuring many of the top supermodels of the 1990s, the campaign captured a striking beauty and elegance that was also inflected with a youthful, playful, unstudied air: a new vision for Chanel for a new generation of women.

ALLURE
CHANEL

Be enriched by the simplest things.

The smell of fresh laundry. The lick of a dog.

The lazy freedom of a cotton sweater.

The sound of rain. The old movie you saw last night

when you could not sleep. Life is made of experience.

Weave yourself a lifetime.

BANANA REPUBLIC

BANANA REPUBLIC

1995

Banana Republic was originally built around a travel and safari-themed brand story, selling products that were durable, functional, and decidedly utilitarian. Stores were filled with props like palm fronds, elephant tusks, army jeeps, tin shack roofs, and life-sized model giraffes, and while this approach represented a highly distinctive brand vision, it tended to feel gimmicky, and arguably detracted from the classic, functional, high-quality clothing that the retailer had staked its reputation on.

After being purchased by The Gap—then headed by the legendary fashion marketer Mickey Drexler—the brand's relevance was waning, and it needed to combat perceptions of being niche and find ways to appeal to a far broader range of customers. To accomplish this, Banana Republic was completely reimagined and repositioned as an elegant and contemporary lifestyle brand. "True outfitters of modern American style" became the brand's guiding ethos, and this new style immediately made Banana Republic freshly relevant to broad swaths of shoppers who wanted beautiful, timeless, high-quality wardrobe staples. To reintroduce the radically transformed brand, a strategy was needed to bridge the gap between these two seemingly disparate chapters in Banana Republic's history. Ultimately, the through-line that emerged was simple, intuitive, and completely authentic to the brand: the spirit of adventure. In its previous incarnation, "adventure" had more literal connotations: far-off frontiers, distinct horizons, and wild terrain. Now, the sense of adventure that the brand represented was more urbane, more subtle, and more poetic: exploring a new city; starting a new job; falling in love.

The communications strategy meanwhile was built around the idea of "urban adventure"—a conceptual bridge between the brand's new attitude and its original essence that maintained the core values of the Banana Republic brand while allowing the spirit of adventure to come alive in a new way. Simple imperatives like "Live an extraordinary life" and aphorisms like "Anything can happen in a white shirt" encouraged customers to transcend the everyday and open their eyes to the beauty, joy, and sense of adventure that existed all around them.

To complete Banana Republic's transformation to a lifestyle brand, a Bath & Body line was subsequently launched with a series of elegant, black-and-white communications that featured the clever, tongue-in-cheek directive to "Just Add Water."

Anything can happen
in a white shirt.
BANANA REPUBLIC

BANANA REPUBLIC

Live an extraordinary life.
BANANA REPUBLIC

All you need for adve ure
is the desire to have o
BANANA REPUBLIC

BATH SALTS 12.5 OZ $16.00
CONDITIONER 8 OZ $8.00
SOAP 3.5 OZ $3.50 7.5 OZ $6.50
BODY LOTION 6.75 OZ $14.00 13.5 OZ $22.00
SHAMPOO 8 OZ $8.00
PUMICE STONE $8.00, SEA WOOL SPONGE $16.00, BODY BRUSH $28.00
SHOWER AND BATH GEL 8.5 OZ $14.00
CREME 6 OZ $16.00
$12.00 13.5 OZ
Just Add Water

Just Add Water
BANANA REPUBLIC

RAY-BAN

1995

Ray-Ban, the American eyewear brand created by Bausch + Lomb in 1936, had for decades been a fashion staple, with its iconic styles (Wayfarers and Aviators in particular) seeming to remain perennially in vogue. But by the late 1990s, they were beginning to lose ground to other brands that were seen as younger, more fashionable, and higher-end.

To combat this perception, a comprehensive creative and marketing strategy was developed, along with a corresponding series of print communications featuring stylish young people sporting their Ray-Bans. Photographed and art directed by Peter personally, the ads urged consumers to "Take a Good Look," a sly double entendre that quietly acknowledged the fact that many consumers at the time might have overlooked Ray-Ban, while also endowing the brand with a bold, brazen, sexy attitude.

This new approach helped to turn Ray-Ban's fortunes around, and catapulted the brand back to the forefront of fashion and relevancy.

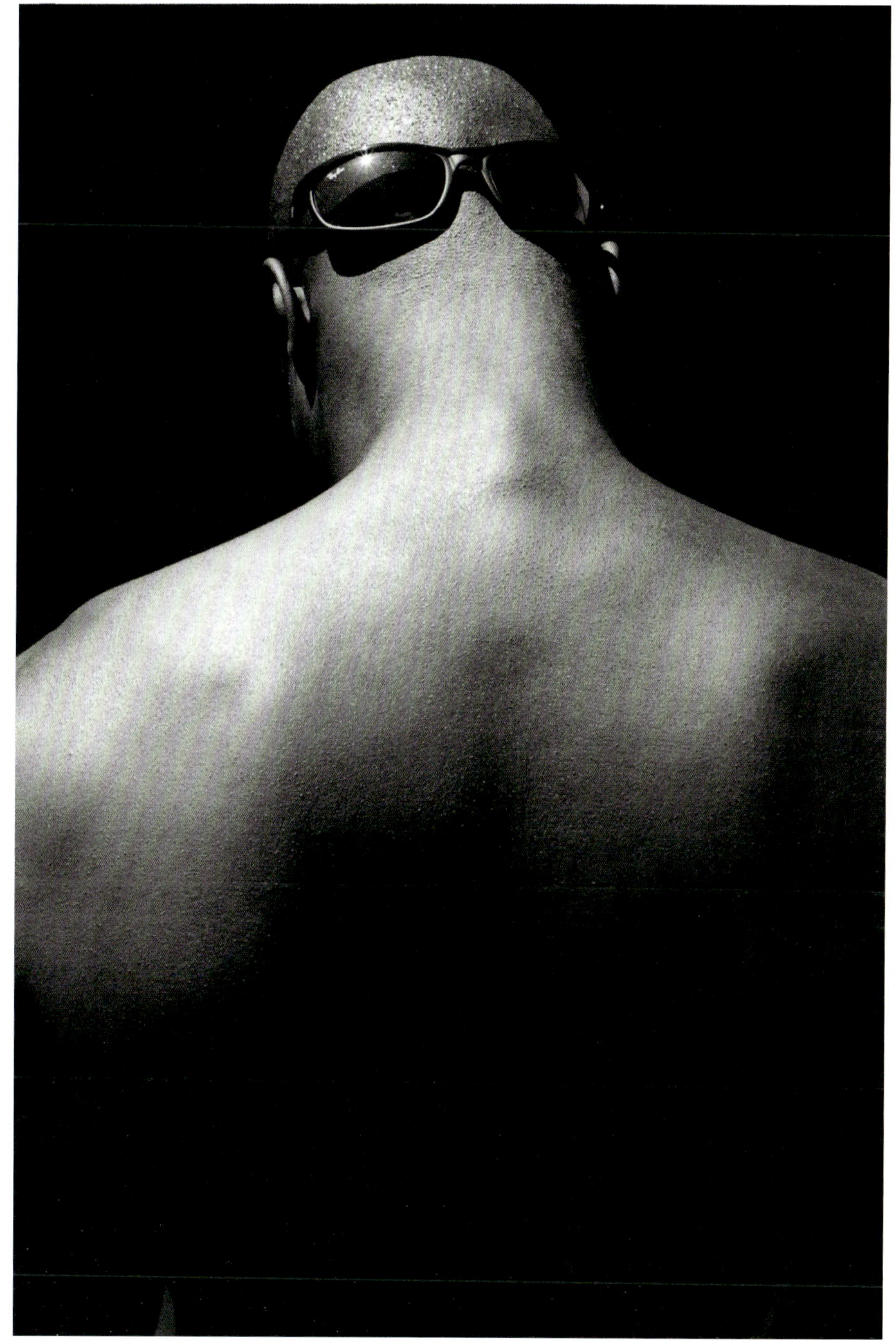

HANES

1996

By the mid-1990s, Hanes understood that in order to connect with the modern woman and make their hosiery products appealing and relevant, the conversation around hosiery had to be entirely reengineered. Rather than focusing on the usual tropes of support, control, and the erasure of perceived "flaws," their new range of hosiery would be called Resilience—a testament to the strength of the product, and of the women who wore it. To launch this new line, a comprehensive marketing strategy was built around a true icon who was herself resilience personified: Tina Turner.

A monument to perseverance, and a living testament to overcoming the direst of circumstances, Tina glowed with a love of life. Working closely with Cathy Volker, the president of Hanes Hosiery, Roger Davies, Tina's agent, and Erwin Bach, Tina's husband and a music executive himself, a total brand activation was scripted out, with Tina at the center as the living, breathing embodiment of strength and resilience for women of all ages and backgrounds.

In a sweeping TV spot created to launch the new product range, Tina danced and strutted through the streets, radiating joy and confidence, her world-famous legs front and center. It was all about getting out of the way and letting Tina shine: her energy, her vitality, her wisdom, her grace, and her magic poured out before the camera. And when, during filming, she made the offhand remark "Resilience: It's all about strength and beauty," in an instant, she captured the true essence of the product because she was speaking from a place of true conviction and profound lived experience.

As Tina subsequently embarked upon her Hanes-sponsored "Wildest Dreams" concert tour, she partnered with Oprah Winfrey to create the Wildest Dreams initiative: a charitable endeavor that donated funds to women all over the country who, like Tina, had beaten the odds and fought back against life's toughest challenges. To celebrate the strength and resilience of these extraordinary women, Oprah and Tina traveled all across the country to make their "wildest dreams" come true.

This collaboration also marked the beginning of a cherished decades-long friendship between Tina and Peter: an unlikely pair who are nonetheless united by their zest for life, relentless determination, and utterly indomitable spirits.

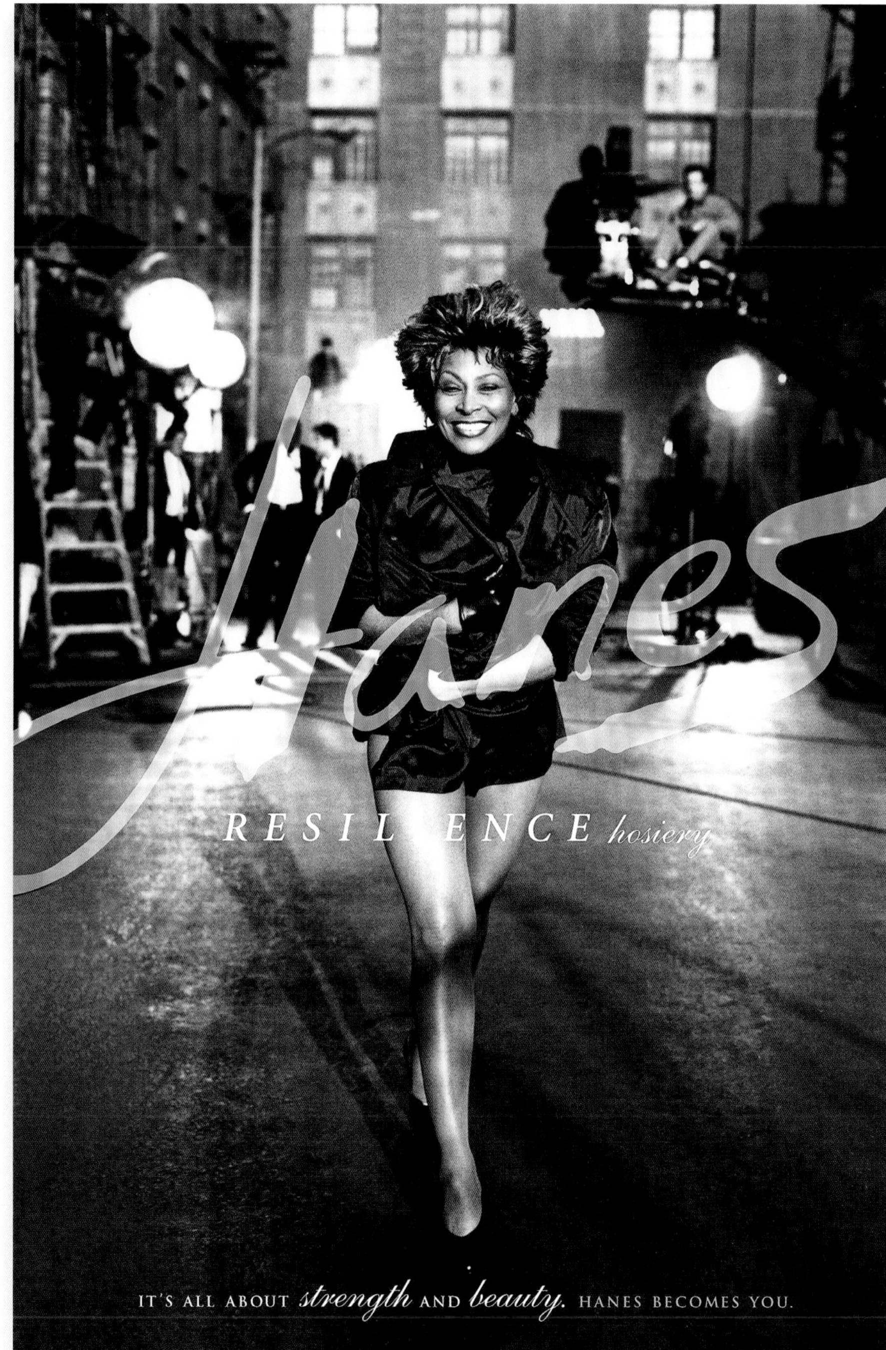

Simply the best, better than all the rest *Hanes*

SAMSUNG
microwaves
vegetables cooked in
a microwave retain more
vitamins than with
any other form of cooking.
1 800 so simple.
simply samsung.

SAMSUNG

1996

In the mid-1990s, Samsung embarked upon a massive global expansion with their range of well-designed, affordable consumer electronics and home appliances. Up until that point, Samsung had been a marginal player in these spaces, selling solid but unremarkable items with generic designs. They needed to make a fresh statement about who they were as a brand in order to reach a mass consumer base for the first time, and they needed customers to see their product through new eyes. The Simply Samsung campaign was developed to introduce (or reintroduce) Samsung as an exciting and dynamic consumer-facing brand.

Samsung's new range of products were highlighted with striking communications that showed a more human and approachable side of these products that were an indispensable part of modern life. Provocative, witty, elevated, and even sexy, these images alluded to the unexpected intimacy that exists between people and the functional objects that they rely on in their daily routines. Research had revealed that the gift of choice from parents to their college-bound kids was a microwave, but to gain the attention of this demographic, you had to speak their language: which, in the context of the mid-1990s zeitgeist, was unbridled sex appeal.The so-called Microwave Man—an extra-fit young man with a Samsung microwave tucked under his arm—transformed the utilitarian appliance into a genuine object of desire, albeit with a wink and a nod. It was unexpected, subversive, and instantly memorable. At the bottom of the now-iconic billboard, there were just two words: "Simply Samsung." Nothing more needed to be said.

In a market segment filled with needlessly complex and frustrating products, the "Simply Samsung" tagline promised ease, simplicity, and intelligent solutions to the problems that consumers most urgently needed solved. Following the success of the Microwave Man ads, a blank check was given to engineer similarly bold activations for the remainder of Samsung's consumer electronics division: TVs, VCRs, video cameras, computers, printers, fax machines, and cell phones—the latter of which in particular marked the beginning of Samsung's extraordinary ascent. By explicitly showing technology as something that's deeply personal, the Simply Samsung campaign anticipated the ever-growing degree to which technology would become embedded in the most intimate aspects of our lives, and sought to make it more familiar, more approachable, more desirable, and more human.

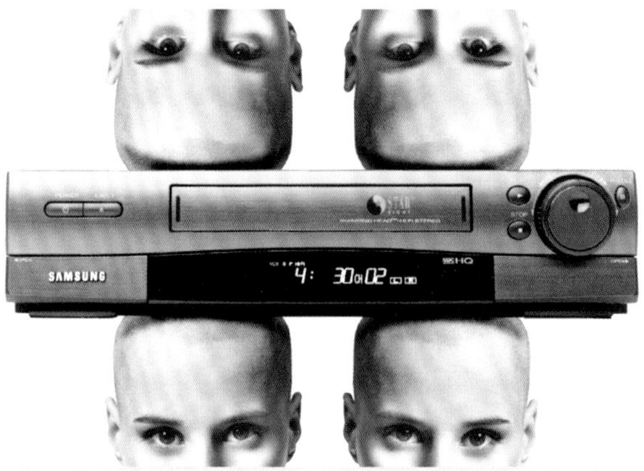
SAMSUNG

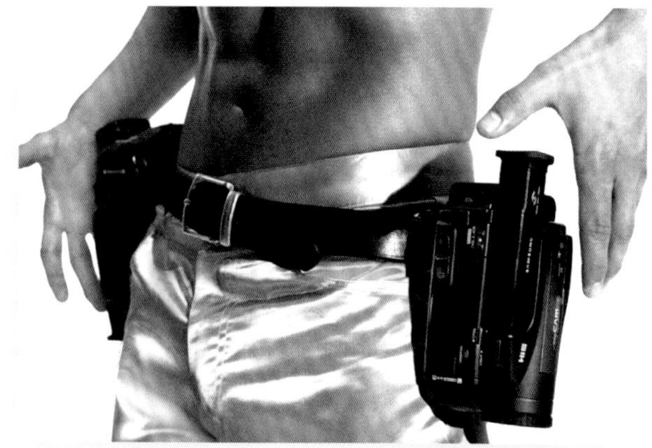

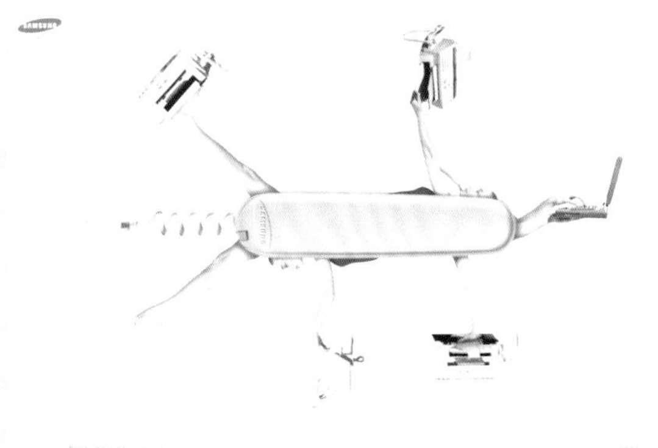

S
im
ply
perfect
vision

If The Phone Fits, Wear It
simply the perfect accessory. Introducing Samsung's newest digital wireless phone with voice activated dialing. You say the word and the phone dials the number, letting every name ring a bell. The SCH-2010 also features voice memo, 10 different ring options including a silent vibrate alert, long life battery, and caller ID so you can see who's trying to reach you. Samsung offers solutions to help simplify your busy life. For more information, call 1 800 SAMSUNG or visit our web site at www.samsungtelecom.com simply connected. simply samsung.
SAMSUNG

Home Away From Home
simply the next best thing to being there. Introducing Samsung's newest digital wireless phone with voice activated dialing. You say the word and the phone dials the number, letting every name ring a bell. The SCH-2010 also features voice memo, 10 different ring options including a silent vibrate alert, long life battery, and caller ID so you can see who's trying to reach you. Samsung offers solutions to help simplify your busy life. For more information, call 1 800 SAMSUNG or visit our web site at www.samsungtelecom.com simply connected. simply samsung.
SAMSUNG

For Stepping Out
simply the way to go. Introducing Samsung's newest digital wireless phone with voice-activated dialing. You say the word and the phone dials the number, letting every name ring a bell. The SCH-2010 also features voice memo, 10 different ring options including a silent vibrate alert, long life battery, and caller ID so you can see who's trying to reach you. Samsung offers solutions to help simplify your busy life. For more information, call 1 800 SAMSUNG or visit our web site at www.samsungtelecom.com simply connected. simply samsung.
SAMSUNG

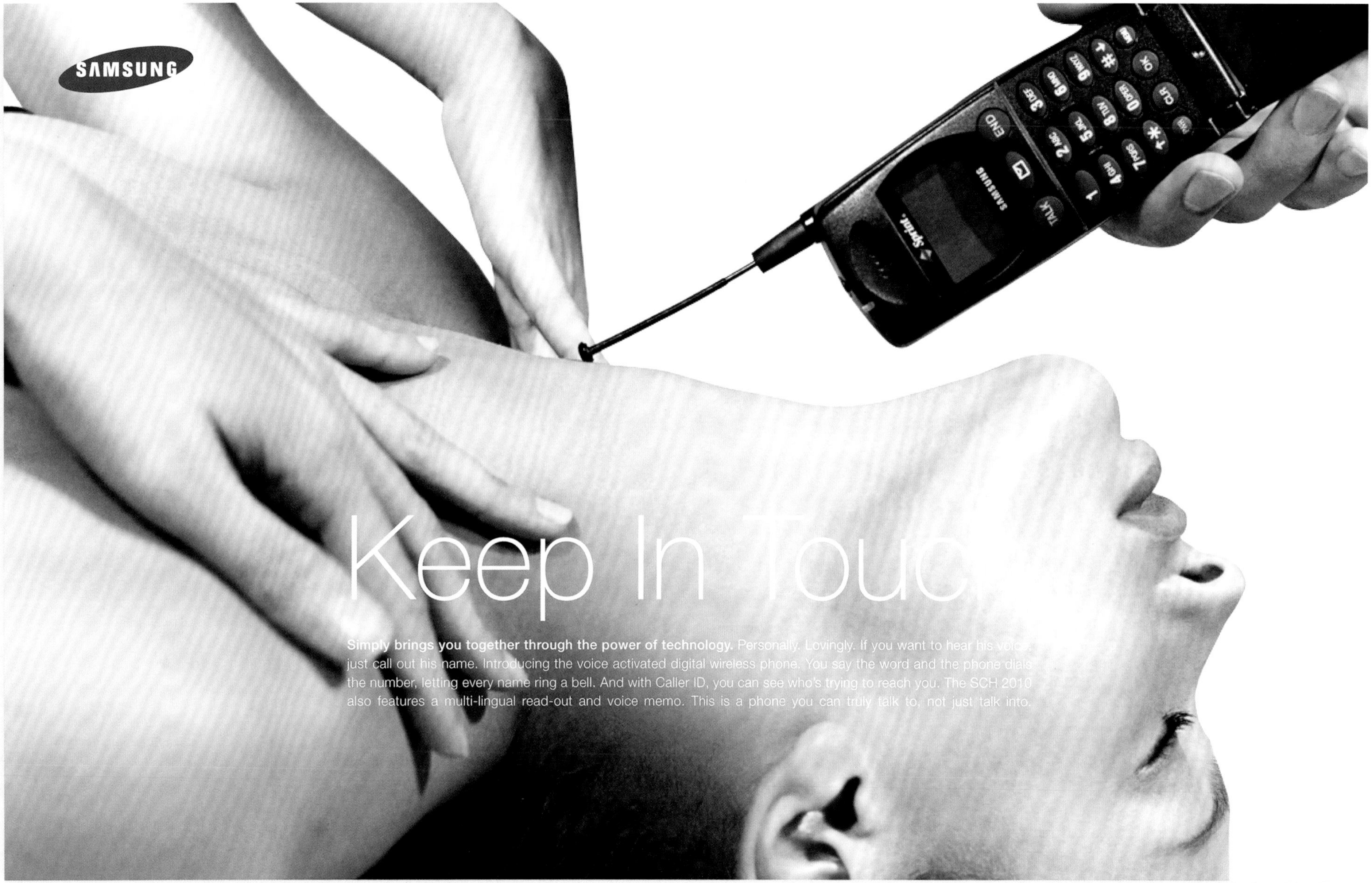
SAMSUNG
Keep In Touc
Simply brings you together through the power of technology. Personally. Lovingly. If you want to hear his voic
just call out his name. Introducing the voice activated digital wireless phone. You say the word and the phone dials
the number, letting every name ring a bell. And with Caller ID, you can see who's trying to reach you. The SCH 2010
also features a multi-lingual read-out and voice memo. This is a phone you can truly talk to, not just talk into.

SAMSUNG GXTV

1996

The Samsung GXTV was created to meet the specialized needs of the rapidly expanding video gaming market of the mid-1990s. The GXTV, one of the first products designed by Arnell Group's in-house industrial design team, featured full stereo sound with two foldout speakers on either side of the screen for an immersive audio experience aimed directly at the player. The unusual retro-futuristic look of the device not only garnered attention, but also served as confirmation of the adage that form does indeed follow function.

Like the Simply Samsung campaign before it, launch communications for this novel product centered on the human aspect of the technology, showing the feeling and emotion of being fully immersed in a new digital realm, rather than fixating on specifications and arcane technical details.

With the subsequent explosion of Virtual Reality technology in gaming, this early championing of technology built to suit the ergonomics and scale of the gamer has only come to look all the more prescient. And although gaming has long since entered the mainstream, in the mid-1990s, the idea that it could be sleek, cool, and adult was nothing short of radical.

NESPRESSO

NESPRESSO

1996

Nespresso, Nestlé's single-serving capsules, were developed to provide a premium, barista-prepared level of quality with the ease, convenience, and consistency of an automated coffee machine. Yet, unlike those systems—which largely produced low-quality, watery brewed coffee—Nespresso was a true premium product, and it needed a comprehensive strategy to introduce itself as a new paradigm in coffee-making.

To introduce Nespresso to the North American market, a campaign was devised that focused on the product's most important point of differentiation: its rich, caramel-colored, foamy topping (or crema) that previously could only be produced by a highly trained barista using a traditional hand-pulled espresso machine. This beautiful crema was prominently featured in communications to capture the richness of a sensory experience that only Nespresso could provide. The color-saturated close-up images evoked the distinctive experience of drinking a perfect cup of coffee and immediately announced Nespresso as a true premium product unlike any other. A series of collages utilizing warm, natural, highly tactile materials were also created as a way of further emphasizing the exceptionally high-quality product that could be had with just the single push of a button.

To this day, the rich, swirling, umber-colored crema is the centerpiece of most Nespresso communications, and has become, along with the thimble-sized metallic pods, an iconic expression of this truly singular product.

Peter

This is the one

Paul

Thanks for the Michelangelo
and everything else.
Keep in touch.

PAUL RAND & APPLE

1997

For Apple's iconic Think Different campaign, conceived by advertising legend Lee Clow and his team at Chiat\Day, a photograph that Peter had taken of the late Paul Rand—the godfather of modern logo and brand identity design—was selected for inclusion in the campaign.

Jobs considered Rand a personal design hero, and revered the sleek, spare, modernist logos Rand had designed for companies such as IBM, UPS, and ABC—he had even commissioned him to design a new logo for NeXT, the computer and software company that he had founded in 1985. The emphasis on detail, precision, universality, and the complete elimination of superfluous details that characterized Rand's graphic works are clearly echoed in every consumer product that Jobs developed over the course of his singular career.

What was particularly interesting about Rand's inclusion in the Think Different campaign was that he was one of the few figures selected who didn't enjoy near-universal name recognition: in the company of fellow iconoclasts and rule-breakers like Gandhi, Mother Teresa, Albert Einstein, Muhammad Ali, Alfred Hitchcock, and Picasso, Rand was a decidedly niche personality who was not widely known outside of the design world, and yet Jobs felt passionately that he deserved to be in the company of other legendary figures for his peerless contributions to the visual culture of postwar America.

Rand's widow Marion granted access to countless portraits taken over the span of Rand's six-decade-long career. Jobs, working with his creative partner Lee Clow, selected a photograph that Peter had shot during informal visits to Rand's Connecticut studio. The image showed the legendary designer looking up from his drafting table directly into the camera, with a twinkle in his eye and a mischievous grin spreading across his face.

In a subtle twist of irony that Rand himself no doubt would have enjoyed, the project that he is shown working on in the image selected for the Think Different campaign is actually packaging designs for IBM—Apple's longtime arch-rival.

inside every woman is a star

IS**C BY ISAAC MIZRAHI

1997

After presenting his first collection in 1987 at a Bergdorf Goodman trunk show, Isaac Mizrahi was catapulted into the global spotlight as one of the most notable young designers of his generation. His namesake brand was highly sought after for a signature blend of glamour, elegance, and edginess, a reputation that helped him become the quintessential New York City designer for an entire generation. His designs provided elevated, impeccable basics, but with a younger twist.

Following the partial acquisition of Mizrahi's brand by Chanel in 1992, the time was ripe to build on the brand's breakthrough success with the introduction of a more accessible secondary line. Having previously worked together on Chanel's No. 5 and Allure fragrances, and being a great admirer of his successes with Donna Karan and DKNY, Arie Kopelman enthusiastically recommended Peter to Mizrahi to help shape and launch his new brand.

The challenge facing Mizrahi was the same one that faced every so-called diffusion line: how to translate the DNA of the main line to a more mass-market context without diluting the brand's essence or diminishing its exclusivity. The solution to this conundrum was a simple but counterintuitive gesture: a self-inflicted act of brand effacement. The new line would be called IS**C—the designer's name with the two "As'" struck out in the manner of an expletive. By preemptively defacing the brand name, IS**C immediately announced itself as distinct from its parent brand: a younger, edgier, more irreverent take on Mizrahi's signature style.

The accompanying tagline "Inside Every Woman Is a Star" affirmed that the new brand was designed for women from all walks of life and encouraged them to lay claim to their own unique beauty and vibrancy—on their own terms.

IS★★C

U BY EMMANUEL UNGARO

1997

Throughout the 1980s and 1990s, Emmanuel Ungaro had built his reputation as a designer who created clothing that accentuates women's beauty, femininity, and sensuality. While the ideas of seduction and attraction were integral to the Ungaro mystique, in considering the potential direction of a more accessible secondary brand, these core brand tenets needed to be gently evolved to meet the needs of a wider range of customers.

Following several collaborations on Ungaro's namesake brand, a close professional as well as personal relationship had emerged, and the visionary designer entrusted Peter to create a compelling concept for a this new line. U by Emmanuel Ungaro was developed to hold up a mirror to the woman wearing the clothes: reflecting back her style, her character, and her priorities. What Ungaro had always instinctively understood was that a woman's clothing should be a means to enhance her own beauty, personality, and identity.

As a symbol, "U" clearly represented balance, harmony, and openness; but it also established an intimate and personal connection with the woman wearing the clothes. The new brand maintained the intimacy and mystique that Ungaro had carefully constructed over the course of his career, while placing the emphasis more firmly on personal expression.

PEACE

ESPRI

ESPRIT

1997

During the 1980s and early 1990s, clothing retailer Esprit was among the most sought-after brands in the world. Offering bold, youthful designs at an affordable price point, Esprit brought an elevated sense of design to a mass audience, not only through its fashions, but also with its now-iconic retail environments created by Italian architect Ettore Sottsass.

But by the late 1990s however, the brand had stagnated, and its bright, bold, branded designs had lost market share to other retailers whose more casual and classic "anti-branded" designs spoke to the spirit of informality and ease that many young consumers were seeking out.

Under the direction of president Cathy Volker, who had recently moved to Esprit from Hanes to help the brand regain its relevance with young consumers, a new campaign was devised that shifted the emphasis away from the clothes themselves and refocused attention on the joyful spirit that Esprit had always represented. Atop these images were simple and evocative words like "Love," "Shine," "Breathe," "Peace," and "Imagine"—with the "E" in each word shown as Esprit's instantly recognizable three-stroke "E" from its iconic logo.

The campaign helped Esprit to inaugurate a new phase in its storied history by showing a softer, simpler, more casual and lived-in side of the brand that would prove to be freshly relevant to consumers in the 1990s and beyond.

ESPRIT

LOVE

ESPRIT

LOEWE

1997

Renowned for their traditional handcrafted leather goods and meticulous attention to detail, venerable Spanish luxury house Loewe had long been a distinctive niche brand beloved by a small but devoted core customer base. And while the story of their quality and craft was well known and well established, following its acquisition by LVMH in 1996, an injection of new energy was needed in order to broaden the brand's appeal and reach a wider audience as a truly global luxury house.

A series of evocative print communications was developed to help tell a different story about Loewe: a story about desire, yearning, and intimacy. These quiet but powerful images painted the venerable brand in a new and more sensual light, and created narrative, theater, and fantasy to elevate the appeal of the brand's exquisitely well-made products.

LOEWE
MADRID

LOEWE
MADRID

HENRI BENDEL

1997

For over a century, Henri Bendel, the veteran New York City luxury retailer, had been revered by discerning New Yorkers for its exquisitely well-curated range of luxury fashion brands, along with its own famously high-quality private label range.

Renowned for its cashmere sweaters and coats, the Henri Bendel Collection had long been a staple in the wardrobes of well-heeled New Yorkers, but over time, the venerable retailer started to take notice of the many younger customers coming into the store looking for the simple, elegant, high-quality knitwear pieces that had become an open secret among the city's fashion crowd.

To more fully embrace and appeal to this expanded audience, a new brand identity, retail design, merchandising strategy, and communications were developed to show the Henri Bendel Collection in a fresher, younger, more casual light—while still emphasizing the exceptionally high quality and understated beauty of the garments themselves.

The campaign showed a different side of Henri Bendel, while also challenging prevailing ideas of what true luxury is. By showing women in real-life scenarios, relaxed and unguarded, customers were invited to imagine a new definition of luxury, one that was rooted in simplicity, ease, and comfort.

HENRI BENDEL
NEW YORK

MISTY HARBOR

1997

To catch the attention of consumers and distinguish the brand from its more established competitors, a communications campaign was developed for the Misty Harbor line of rainwear that provocatively reframed a rainy day as something to be looked forward to.

Rather than focusing on protection from gray-skies gloom, this campaign chose instead to focus on the peaceful, romantic, and even restorative qualities of rain. With the unexpected directive to "Pray for Rain," these ads cleverly subverted the ideas and expectations around inclement weather. Meditative, poetic, and serene images of rain-streaked windows and umbrella-choked sidewalks reminded the world of the special charms of a rainy day, transforming feelings of dread into feelings of anticipation.

Having pioneered the genre of product-less advertising with his work for Donna Karan and DKNY, Peter understood that the appeal of a brand had to be rooted in something more than the wares it was selling: it had to evoke a mood, an aura, an entire state of being that could take root in the imaginations of customers, living on beyond any one product or campaign. For Misty Harbor—a brand that existed to solve an unpleasant problem—this was doubly true, and, as such, the success or failure of the brand depended not on crafting an appealing image, or presenting a provocative idea, but rather, on engineering an enduring emotional state.

MISTY HARBOR
ORIGINAL

FILA
FILA
2
BASEBALL

FILA

1997

Founded in Italy in 1911, Fila had been a premier sportswear brand for decades, reaching its peak with the tennis craze of the 1970s. But by the 1990s, Fila had become largely overlooked in favor of its larger competitors, especially in the US market.

To effectively reintroduce Fila to a broad American audience, a spectacular series of large-scale outdoor murals were installed across New York City, featuring a roster of contemporary American sports heroes striking aggressive, heroic poses. The painstakingly hand-painted murals—a quintessentially New York public art form—featured athletes like Grant Hill, Lee Janzen, Derek Jeter, Mike Powell, and Jerry Stackhouse: a pantheon of modern-day greats.

This campaign immediately repositioned Fila as a true performance brand beloved by world-class athletes who demand a balance of functionality and style, both on and off the field.

FILA WOMEN

1997

For Fila's women's line, a campaign was devised that featured some of the most celebrated and accomplished female athletes of the day along with the simple but powerful invitation to "Change the Game." This message immediately announced Fila's new identity, mission, and principles, while also encouraging women everywhere to think about what they might do in their own lives—even beyond the realm of sport—to truly change the game.

Dramatic black-and-white photographs of these incredible women in action were accompanied by the clever and the evocative line "Designed for a Woman's Movement," which drew a clear line between the power, dedication, and accomplishment of these world-class athletes and the broader aims of women's empowerment. The result was an aspirational and empowering campaign that celebrated the unique power of women, not only in sport, but in all areas of their lives.

FILA
CHANGE THE GAME

FILA

PENNY BLACKWELL, RICHMOND, VIRGINIA. THE FILA CRC

URT TOP AND SHORT. THE FILA PUENTE CROSS TRAINER

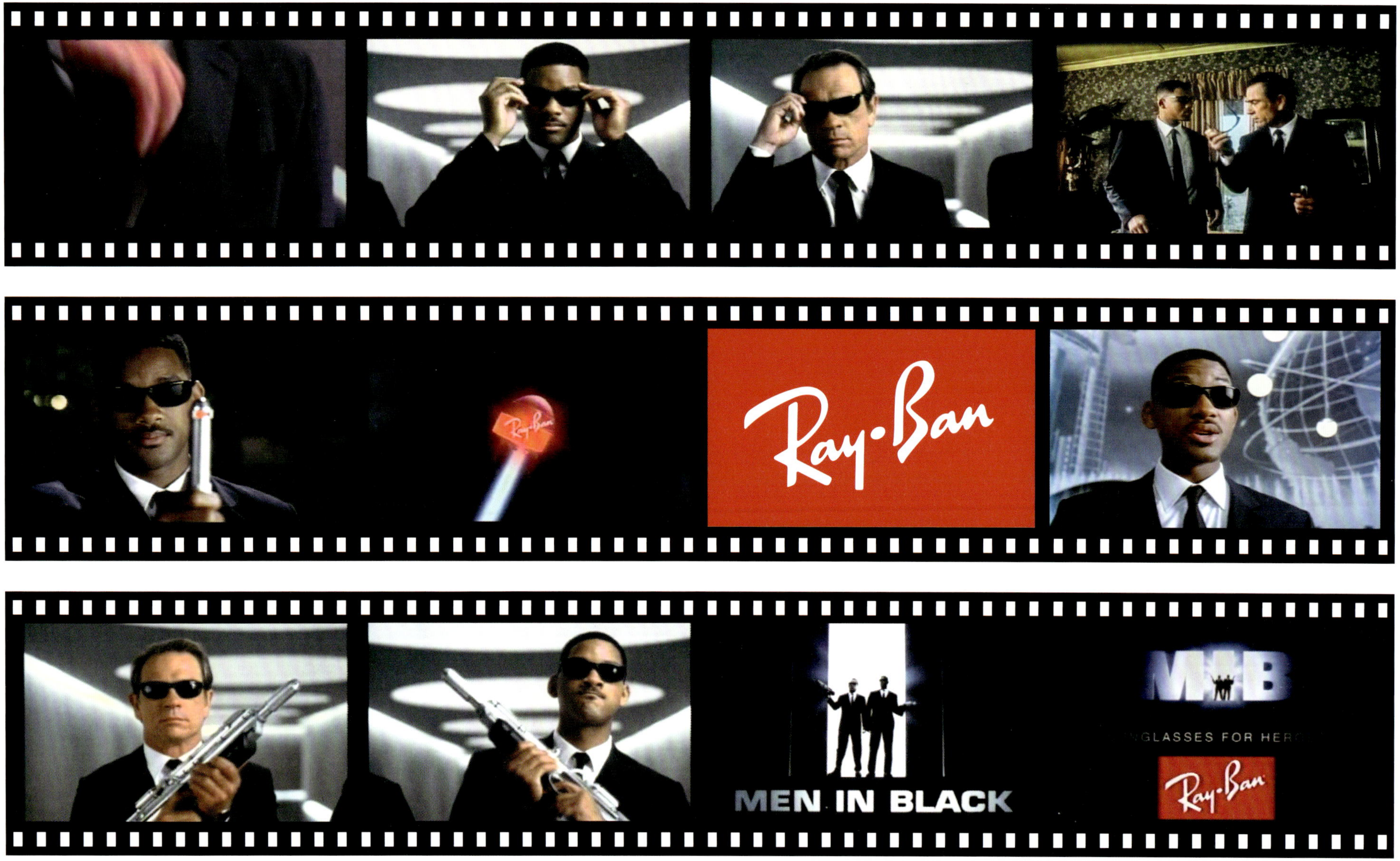
Ray-Ban
MEN IN BLACK
Ray-Ban

RAY-BAN MEN IN BLACK

1997

The 1997 partnership between Ray-Ban and *Men in Black*—the Will Smith and Tommy Lee Jones smash sci-fi comedy blockbuster—elevated corporate sponsorship and pioneered plot-based product integration by making Ray-Ban glasses an integral plot device that seamlessly fit into the film. While there was a long and established history of successful product placement in movies, the idea of integrating a brand in a way that organically complemented and enhanced the plot, dialogue, and emotional tenor of the film was entirely new. Working in close collaboration with the film's legendary producer Walter Parkes, the presence of Ray-Ban glasses in *Men in Black* was carefully scripted out to be a seamless addition that authentically belonged in the world depicted on-screen.

Prior to the release of the film, the Balorama style worn by Smith and Jones throughout the movie were generally less well-known than Ray-Ban's iconic Wayfarer and Aviator styles, but practically overnight, the sleek futuristic wrap-around style of the Balorama became the most sought after sunglass style of the late 1990s.

Unlike the clumsy, obvious, and unnatural product placement of the past, the glasses played a pivotal role in the plot, and had authentic meaning in the world of the film as both a part of the agents' uniforms and as a point of contention between the main characters, with Will Smith's character famously taunting his partner by declaring (of his Ray-Bans): "I make this look good."

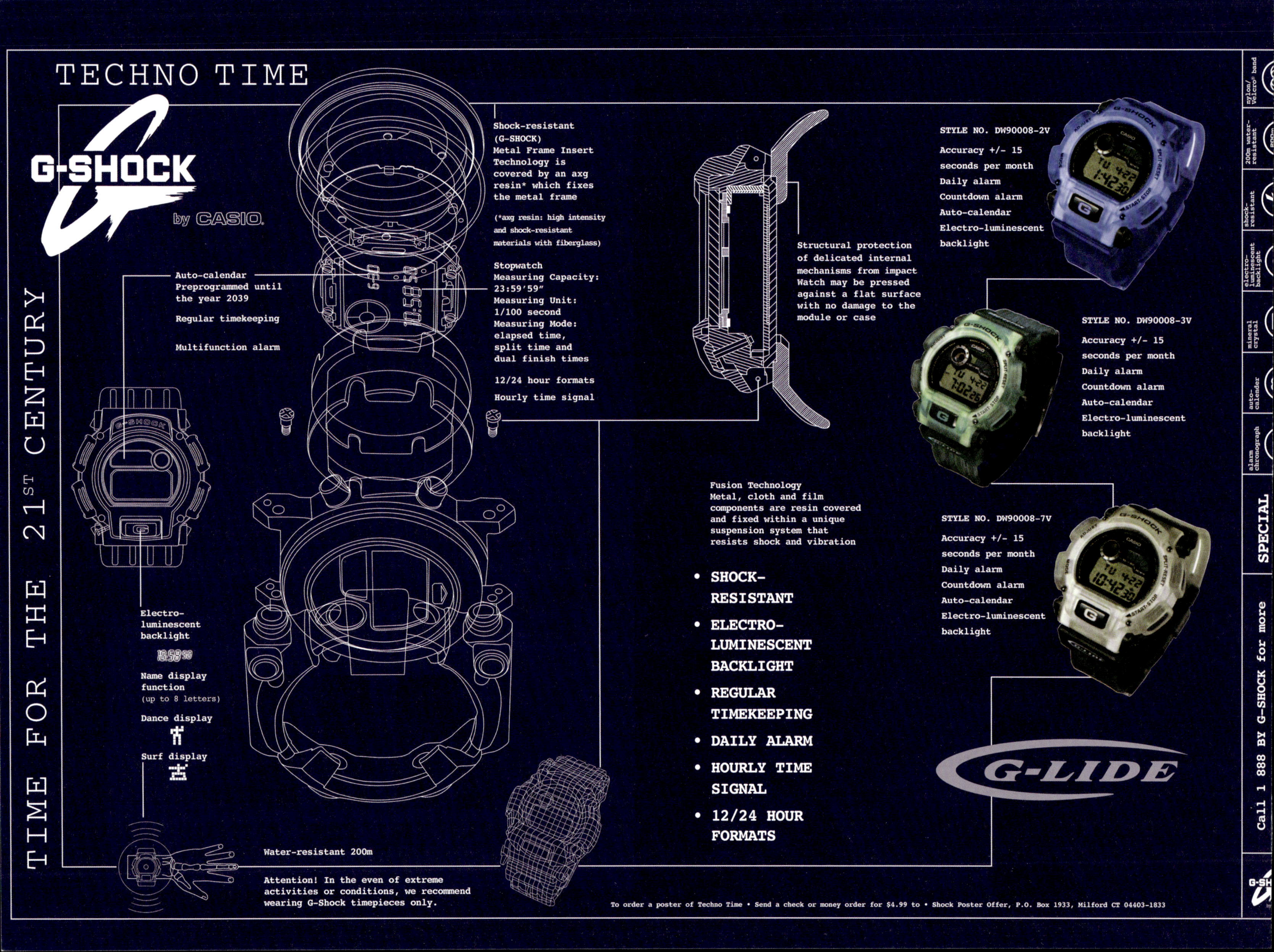

TECHNO TIME
G-SHOCK
by CASIO
TIME FOR THE 21ST CENTURY
Auto-calendar
Preprogrammed until the year 2039
Regular timekeeping
Multifunction alarm
Electro-luminescent backlight
Name display function
(up to 8 letters)
Dance display
Surf display
Water-resistant 200m
Attention! In the even of extreme activities or conditions, we recommend wearing G-Shock timepieces only.
Shock-resistant (G-SHOCK) Metal Frame Insert Technology is covered by an axg resin* which fixes the metal frame
(*axg resin: high intensity and shock-resistant materials with fiberglass)
Stopwatch
Measuring Capacity: 23:59'59"
Measuring Unit: 1/100 second
Measuring Mode: elapsed time, split time and dual finish times
12/24 hour formats
Hourly time signal
Structural protection of delicated internal mechanisms from impact Watch may be pressed against a flat surface with no damage to the module or case
Fusion Technology
Metal, cloth and film components are resin covered and fixed within a unique suspension system that resists shock and vibration
• SHOCK-RESISTANT
• ELECTRO-LUMINESCENT BACKLIGHT
• REGULAR TIMEKEEPING
• DAILY ALARM
• HOURLY TIME SIGNAL
• 12/24 HOUR FORMATS
STYLE NO. DW90008-2V
Accuracy +/- 15 seconds per month
Daily alarm
Countdown alarm
Auto-calendar
Electro-luminescent backlight
STYLE NO. DW90008-3V
Accuracy +/- 15 seconds per month
Daily alarm
Countdown alarm
Auto-calendar
Electro-luminescent backlight
STYLE NO. DW90008-7V
Accuracy +/- 15 seconds per month
Daily alarm
Countdown alarm
Auto-calendar
Electro-luminescent backlight
G-LIDE
nylon/ Velcro® band
200m water-resistant
shock-resistant
electro-luminescent backlight
mineral crystal
auto-calender
alarm chronograph
SPECIAL
Call 1 888 BY G-SHOCK for more
To order a poster of Techno Time • Send a check or money order for $4.99 to • Shock Poster Offer, P.O. Box 1933, Milford CT 04403-1833

CASIO G-SHOCK

1998

The G-Shock line of watches introduced by Casio in the mid-1990s were an unexpected runaway success from the moment they hit the market. With barely any traditional promotion or advertising, the tough, waterproof, shock-resistant digital timepieces loaded with performance-oriented features unexpectedly found mass appeal despite being designed specifically for high-performance athletes and extreme sports enthusiasts. Consumers of all types flocked to the rugged design, high-tech aesthetic appeal, and set of features not typically available in a mass-market timepiece. The unexpected popularity of the G-Shock products led to a surge of imitators, and with that, Casio saw an opportunity to solidify their dominance in the sector by emphasizing that their G-Shock watches were first and foremost true performance devices.

Striking black-and-white photographs of athletes sporting G-Shocks were tinted bright and bold primary colors (like the watches themselves) to create an eye-catching and indelible brand image, one that was instantly distinguishable from all the competitors. Whether running, biking, swimming, or weightlifting, the G-Shock was shown to be an indispensable tool to boost performance while also being a sleek, sexy, fashionable accessory for everyone—athlete or not.

Under the new mantra of "Time for the 21st Century," an accompanying print campaign was developed that showed the G-Shock in schematic blueprint form, breaking down the individual components of the watch to give consumers an inside view of the cutting-edge technology inside the watches. These communications affirmed that G-Shocks were the real deal and had the technological sophistication to match—and even exceed—any other performance timepiece on the market.

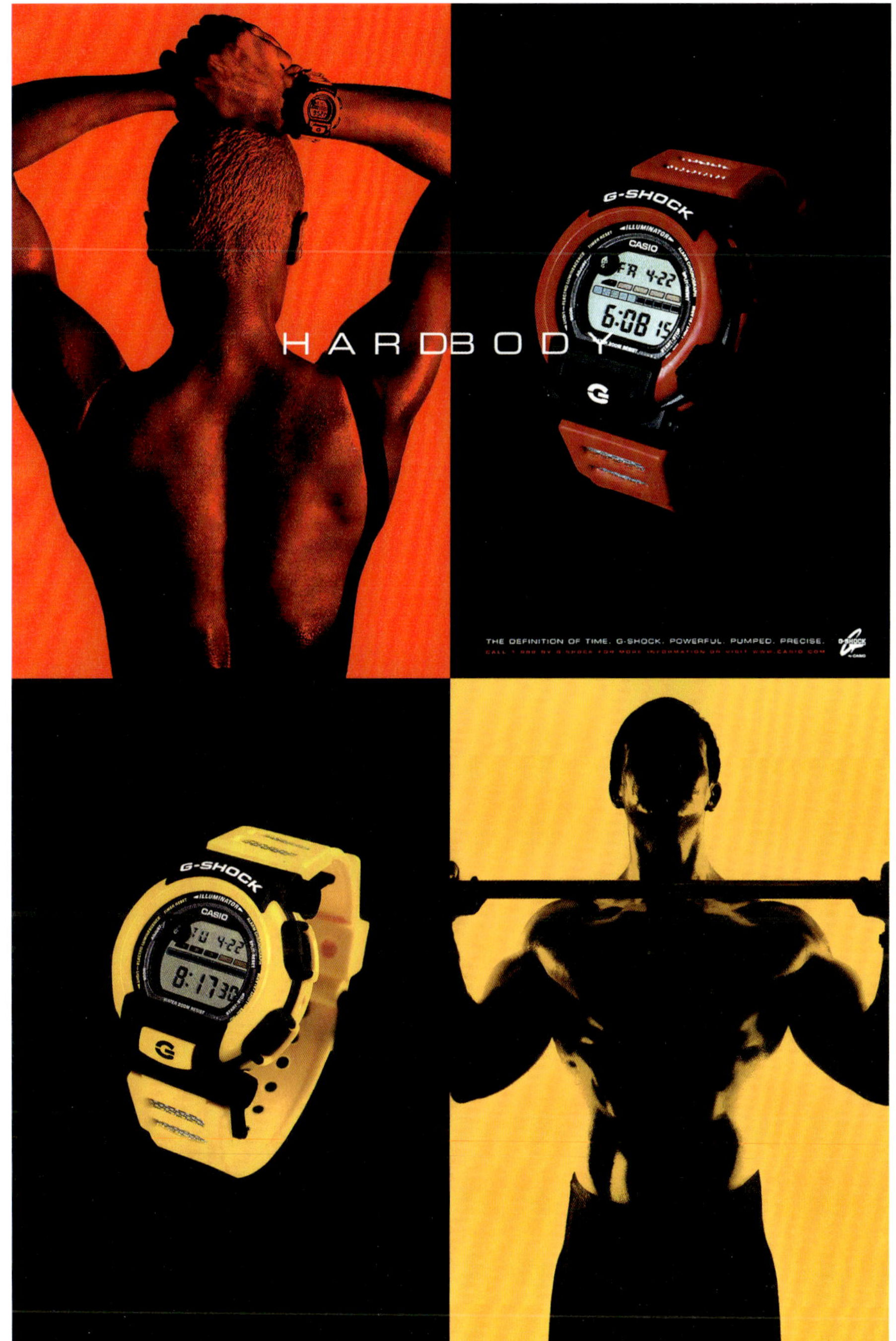

HARD BODY
G-SHOCK
ILLUMINATOR
CASIO
THE DEFINITION OF TIME. G-SHOCK. POWERFUL. PUMPED. PRECISE.
CALL 1 888 BY G-SHOCK FOR MORE INFORMATION OR VISIT WWW.CASIO.COM
G-SHOCK
by CASIO

G-SHOCK
CASIO
ONE WAY
ALL TRAFFIC

CASIO BABY-G

1998

When Casio's G-Shock brand was first launched, it featured rugged performance timepiece that were aimed squarely at the male market. But it soon became clear that in addition to this target customer base, huge numbers of women were also borrowing their boyfriend's G-Shocks, or simply buying them for themselves. This represented an opportunity to exponentially expand the potential customer base by targeting women who were drawn to the utilitarian appeal of a durable, functional timepiece.

The Baby-G line of watches, with its cheeky mantra of "secs appeal," dramatically opened up the customer base for Casio's watches, and encouraged consumers to see what had previously been a purely functional item as something stylish, fashionable, and, of course, sexy.

The accompanying campaign featured women sporting G-Shock watches as a true fashion and lifestyle accessory: on the waist … around the ankle … anywhere but on the wrist.

Baby-G
CASIO
secs appeal
IRRESISTIBLY ATTRACTIVE
ALL 1 888 BY G-SHOCK FOR MORE INFORMATION OR VISIT WWW.CASIO.COM

Baby-G
CASIO
24 hours of secs
ANYTIME. ANYWHERE FROM G-SHOCK. SMALLER SIZES. BIGGER COLORS.

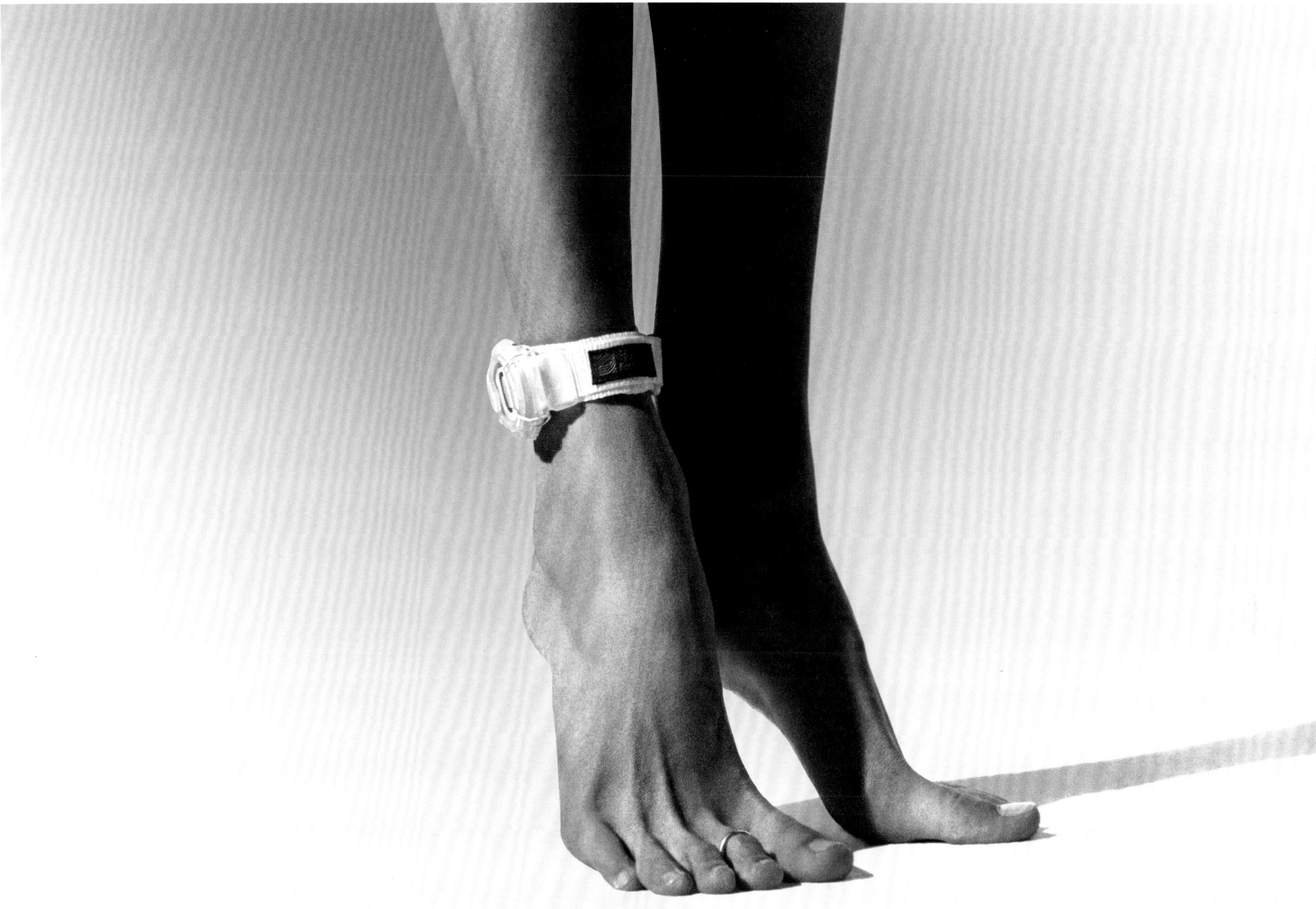

tom·my hil·fig·er **1 :** designer: AMERICAN **2 :** quality tailored clothing, sportswear, jeans and athletic apparel for men and women *adj* **3 :** fun, free, spirited, e.g., **a desirable addition to your wardrobe** **4 :** *chiefly* a group of people who demand distinctive style < ~ customers have an eye for the details> *v.* **5 :** to twist tradition, to re-invent the classics: *often attrib.* FASHION FREEDOM. **6 :** to chuck convention, but with panache ***syn*** see Duke and Duchess of Windsor. **7 :** Whoever you are, whatever you wear, TOMMY IS YOU

TOMMY HILFIGER

1998

Throughout the 1990s Tommy Hilfiger was unquestionably one of the most recognizable and influential brands in the world, and its fashions became a ubiquitous staple in the wardrobes of young people everywhere. By the middle of the decade, Tommy Hilfiger's fashions had become synonymous with both a more classic American preppy look, but it had also become a staple of hip-hop fashion. As arguably the first major brand to bring together these two distinct subcultures, the new hybrid style that Tommy pioneered brilliantly captured a cross-pollination of styles and cultures that was young, fresh, and distinctly American.

A lifelong fan of rock and roll, Hilfiger's collections had often been inspired by the fashions of musicians and the distinctive looks that emerged from musical subcultures. And throughout the 1990s Tommy Hilfiger had become a sponsor for the tours of some of pop, rock, and R&B's biggest stars such as Sheryl Crow, Pete Townshend, and Aaliyah. Building upon these successful creative partnerships, the Tommy Rocks campaign was created in collaboration with some of the day's biggest music and movie superstars as a way of further solidifying Tommy Hilfiger's position firmly at the center of the youth zeitgeist.

The holistic communications campaign for Hilfiger's women's and men's sportswear collections bolstered the brand's association with the worlds of music and film through total brand integration activations, like in the star-studded movie *The Faculty*, and through innovative, event-based marketing activations, like the multi-pronged partnership with the Rolling Stones for their 1999 "No Security" tour.

Featuring other famous faces such as Britney Spears, Usher, and Lenny Kravitz, these campaigns deftly merged the worlds of celebrity and style, and cemented the position of Tommy Hilfiger (both the brand and the man) at the forefront of popular culture, with Tommy himself later commenting that Peter "made me a great American designer."

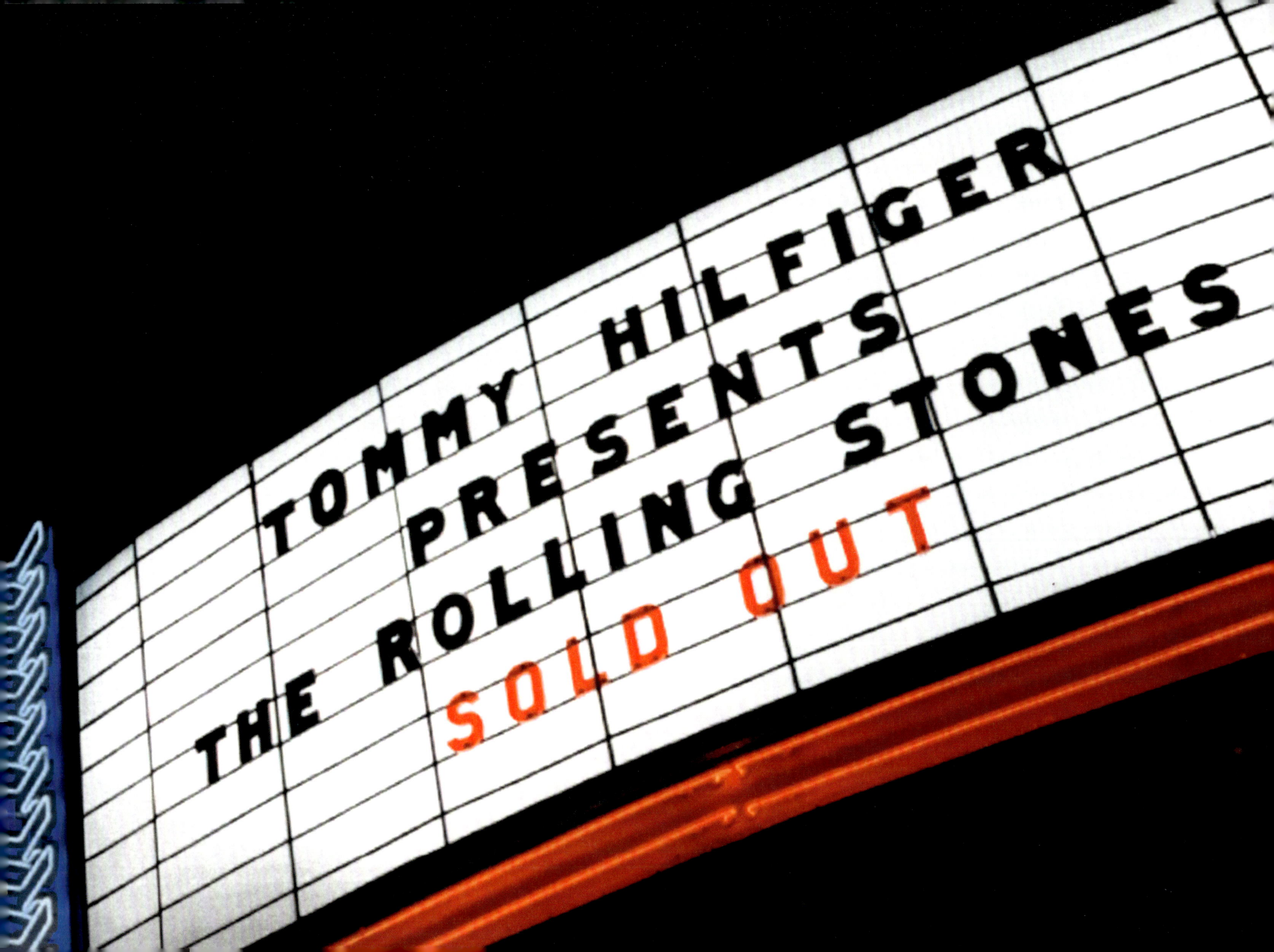
TOMMY HILFIGER
PRESENTS
THE ROLLING STONES
SOLD OUT

TOMMY HILFIGER PRESENTS
THE ROLLING STONES
NO SECURITY TOUR 99
©1999 TOMMY HILFIGER LICENSING INC. TOMMY HILFIGER AND ARE REGISTERED TRADEMARKS OF TOMMY HILFIGER LICENSING, INC.
"ROLLING STONES" AND TONGUE AND LIP DESIGN ARE TRADEMARKS OF MUSIDOR B.V.

TOMMY ROCKS
Tommy Hilfiger presents the Lenny Kravitz Freedom Tour
TOMMY HILFIGER

tommy
jeans

TOMMY HILFIGER PRESENTS THE ROLLING STONES
NO SECURITY TOUR

TOMMY ROCKS
Tommy Hilfiger presents Jewel in concert
TOMMY HILFIGER

TOMMY HILFIGER

TOMMY HILFIGER PRESENTS
SUNGLASSES FOR WOMEN

TOMMY HILFIGER
PRESENTS
THE ROLLING STONES
NO SECURITY TOUR

TOMMY JEANS
introduces michael fredo in concert
TOMMY
JEANS

TOMMY HILFIGER
introducing toddler girls

ICEBERG

1998

Long known for their luxurious knitwear, by the late 1990s Italian brand Iceberg had begun to transition from a niche brand to a fully-fledged luxury fashion house offering a complete range of ready-to-wear clothing and accessories. This expansion of the brand saw Iceberg emerge as an early proponent of fashionable, high-end sportswear with a fresh new look that was luxurious yet approachable, sophisticated yet sexy. For a brand with "coolness" embedded in its name, it only made sense to lean into these associations and embrace a younger, freer, more daring image.

Alongside a refreshed brand identity, a series of print ads and billboards was developed featuring some of the biggest stars of hip-hop. Artists like Mya, Lil' Kim, and The Lox were driving youth culture with their creative vitality, artistic experimentation, adventurous fashions, and urgent social and political messages. These ads injected a fresh, bold, rebellious energy into the brand and helped to position Iceberg as a new sort of designer brand for a new generation. As one of the first major fashion houses to embrace and pay tribute to the hip-hop community, Iceberg established itself as a different kind of luxury brand, one that was open, accessible, and genuinely welcoming of self-expression.

To expand the brand into the lifestyle market and further energize the growth of the company, a new denim line was created, along with a range of luxury fragrances—complete with eye-catching designs that evoked the brand's DNA in new and unexpected ways. For Iceberg's first fragrance, a solid glass brick with a blue void within gave the appearance of an actual iceberg submerged below the surface, while for the brand's second fragrance, a stainless steel bottle with a faceted geometric pattern evoked the angular, fractal character of a single frozen water droplet.

ICEJEANS
ICEBERG

IceJeans
ICEBERG
LIL'KIM ICED
GOLDEN RING
GOLDEN RING

ICEBERG
ICEJEANS
ICEBERG
LOX ICED WE ARE THE STREET

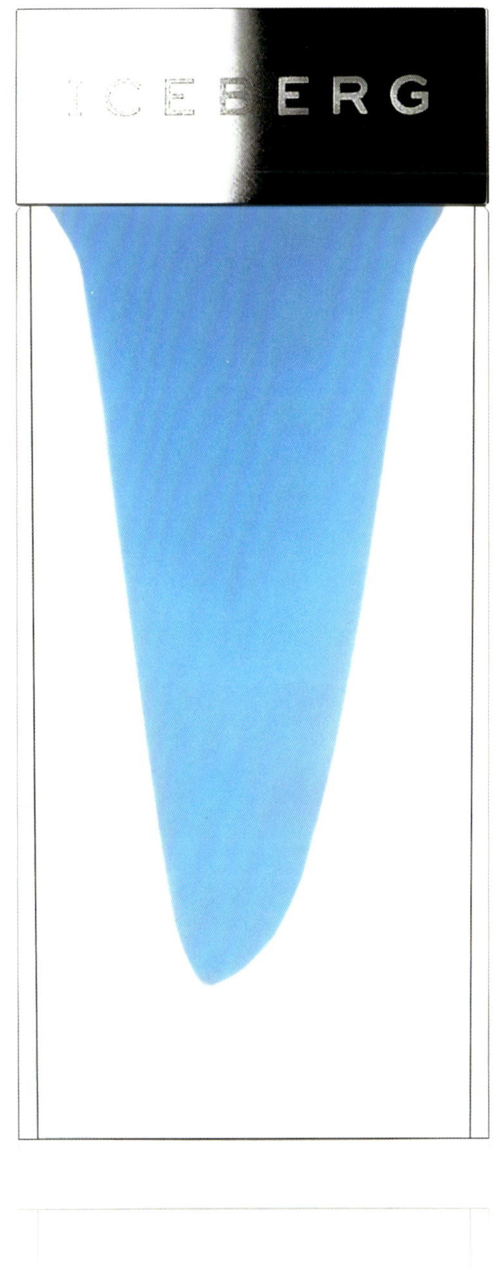
ICEBERG

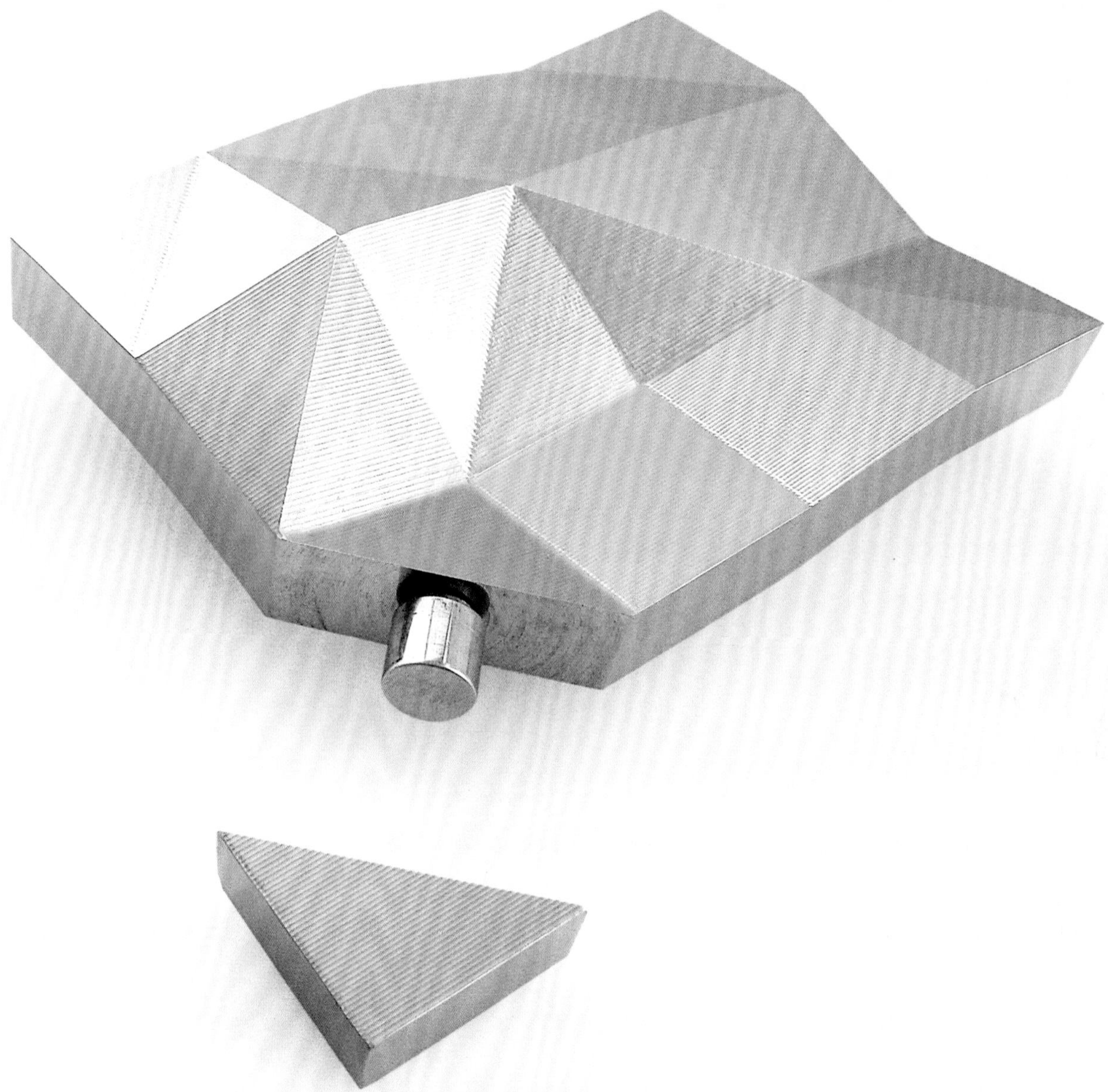

RENA LANGE

1998

Rena Lange was founded in Munich in 1916 as a lingerie label, but by the 1980s the brand had expanded its offerings, showing its first ready-to-wear collection in Milan in 1982. Over the next decade, the brand's sales and reputation steadily grew, and it quickly became renowned for its restrained, minimalist take on luxury.

By the late 1990s the brand's international presence was still expanding, but it largely remained a niche brand. To broaden its appeal and invite new customers into the story of the brand, a younger and more provocative brand image was needed. The resulting campaign, shot in London over the course of a few days at the iconic Savoy Hotel, was unlike anything the brand had done before: the model appeared mature and classically chic in her formal wear, complete with polished hair and makeup. And yet, she was also notably informal in her attitude: unposed, unstudied, and casual, as though she was being captured in between takes (which in many cases, she was).

The images stripped away the formality and artifice that typically accompanied fashion campaigns. What was left was a series of beautiful, intimate, almost voyeuristic shots that showed a woman, who, while poised, polished, and beautiful, was also unguarded, vulnerable, and real.

CLASSIC.
ORIGINAL.
MODERN.
WARM HEART.
COOL PRECISION.
RENA LANGE.
RENA LANGE
M. LANGE & CO. GMBH ROSENHEIMER. STRASSE 139 81671 MÜNCHEN TEL.:0 89/41 86 60

SAVOY

GUGGENHEIM MUSEUM

1998

As the Guggenheim Museum approached its sixth decade and the turn of the millennium, the iconic institution needed to find new ways to challenge cultural gatekeepers, dismantle preexisting ideas about what a cultural institution should look like, and make art relevant and accessible to more people all over the world.

Working closely with Thomas Krens, the visionary director of the Solomon R. Guggenheim Foundation, a comprehensive vision for the future of the museum was devised around the simple but powerful idea that culture is a universal good. Krens, who had led a massive diversification of the Guggenheim's collection and spearheaded a rapid international expansion effort, considered it not just a responsibility, but an obligation of the institution to bring culture—in all its forms—to the broadest possible audience. In his role as director, Krens took an early lead in acknowledging that the guardians of so-called high culture would need to adapt if they were to remain relevant, and find new ways to reach audiences, both within the museum itself, and out in the world at large.

This multi-pronged vision identified a number of potential new initiatives: a task force was convened to help the Guggenheim to continue to building out its collection with diverse contemporary artists; additional museum alliances were identified and then established worldwide; the museum network was expanded to Asia, South America, and the Middle East; new business and investment partnerships with the private sector were explored; and Guggenheim web and TV platforms were launched as the new online homes of documentaries, movies, art films, educational programming, interactive learning, cultural news, and performing arts broadcasting.

As part of an overarching strategy, a refreshed logo was proposed. Built around the Frank Lloyd Wright-designed typeface that spells out the museum's name on its Fifth Avenue façade, the Guggenheim's famous G was dimensionalized and shown at a slightly oblique angle: a completely fresh and modern expression for the digital age.

THE
GUGGENHEIM
IS NOT
A PLACE

THE
GUGGENHEIM
IS A POINT
OF VIEW

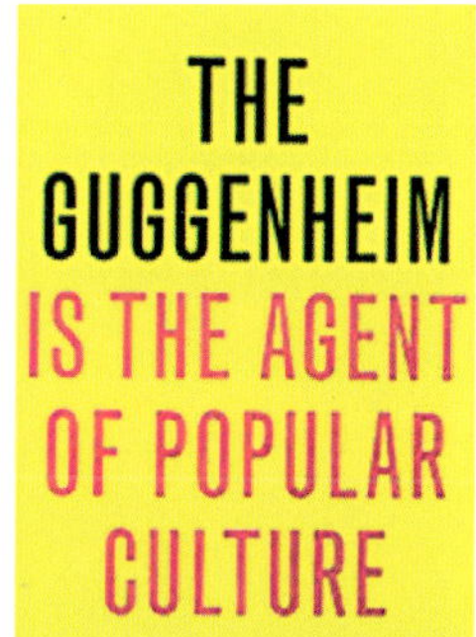
THE
GUGGENHEIM
IS THE AGENT
OF POPULAR
CULTURE

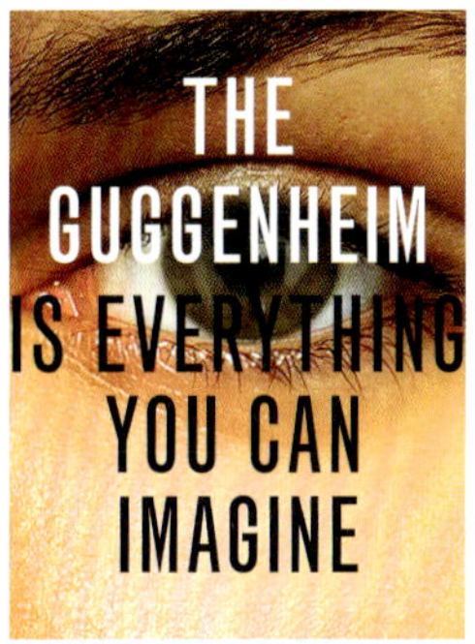
THE
GUGGENHEIM
IS EVERYTHING
YOU CAN
IMAGINE

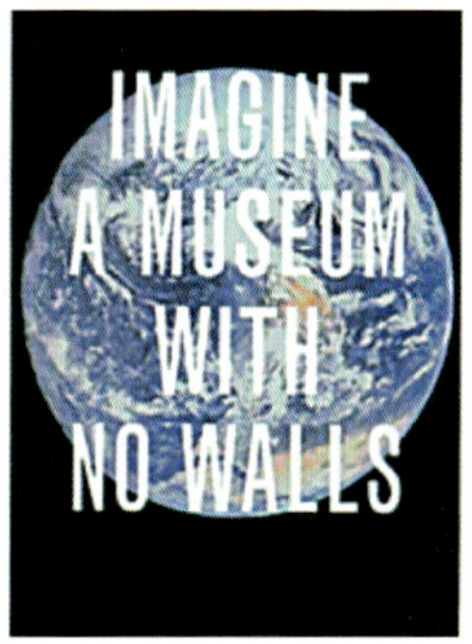
IMAGINE
A MUSEUM
WITH
NO WALLS

IMAGINE
A MUSEUM
IN CONSTANT
MOTION

IMAGINE
MEMORY
WITHOUT
EXPERIENCE

IMAGINE
TIME
AS A SOLID
OBJECT

IMAGINE
WRONG
THAT IS
RIGHT

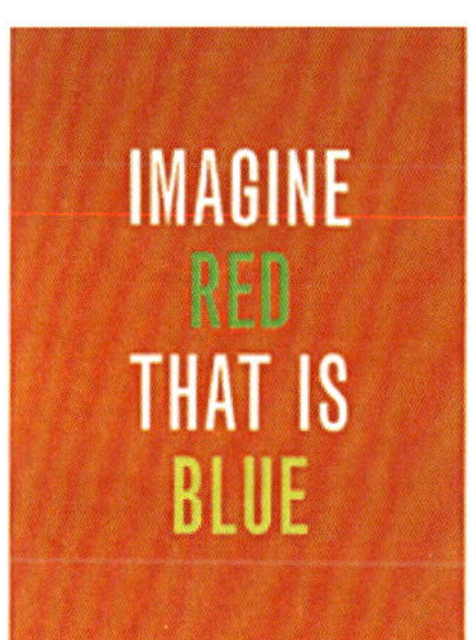
IMAGINE
RED
THAT IS
BLUE

IMAGINE
A VOICE
THAT MAKES
NO SOUND

IMAGINE
ART THAT
CANNOT
BE SEEN

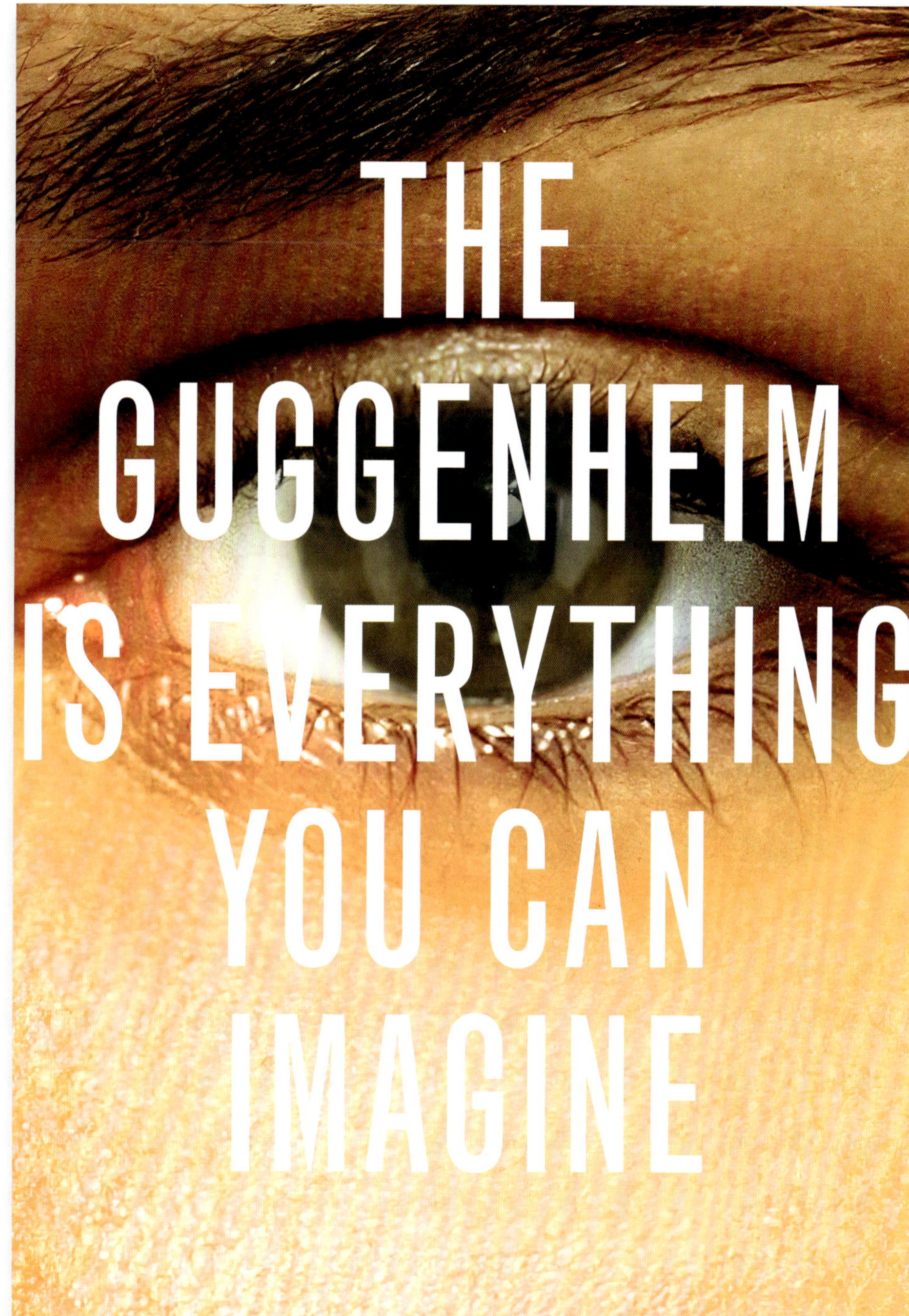
THE
GUGGENHEIM
IS EVERYTHING
YOU CAN
IMAGINE

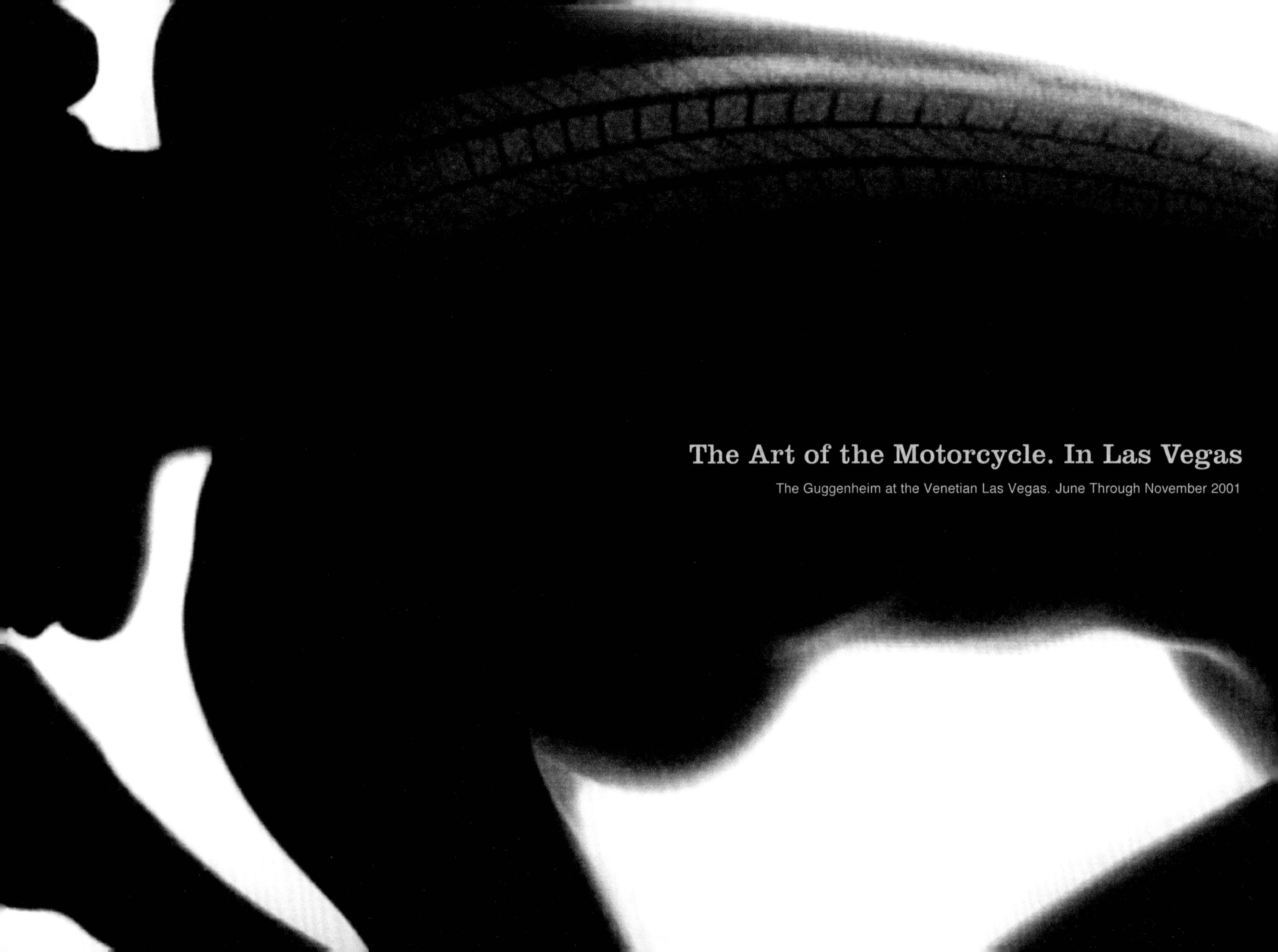
The Art of the Motorcycle. In Las Vegas
The Guggenheim at the Venetian Las Vegas. June Through November 2001

THE ART OF THE MOTORCYCLE

1998

The landmark 1998 *Art of the Motorcycle* exhibition marked a turning point for the Guggenheim Museum in many ways: as the Guggenheim's first large-scale show devoted to mass-produced objects, it signaled the museum's broadened curatorial vision, and its commitment to engaging with audiences outside of the traditional art world. A striking exhibition design by Frank Gehry created a series of dramatic, dynamic tableaus that highlighted the true artistry of these machines, showing them as the incredible design artifacts that they were.

Over the course of the three-month-long show at the Guggenheim's Fifth Avenue home, hundreds of thousands of attendees made the journey to appreciate these spectacular machines, and were encouraged to consider these industrial creations them as true works of art: as potent a cultural artifact as any painting or sculpture. While controversial at the time for its focus on mass-produced consumer goods, by the time of its closing, it had become the most-attended show in the Guggenheim's history.

Following this unprecedented success, the show was remounted at the Guggenheim's Las Vegas outpost in 2001. To promote the expanded version of this landmark exhibition, a series of bold, vibrant, and provocative communications was created, featuring celebrities like Laurence Fishburne, Dennis Hopper, and Lauren Hutton: all motorcycle aficionados, and all associated with the "art of the motorcycle" in the popular imagination. A new curatorial strategy was also devised in collaboration with Thomas Krens that emphasized the historicity of the motorcycles themselves, situating them within the broader cultural and political landscapes of their eras.

Both the New York and Las Vegas shows were enormously successful and heralded the beginning of a new trend in blockbuster, mass-appeal museum exhibitions. Two decades later, the exhibit is still among the most popular and most remembered in the Guggenheim's history, and was instrumental in pushing museum institutions everywhere to radically expand the notion of what an exhibition can and should be.

Indian

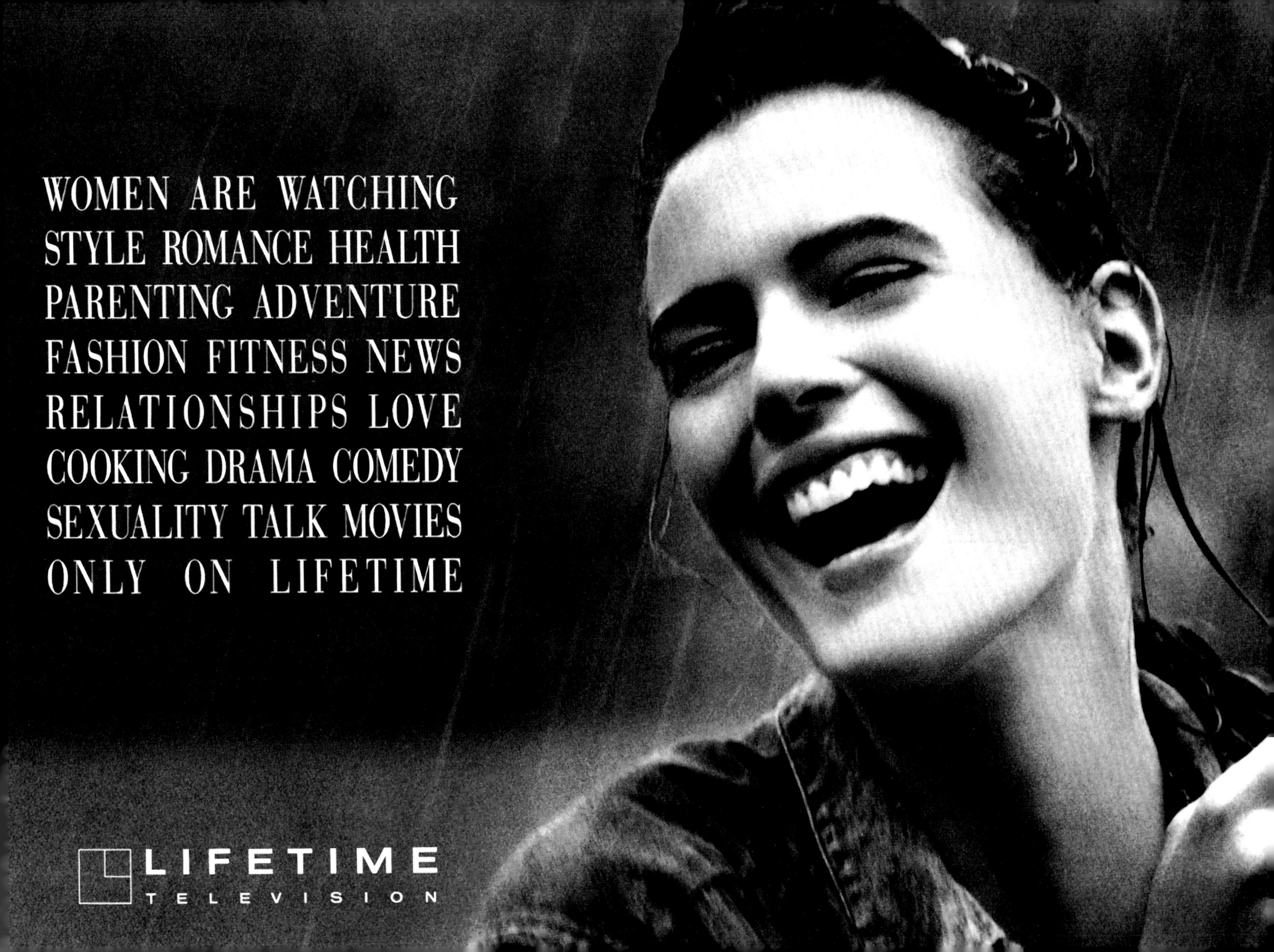
WOMEN ARE WATCHING
STYLE ROMANCE HEALTH
PARENTING ADVENTURE
FASHION FITNESS NEWS
RELATIONSHIPS LOVE
COOKING DRAMA COMEDY
SEXUALITY TALK MOVIES
ONLY ON LIFETIME
LIFETIME
TELEVISION

LIFETIME TELEVISION

1998

Lifetime Television launched in the 1980s as an undistinguished cable network running a broadcast slate of talk shows, call-in programs, and syndicated hits from other networks. But by the early 1990s the network shifted the focus of its signature talk shows from general interest to women's issues, began broadcasting women's sports coverage, and started producing original Lifetime TV movies with targeted appeal to female viewers.

To accompany this new chapter in the network's history, a colorful new brand identity was designed that evoked the colors and forms of the ubiquitous TV test pattern. A four-quadrant square that read as a stylized L, the brand mark also abstractly suggested a clock face with the hands positioned at twelve and three—a nod to Lifetime's 24-7 programming. Rendered in an array of colors, the fresh logo was designed to have no one fixed expression: a celebration of the diversity of women, and an implicit promise of something for everyone.

A new series of communications were developed to address how Lifetime was different, and specifically how it would speak to the unique interests and concerns of its largely female viewership. This new direction was succinctly captured in the simple but indelible tagline that continues to inform the network's identity to this day: "Women Are Watching."

LIFETIME
TELEVISION

PROGRES
OGRESSIVE

PROGRESSIVE INSURANCE

1999

For Progressive Insurance, a sponsor of the Super Bowl XXXIII halftime show, a magical thirty-second spot was created that offered life-saving traffic safety tips from an unlikely spokesperson: E.T. the Extra-Terrestrial. Over the years, E.T. had been carefully guarded by his creator Steven Spielberg, who was reluctant to allow his iconic alien to appear in advertisements for fear of diluting the charm of the beloved character. But upon being approached with the premise for the Progressive spot, Spielberg was impressed, and agreed to collaborate on a Super Bowl ad that featured the famous extraterrestrial as an unlikely advocate for traffic safety.

The spot, produced in collaboration with George Lucas's Industrial Light & Magic and original *E.T.* cinematographer Allen Daviau, finds E.T. back on board his spacecraft regaling his fellow extraterrestrials with stories of his time on Earth. They're all amused by humans' unsophisticated technology, and E.T. confirms that accidents are constantly happening, particularly because of earthlings' reliance on the "primitive" wheel in their vehicles. But there's hope, he assures them, as he summons a hologram of a Progressive vehicle. Distracted by E.T.'s tales, the extraterrestrials almost fail to notice an approaching space shuttle piloted by astronauts Buzz Aldrin and Jim Lovell. At the last minute they see it and make a sharp turn to avoid collision—a near miss that reminds E.T. and his friends that accidents can happen to anyone, anywhere … even in space. The spot ends with the two spacecrafts heading safely back to their home planets, as a voice over promises viewers: "Finally, intelligent life in auto insurance."

The campaign extended beyond this spot with specially designed drivers' education courses, a Safety Club for kids, and a series of communications featuring E.T. with his digit raised and his seat belt buckled, urging others to "buckle up" as well. In his new role as an ambassador for traffic safety, E.T. reminded everyone to do their part in preventing accidents and buckle up. This safety message proved to be an impactful and unforgettable way to harness the magic of this beloved character for a worthy and life-saving cause.

ALDRIN

BIKE ROUTE

They still rely on the wheel
The big ones drive cars - always rushing
Sometimes they have accidents
But there is hope
There are people who can help
PROGRESSIVE
1 800 AUTO PRO
BUCKLE UP
visit progressive.com
or your independent agent

MIKIMOTO

1999

More than a century ago, Kokichi Mikimoto pioneered the process of creating a cultured pearl within the shells of an oyster. In the years since, his namesake brand has become world-renowned for its craftsmanship and fine jewelry design.

To refresh the brand and leave it well-positioned for another century, a new global brandmark was designed that respected the heritage of the brand while still being entirely fresh and contemporary. The elegant simplicity of a single pearl placed symbolically above an M was modern yet timeless: exactly like the brand itself.

To expand the Mikimoto brand outside of jewelry, a flagship line of gift and office items was also developed. These beautiful products maintained the brand's heritage by incorporating Japanese design principles and the related idea of *wabi-sabi:* an aesthetic philosophy which finds beauty in imperfect, incomplete, and impermanent things. This new range of functional design objects respected traditional Japanese notions of aesthetics while also paying tribute to the spirit of distinctive but refined elegance that the brand had always embodied.

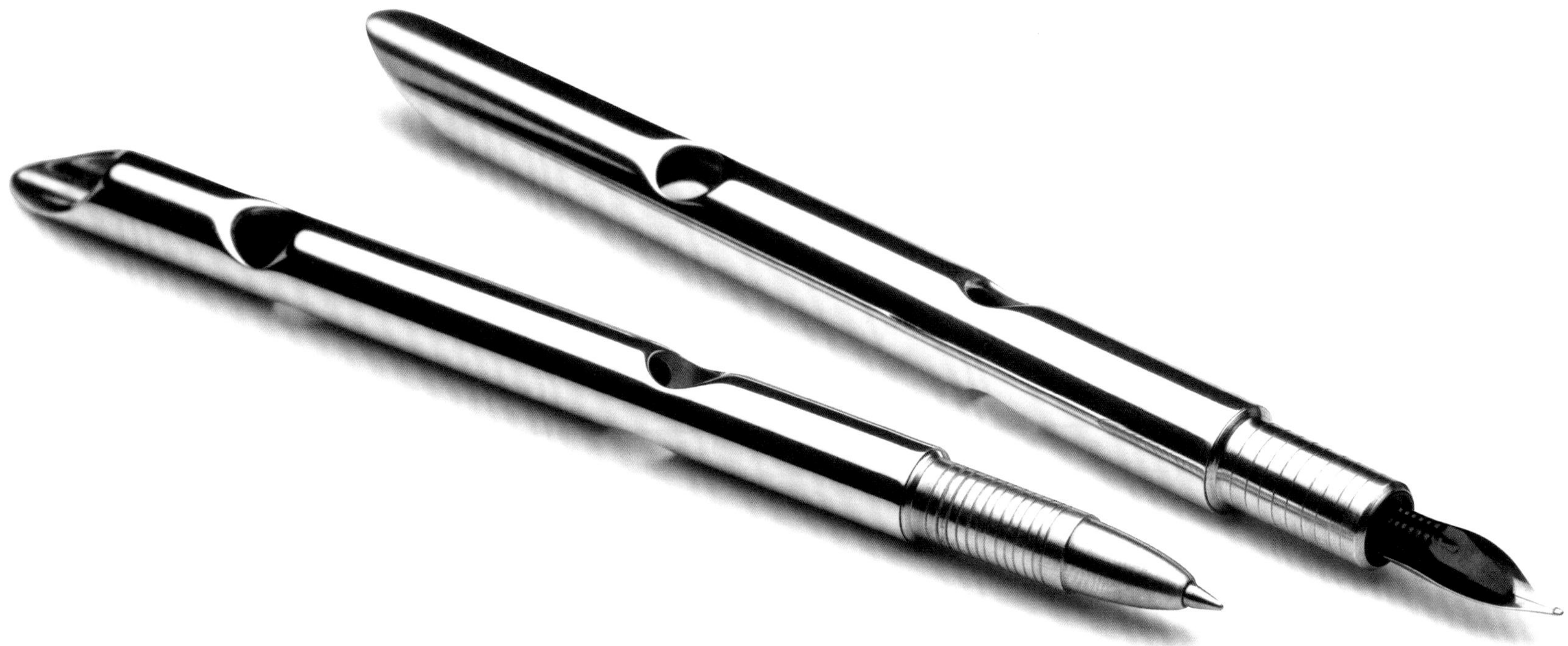

MARUI

1999

For Marui, the largest department store chain in Japan, a comprehensive brand overhaul was engineered to help the retailer modernize and remain competitive with the many global brands that were now vying for their customers' attention.

To achieve this, Marui's range of store brands had to be transformed from sensible, reliable private labels into a family of global power-brands: distinct entities with unique personalities that had strong links to the parent brand. To this end, a number of new in-house brands, product lines, and branded intellectual property were developed, including a Marui Eyewear Line, The OIMEN and OIMODEL men's and women's clothing brands, and an adorable new children's brand mascot, the joyful raconteur MIMI.

Candid, emotional, and intimate imagery was produced for use in-store and in global advertising campaigns to help reinforce Marui's new identity as a lively, energetic, globally aware brand. Every brand touchpoint was completely redesigned, including product packaging, wayfinding, store fixtures, employee uniforms, hang tags, and corporate collateral.

A new brand mark was also designed for the retailer: two perfect circles and two vertical lines formed a glyph that seemed to spell out "OIOI"—a transliteration of Marui's name for native Japanese speakers. In the Japanese language, the symbol "O" is read "maru," while he symbol "I" represents the numeral "1," which can be read as a long "i" sound in Japanese: "maru" plus "i" yielded "Marui." The new OIOI brand mark thus acted as both a word and a logo mark: a clever, welcoming, and instantly recognizable symbol in every language.

0101

OICITY

9月20日(金)
MEET ME AT
OIOI

OIOI
MEET ME AT OICITY
MEET ME AT OICITY
MEET ME AT OICITY
FUJI BANK
富士銀行
YOUNG
OIOI
OIOI
MITSUMINE
イタリアン・トマト
3F SUIT

MEET ME
0101

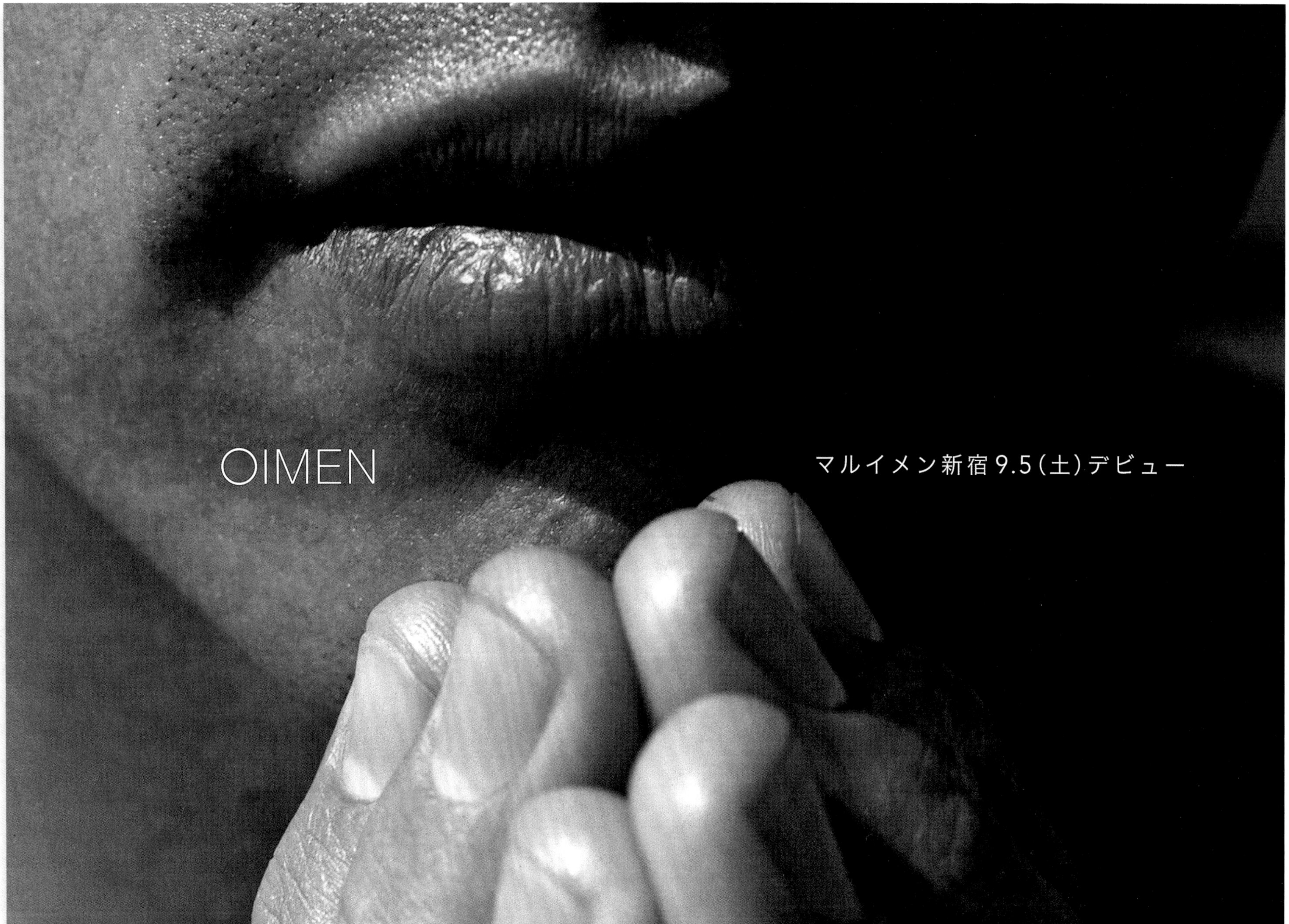
OIMEN
マルイメン新宿9.5(土)デビュー

OIMEN
shirts

OIMEN
watches

OIMEN
sunglasses
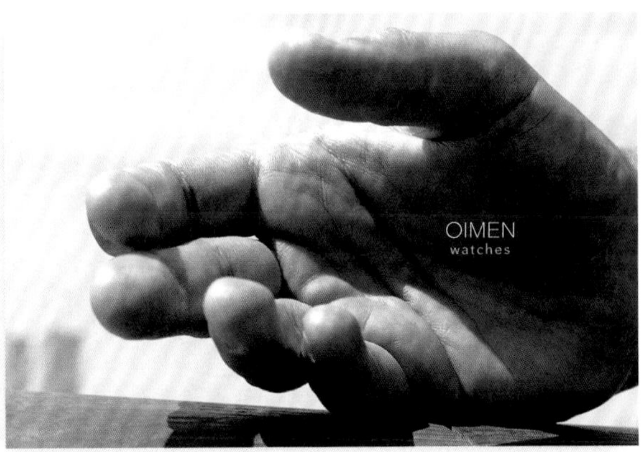
OIMEN
watches
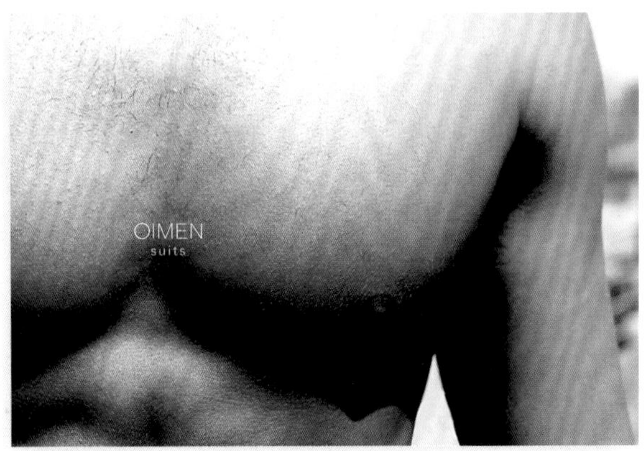
OIMEN
suits
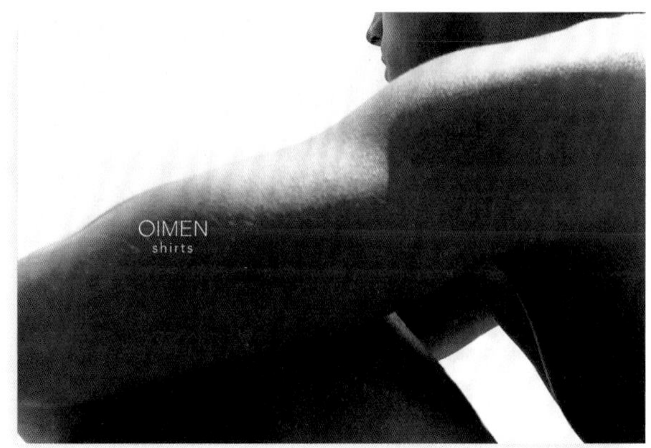
OIMEN
shirts

O|MODEL
FIT FOR A NEW GENERATION
モデルサイズ、始めます。
マルイ各店　マルイシティ新宿・渋谷・池袋・上野ではサイズのメガショップがオープン。
WEAR 13–19号&トールサイズ　SHOES 24.5–26.0cm　FOUNDATION F–Hカップ アンダーバスト 80–100cml
お問い合わせ　マルイプレスルーム 03(5343)6666

OIOI FAMILY

みんな家族がだいすき。

Mimi at Marui
ミミアットマルイ

JIMMY AND
FARMCLUB
DOUG'S
.COM

FARM CLUB

1999

In 1999, Universal Music Group became one of the first major players to understand the unique potential of the Internet to distribute new forms of entertainment and develop new kinds of communities with the launch of its online Farm Club Record Label. Farm Club was envisioned as both a pioneering digital music distribution platform, and a vehicle for aspiring artists to get their first big break: in effect, a precursor to and amalgamation of iTunes, YouTube, and *American Idol*.

With its "headquarters" on the Internet at FarmClub.com, it instantly gave every aspiring artist in America access to some of the most legendary power players in the music industry and offered them a path to the ultimate dream of scoring a major record label deal. The unsigned artists uploaded their music, and the most popular artists—selected from votes through the FarmClub.com website—were then offered record deals and partnerships with Interscope, Def Jam, and other major labels under Universal management.

Working alongside UMG chairman and CEO Doug Morris, and Jimmy Iovine, founder of Interscope Records, a complete brand strategy was built from the ground up, including identity, branding, website, marketing, and communications for this groundbreaking new multimedia business. As a first of its kind, the Farm Club multi-platform endeavor both anticipated and precipitated the models of distribution that define the music industry today: a more democratic and accessible system, where anyone, anywhere, can instantly share their music with the world.

FARMCLUB.COM

SONY

media experience

SONY MEDIA EXPERIENCE

2000

Sony Media Experience was a proposal for an ideation, marketing, and media R&D group at Sony dedicated to the development of emerging media and consumer technologies. It was conceived of as a conduit to provide rapid prototyping capabilities across all Sony brands, business groups, and divisions, with the overarching aim of giving media new meaning in light of rapidly changing conditions across all mediums.

A series of conceptual design studies for so-called Sony Spaces was developed that envisioned new environments in which the physical and digital merged into something entirely new. An attempt to capture the full sensory experience of life with Sony, these immersive multimedia brand spaces—vast three-dimensional and inhabitable versions of Sony's amoeboid logo—highlighted different Sony products, platforms, and experiences, allowing the public to experience firsthand the many transformational digital technologies that Sony had created.

All of these activations were designed to clearly define and instantly communicate the spirit of "Sony-ness"—a global brand experience that shared Sony's cutting-edge, creative, and impactful innovations in media and communications with the world.

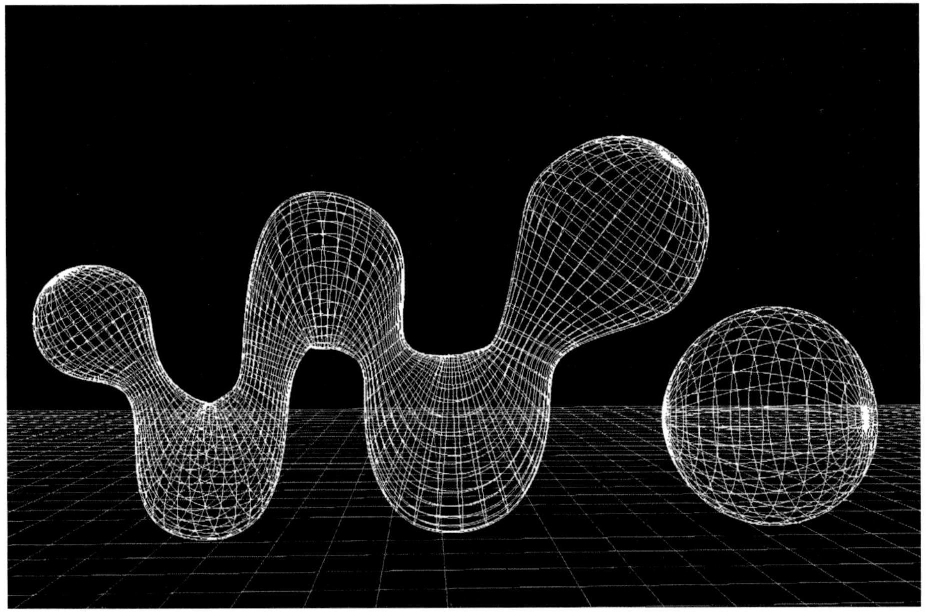

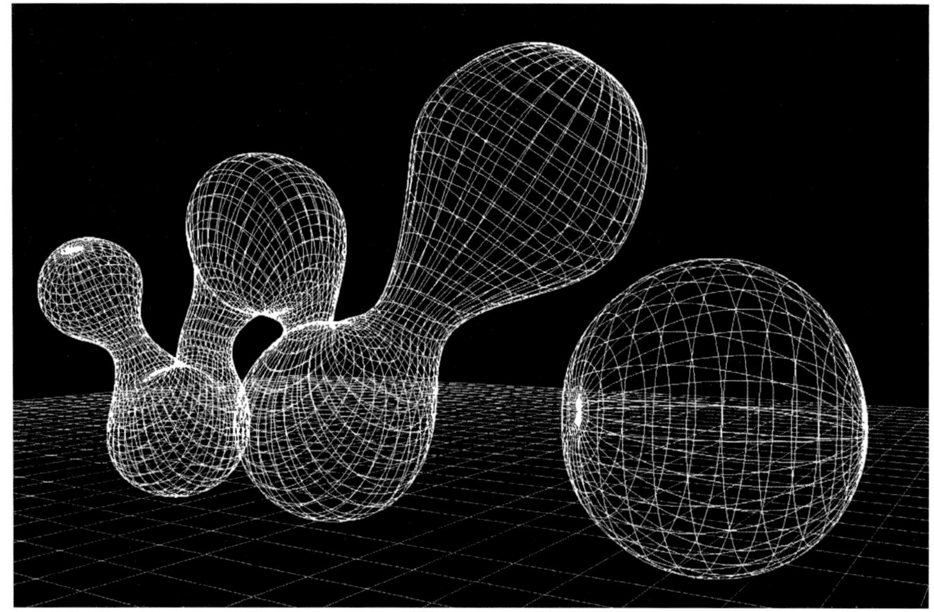

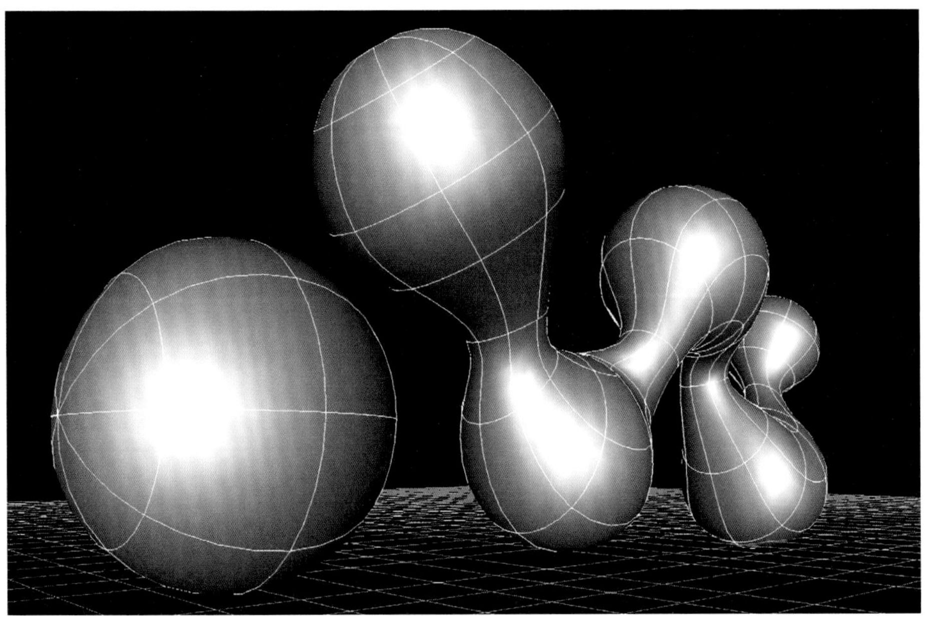

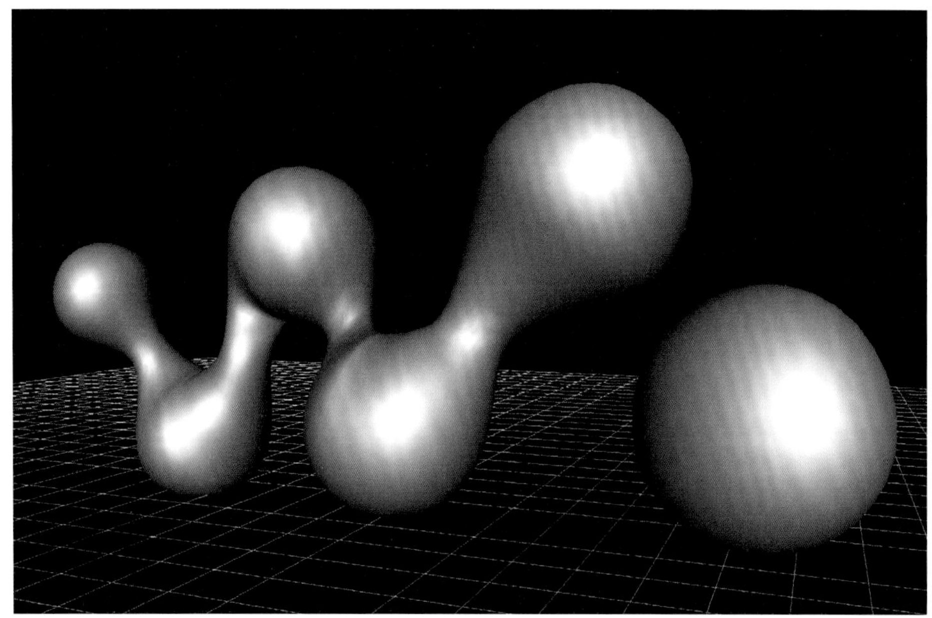

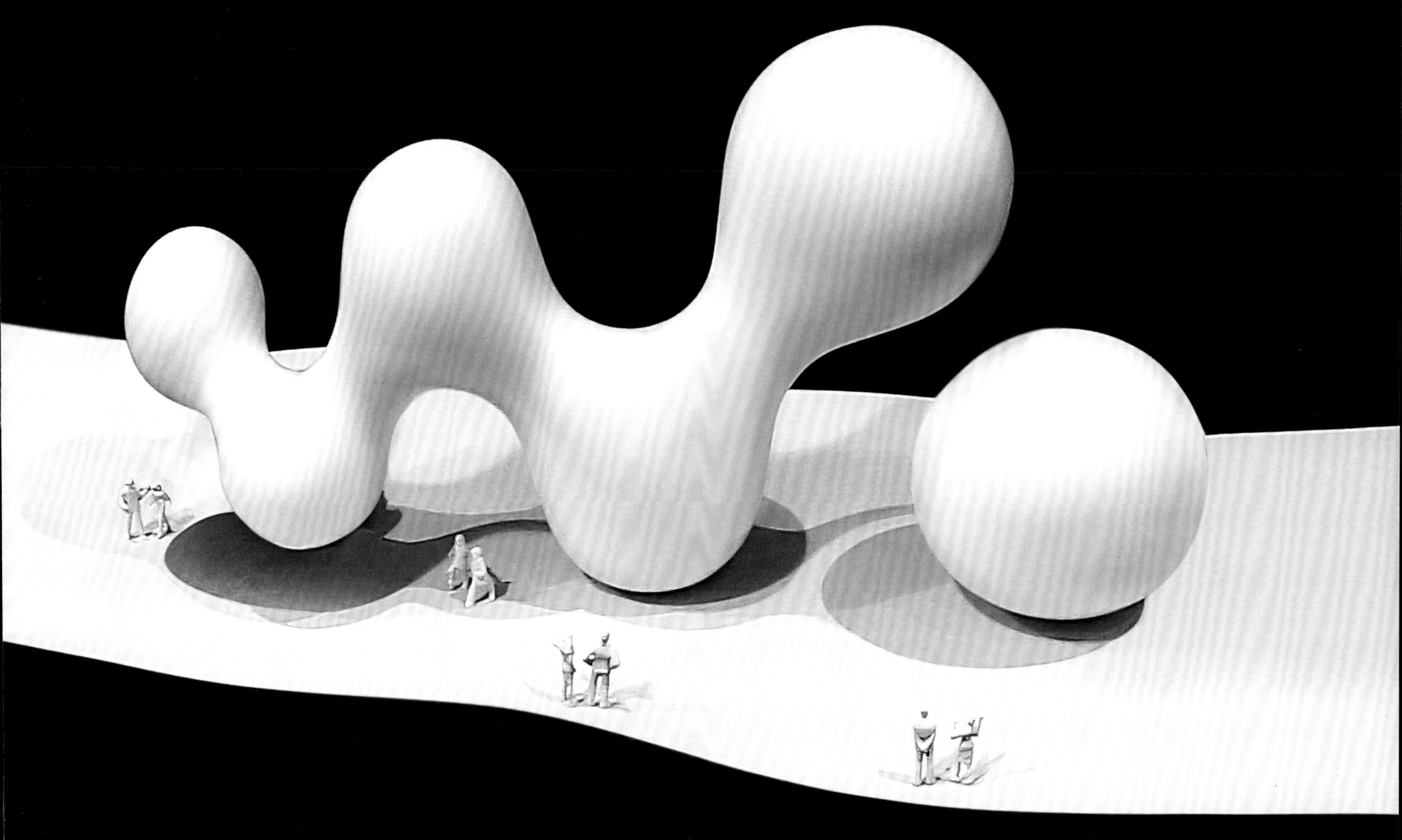

KID ROBOT

2000

For years, Kid Robot has been renowned around the world as a highly influential creator and seller of collectible designer art toys.

When Kid Robot launched its now-iconic Dunny toy—a cute, rabbit-like creature—it recruited artists and designers from around the globe and across all different types of creative backgrounds to truly explore the limits of their imagination with limited-edition Dunny designs.

For the occasion, Peter designed both the Mummy Dunny, as well as the fully functioning fish tank Dunny, called "Jacques Dunnai," in honor of the legendary aquatic explorer, Jacques Cousteau. Both pieces instantly sold out and have since become highly sought-after collectors' items that perfectly encapsulate the unique spirit of irreverence at the heart of everything Kid Robot creates.

nourish your senses
indulge.com

INDULGE.COM

2000

Indulge.com was a groundbreaking early entry into e-commerce that pioneered a more experiential, editorial, and personalized approach to online shopping.

Launched in 2000, the site was developed as an online destination for women to find holistic beauty and wellness solutions for every aspect of their lives. The editorial content for the site was composed of intimate, sensual images of women engaging in their personal beauty routines: private moments of personal care entirely removed from the demands and concerns of daily life. To bolster this sense of escapism, and to heighten emotional resonance, specific products were never explicitly featured; instead, the images were designed to evoke sensations that the products would provide: relaxation, rejuvenation, and true indulgence.

By framing indulgence as an essential act of self-care, women everywhere were encouraged to think of their beauty routine—whatever it consisted of—not as an extravagance, but as an indispensable part of overall well-being.

indulge.com

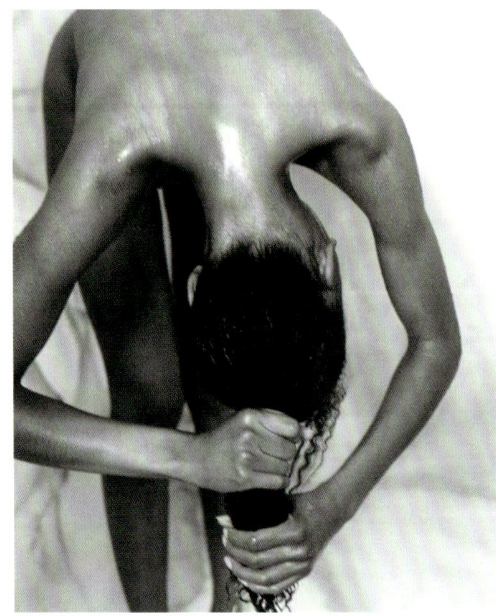

want to indulge?

It's sparkling gems for your hair
It's wearing pashmina pajamas for the softest, most intimate comfort
It's a five-minute facial
It's an aromatherapy head trip
consider the possibilities
It's the softest, warmest cashmere wrap
It's a shower that's revved up with lemongrass and jasmine body wash
It's all-over moisture
It's a new shade of lipstick

THALÍA

2000

In 2000, pop star Thalía was already an established star and major celebrity in her native Mexico, as well as in Latin America and around the world. Tommy Mottola, the legendary music executive who had helped to build Thalía's career, saw an opportunity to leverage her fame to reach a greater audience through a line of cosmetics and lifestyle products.

For this new brand, a bold visual identity and brand mark was developed: an elegant, serif T symbol and motif that recalled the monograms of many of the great European luxury houses, updated for a fresh, young clientele. A striking perfume bottle was also designed for the brand's signature scent: a curvaceous, sculptural form that was modeled after the singer's famously supple lips.

At a time when it was still rare for celebrities to build their own brands from the ground up, Thalía's namesake brand anticipated our current cultural moment where global, billion-dollar celebrity brands are the new normal.

be late
CONCORD
Time is a luxury.

CONCORD

2000

To introduce the Movado Group's Concord brand of ultra-premium timepieces, something bold and unusual was needed to stand out from the crowd. Every brand operating in this highest echelon of the market had essentially the same approach to marketing their timepieces: a focus on precision, superior materials, Swiss heritage, expert engineering, and premium materials. And while Concord's luxury timepieces shared in that pedigree, it wasn't enough to distinguish them in an increasingly crowded market.

The breakthrough came in the form of a radical rethinking of where the utility of a luxury timepiece actually lies. If a customer could afford to buy one of these watches, so the thinking went, they could also afford to take control of their own schedule: showing up where they wanted, when they wanted—and perhaps not even showing up at all.

"Be Late," the communications boldly instructed, as a man and a woman are shown (separately) lounging in bed. They're in control of their life, and in control of their time. They can choose to be late if they want to, and they answer to no one. In their world, lateness is not a source of stress, bur rather a pleasure that they can afford to indulge in.

By counter-intuitively introducing the concept of lateness to the conversation around a timepiece—the very thing it's meant to prevent—customers were challenged to reflect on what really constitutes an accomplishment and success. And by encouraging them to "Be Late," Concord gave its customers permission to take a moment for themselves and indulge in the one of most elusive luxuries of all: time.

parmalat
milk
homogenized
pasteurized
vitamin D added
REAL
(946 mL) QUART

PARMALAT

2000

Since 1961, Parmalat has been well-known across Europe for its shelf-stable milk that combined ultra-pasteurization processes with special sterile containers—a pantry staple for millions. When the Italian company was ready to launch its new fresh milk brand in the North American market, a campaign was created to help launch this new product in a way that would immediately distinguish it from its many competitors in the dairy cooler.

A witty, attention-grabbing communications campaign featured images of beautiful models crying tears of creamy white milk while proclaiming Parmalat to be "milk worth crying over."

Packaging and the overall corporate identity were entirely rethought as well: sleek all-white bottles with bold branding, minimal type, and a distinctive cap immediately distinguished Parmalat from the usual colorful paper cartons and plastic jugs in the supermarket dairy cooler and spoke to the purity and freshness of the milk inside.

FAMILY OWNED SINCE 1795
100% JOSE CUERVO TEQUILA
ALL WORK AND NO PLAY IS TOTALLY MISSING THE POINT.
Vive responsibly. Drink responsibly.
Jose Cuervo Especial ©2000 Imported and Bottled by Heublein, Stamford, CT under license from the trademark owners. Tequila, 40% Alc. by Vol.
Premium
Jose Cuervo Especial
Hecho en Mexico
VIVE Cuervo®
Hecho en México. Desde 1795.

JOSE CUERVO

2000

For many years, Jose Cuervo tequila had been by far the best-selling tequila brand in the world, successfully trading on its association with escape, indulgence, and festivity. While hardly languishing, there was a vast untapped market of consumers, specifically those over twenty-five years old, who were still young, vibrant, and full of life, but who didn't necessarily respond to the wild, hard-partying lifestyle that tequila drinking was often associated with.

To help gently evolve the spirit and culture of Cuervo into something that would appeal to a younger and more cosmopolitan generation, a more elevated and contemporary new global campaign was created to elevate the brand's image from a party catalyst to something sophisticated, approachable, and fun.

The resulting campaign—and brand mantra—Vive Cuervo featured photographs of beautiful, but still normal and approachable-looking young people enjoying life, accompanied by spirited and irreverent declarations that focused on letting go and enjoying life. As part of the coordinated effort to broaden the brand's appeal among those who may not have been regular tequila drinkers, ads were custom-designed to reach certain target demographics—namely, women and the LGBT community. This pointedly inclusive approach reflected the diversity of the audiences that Cuervo wanted to reach, emphatically welcoming in and celebrating the individuals and communities that were often overlooked by marketers at the time.

These new communications elevated Jose Cuervo's signature casual and carefree spirit, lending the brand a decidedly polished and modern twist that perfectly captured the pulse of a new generation.

VIVE Cuervo
Hecho en México. Desde 1795.

FAMILY OWNED SINCE 1795
100% JOSE CUERVO TEQUILA
Vive responsibly. Drink responsibly.
LIVE.
LAUGH.
LOVE.
SLEEP LATER.
Premium
JOSE CUERVO
Jose Cuervo
Especial
Hecho en Mexico
VIVE Cuervo
Hecho en México. Desde 1795.

FAMILY OWNED SINCE 1795
100% JOSE CUERVO TEQUILA
WE'RE HERE.
WE'RE CUERVO.
Vive responsibly. Drink responsibly.
Jose Cuervo
Especial
VIVE Cuervo®
Hecho en México. Desde 1795.

1800

TEQUILA

SUPER PREMIUM

1800 TEQUILA

2000

For decades, 1800 Tequila has produced some of the finest sipping tequilas, meticulously maintaining its original formula and traditional small-batch distillation process that uses only the highest quality 100% blue agave. The necessary secrecy around their proprietary recipe, along with the limited distribution of their signature product, had organically created a certain mystique around the brand that begged to be leveraged.

The Tell No One strategy—conceived of as a series of secret private events in exclusive locales around the world—leaned into 1800's reputation as a true insider's drink, preserving its air of exclusivity while also carefully, quietly, and discreetly raising its profile.

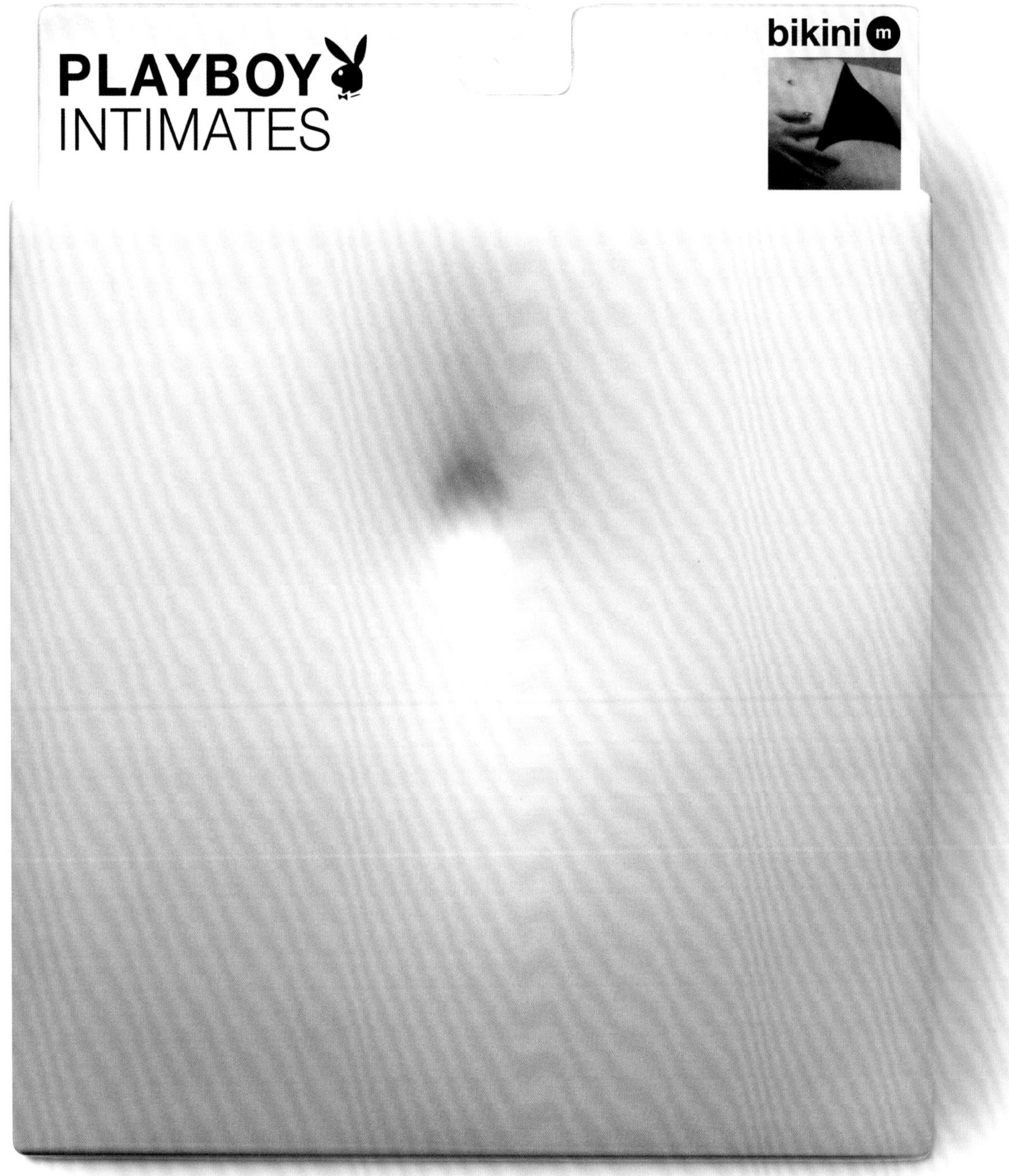
PLAYBOY
INTIMATES
bikini m

PLAYBOY INTIMATES

2000

For *Playboy*, the famed men's magazine founded by Hugh Hefner in 1953, a new brand concept was developed in the form of a line of men's and women's undergarments—fittingly named Playboy Intimates.

The packaging featured hyper-detailed images of vacuum formed plastic molds of male and female torsos that had been photographed at one-to-one scale. The smooth, sleek, highly realistic but also artificial-looking body parts on its packaging immediately distinguished Playboy Intimates from every other brand operating in this sector, most of whom tended to opt for beautiful but interchangeable models and vivid, full-color product shots. Rendered in a smooth matte gray finish, these toned and chiseled bodies took on an uncanny new aspect that was less about sex and more about style.

Attention-getting, yet chic and clever, Playboy Intimates was an ideal and natural extension for a brand renowned for being sexy, provocative, and thought-provoking in equal measure.

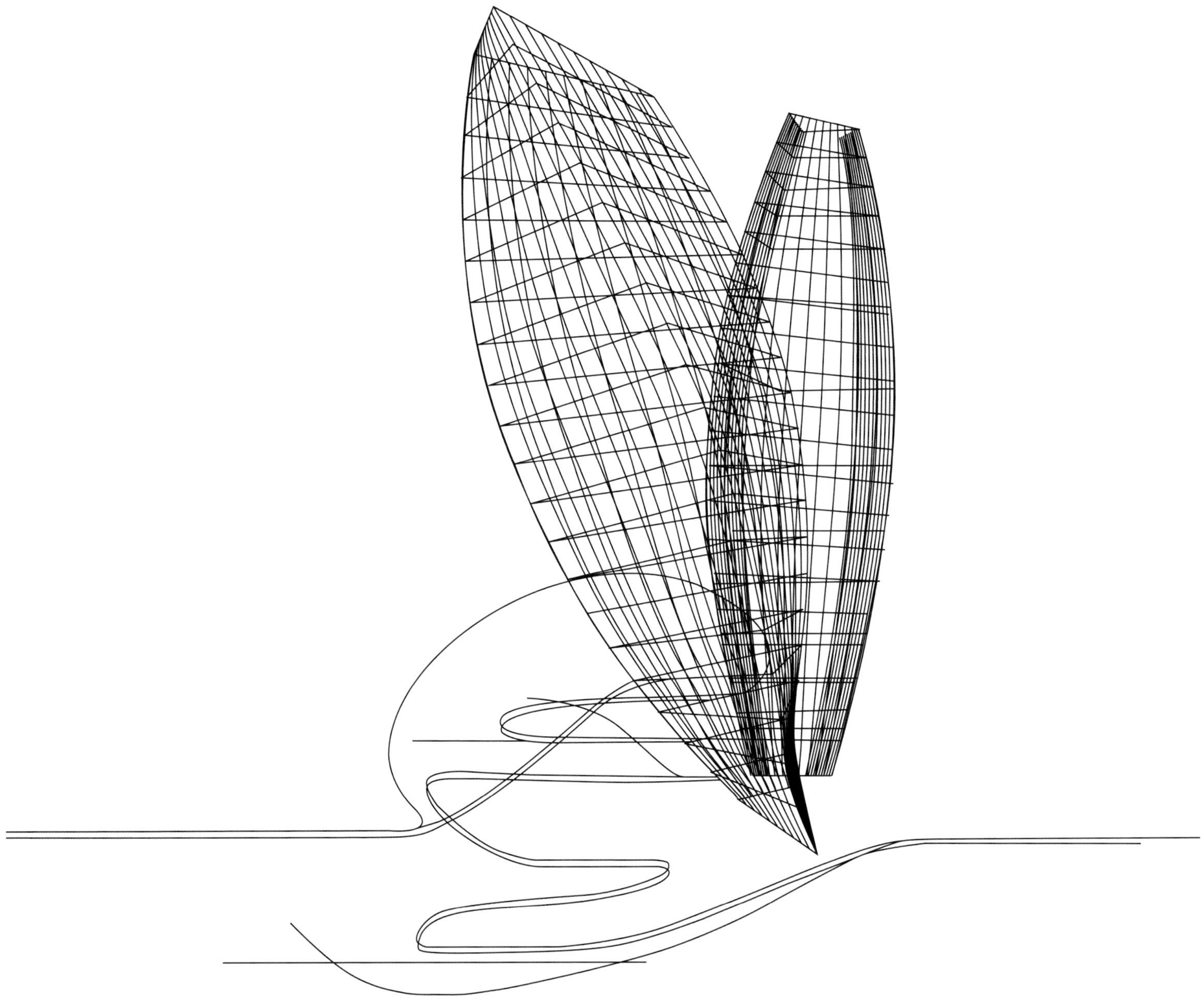

VIRGIN PLAYBOY RESORT

2000

The Virgin Playboy Resort Casino was a design concept developed to explore the opportunity of bringing together two iconic feel-good brands, Virgin and *Playboy*, to make the world's biggest adult playground: a new super-resort in Las Vegas.

The design of the resort's two towers quoted the ubiquitous *Playboy* Bunny logo, dimensionalizing the iconic ovoid shapes of the bunny's ears to create two bold sculptural masses anchored below by a sensuous, fluid, undulating multimedia exterior surface that transformed the ground plane into a dazzling, multisensory wonderland.

The design of other resort elements, including an innovative virtual card table, and the Digicard—an RFID-enabled all-access pass for guests—used cutting-edge technology in order to create an immersive and interactive environment that was truly one of a kind.

A new brand identity was also designed for this landmark joint venture: a stylized serif "VP" letterform topped with *Playboy's* iconic Bunny logo that could be read as "VIP"—a nod to *Playboy's* origins as a private members club, and an implicit promise of a truly exclusive experience.

VP

Virgin Playboy

VP
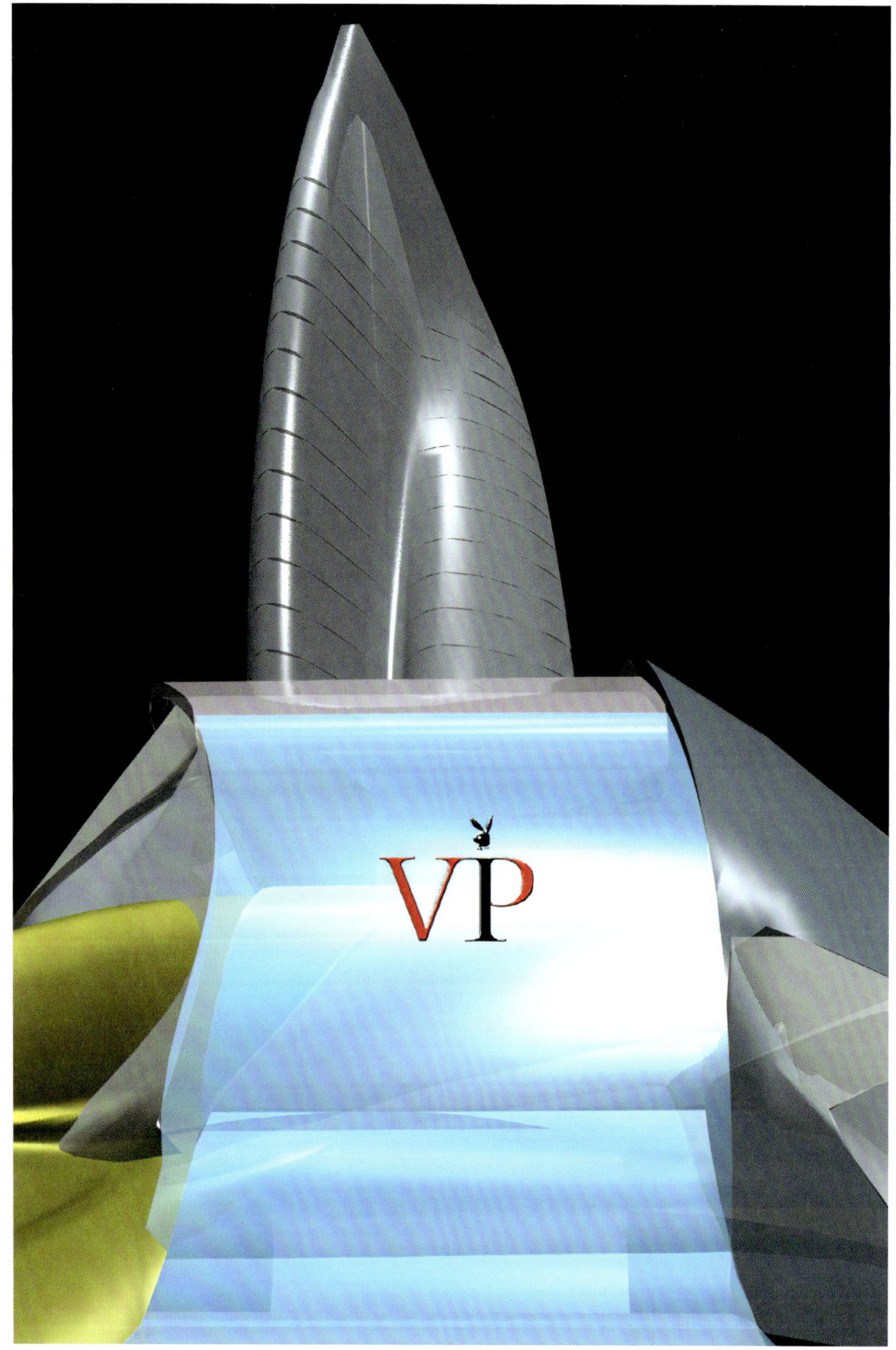
VP

Virgin
Virgin

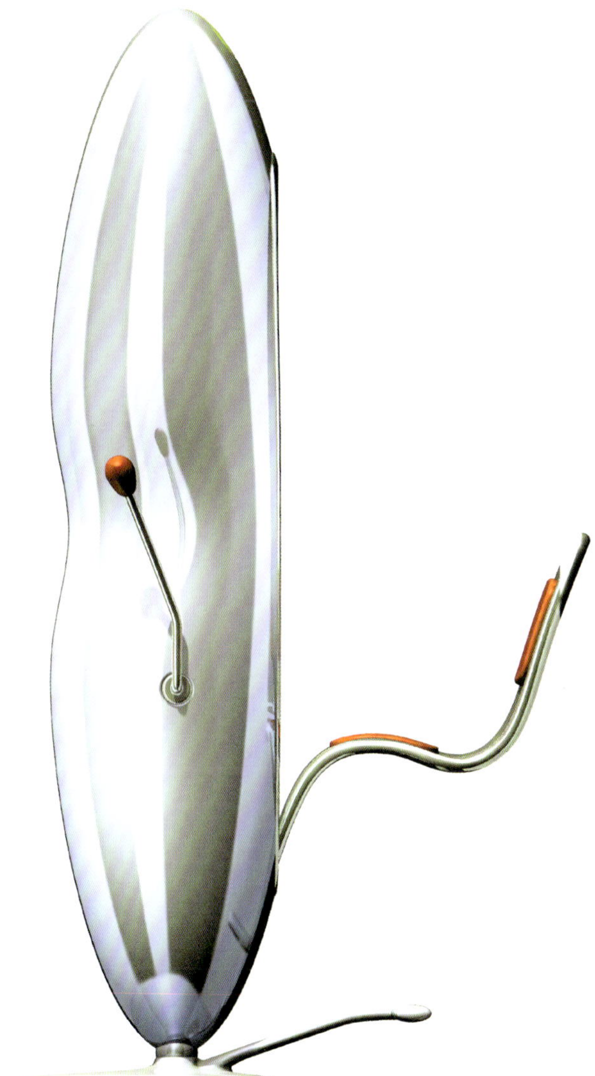

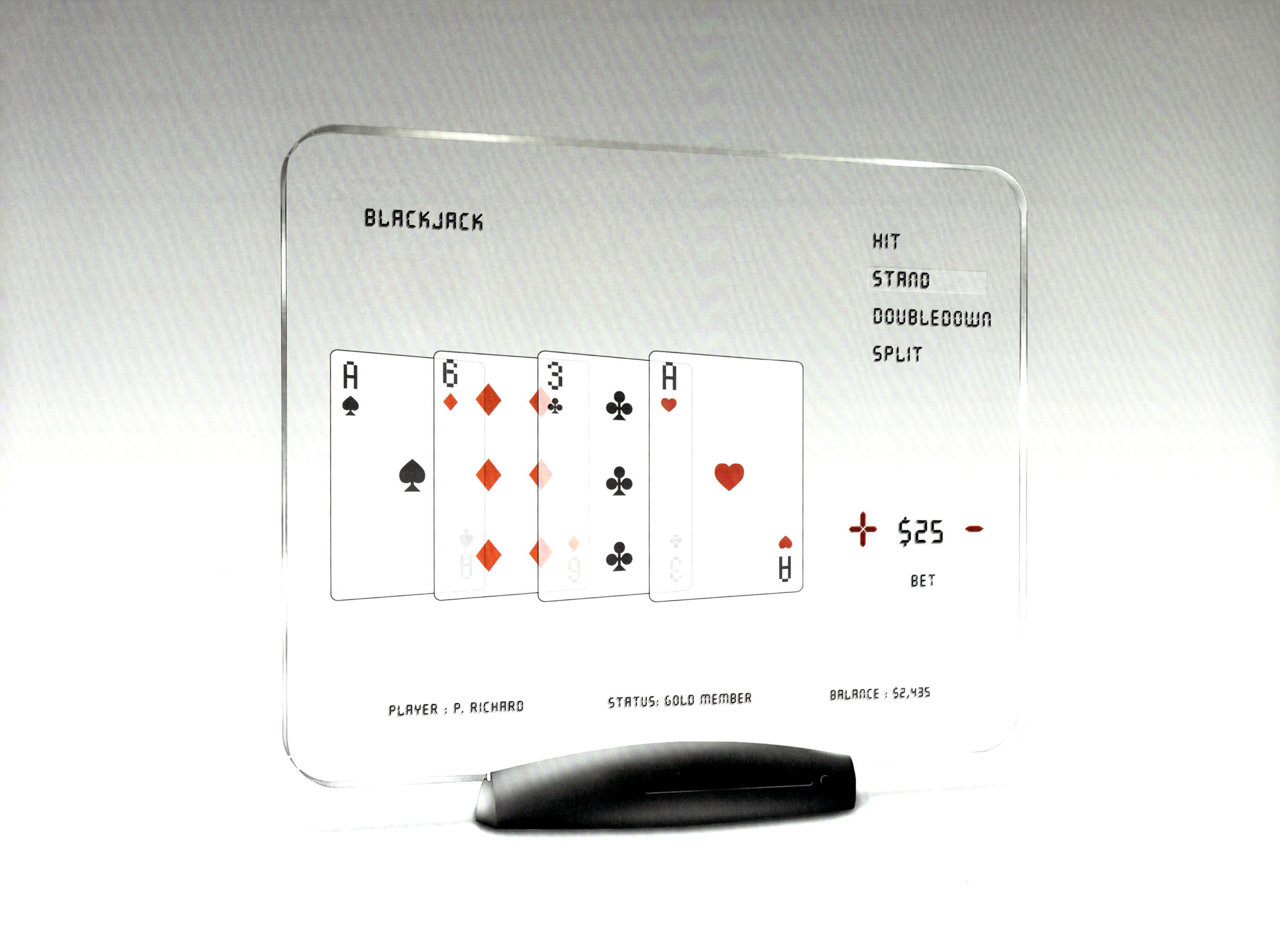
BLACKJACK
HIT
STAND
DOUBLEDOWN
SPLIT
$25
BET
PLAYER : P. RICHARD
STATUS: GOLD MEMBER
BALANCE : $2,435

THE DIGICARD

THE DIGICARD

the digicard

the digicard

the digicard

THE DIGICARD

the digicard

the digicard

THE DIGICARD

THE DIGICARD

the digicard

the digicard

con Edison

CON EDISON

2000

In the late 1990s, facing an increasingly competitive landcape following the deregulation of the utility industry in New York State, Con Edison needed to distinguish itself from its new competitors while offering a sense of trust, reassurance, and dependability to its many customers. Working closely with chairman Eugene McGrath, along with Mary McCarthy, director of corporate communications, and Frances Resheske, senior vice president of corporate affairs, a sweeping overhaul of the company's image, voice, positioning, and mission was engineered.

To replace the previous logo that had been in use for over thirty years, a new brand mark was designed: a C (for Consolidated) and an E (for Edison) fit into one coiled unit reminiscent of an electrical plug. Inspired by the iconic mid-century corporate identity designs for companies like IBM, UPS, and ABC, ConEd's new logo was immediately rolled out and applied across every brand touchpoint: employee uniforms, service vehicles, consumer materials, and even manhole covers and power plants. In the years since, Con Edison's brand mark has become an indelible part of the landscape of the city—the ubiquitous cast iron manhole cover emblazoned with the logo is now even part of the permanent collection of the Museum of Modern Art.

A new campaign was also devised that featured the concise but confident tagline "On It." An honest, straightforward, and proactive statement of intention, "On It" was aspirational without being grandiose, and it clearly promised swift resolution to whatever issue a customer might be facing, whether for a routine repair or in the event of a citywide emergency.

To reinforce this spirit of "On It," the men and women of Con Edison were featured in the accompanying communications: the workers who keep New York powered up portrayed as the everyday heroes that they were. During one photo shoot, a worker wearing a thick work glove gave a casual thumbs up to the camera, and this image quickly became the defining image not only of the campaign, but of the Con Edison brand as a whole. To this day, the iconic thumbs up glove image can be seen on the side of Con Edison service vehicles all over New York: a ubiquitous brand "billboard" that offers an approachable, human face to Con Edison's guarantee that they're on it.

conEdison
www.conEd.com

conEdison

conEdison

JOIN US
212-RECRUIT
www.ci.nyc.ny.us/nypd/

NYPD

2000

For the NYPD, a series of evocative and deeply personal communications was developed to help with diversifying recruitment to the police force. Working alongside commissioner Howard Safir, and deputy commissioner of community affairs Yolanda Jiminez, a series of massive outdoor billboards and accompanying print communications was designed that featured real New Yorkers and the officers who had helped them—or even saved their life—describing in their own words what the NYPD had meant to them.

The resulting ads lent a human face to the NYPD, and functioned not only as a recruitment ad aimed at prospective employees, but as a citywide tribute to the courage, conviction, and hard work of the many men and women who serve and protect New York City.

FDNY

2000

For over two decades, Peter has maintained a close partnership with the FDNY and the FDNY Foundation, the official not-for-profit of FDNY whose mission is to promote fire and life safety education and help the FDNY better protect New York. First invited by Fire Commissioner Tom Van Essen and Deputy Fire Commissioner Lynn Tierney to craft a fire and life safety campaign in 2000, Peter has gone on to apply his design and branding expertise to numerous public awareness campaigns, among many other initiatives.

Having served both on the board of the FDNY Foundation and as creative director of FDNY, and as one of only a handful honorary fire commissioners in the history of the 155-year history of the FDNY, Peter's relationship with the department over the past twenty years has generated countless projects, exhibitions, campaigns, and initiatives that have helped New York's bravest continue to do their vital and life-saving work.

This long-standing partnership, ongoing to this day, has allowed the FDNY and its affiliates to expand their appeal, broaden their mandate, and reach even more New Yorkers with their essential messages of fire and life safety preparedness.

MARTHA
STEWART
everyday
clean
sweep
Whistle while you work. With Martha Stewart Everyday, you'll find everything you need to keep everything in its place and make life beautiful. Get to Kmart and Get Organized.
Available at KMart, bluelight.com and through marthastewart.com®
Kmart

MARTHA
STEWART
everyday
good mornings
every
Rise and shine. With Martha Stewart Everyday, you'll find everything you need to start the day right. Get to Kmart and Get Cooking.
Available at KMart, bluelight.com and through marthastewart.com®
Kmart

MARTHA
STEWART
everyday
good mornings
everyday
From left to right: 18/10 Stainless Steel Tri-Ply 7-Piece Set, $159.99, Glass Ovenware and Storage 14-Piece Set, $19.99, Stainless steel mixing bowls, $7.99 - $12.99, Seven Everyday Kitchen Towels: Set of 7: $9.99.
Rise and shine. With Martha Stewart Everyday, you'll find everything you need to start the day right. Get to Kmart and Get Cooking.

MARTHA STEWART EVERYDAY

2001

Martha Stewart created her Martha Stewart Everyday brand with Kmart to provide mass-market access to the elegant and impeccable, yet still approachable, sense of style that made her into a household name. Millions of consumers around the world trusted Martha to recommend tried-and-true household solutions that promised to make daily life easier and more delightful, so creating a product line at an attractive price point was the next logical step for her rapidly growing empire.

A series of broadcast and print communications campaigns was created to showcase all of the different product categories that Martha Stewart Everyday offered—from bedding to dishes to outdoor furniture to home storage to cookware. Each campaign was unique to the category but consistent with the brand's overarching standards of taste and quality, and of course they all featured Martha herself as the confident, poised, relatable authority that so many people looked to for guidance and inspiration on all things home-related.

MARTHA
STEWART
everyday

MICHAEL JACKSON
INVINCIBLE
MICHAEL JACKSON
INVINCIBLE
MICHAEL JACKSON
INVINCIBLE
MICHAEL JACKSON
INVINCIBLE

MICHAEL JACKSON'S INVINCIBLE

2001

For Michael Jackson's tenth studio album *Invincible,* an overall visual identity was created that showed the world's biggest star in an entirely new light. Jackson, a longtime fan of the Pop Art movement, loved the idea of adapting the visual language of this style for his new album—a knowing nod to his undisputed status as the "King of Pop."

The resulting album cover—a high contrast monochrome photograph by legendary photographer Albert Watson—was overlaid with vivid color fields in a clear evocation of other iconic images from pop culture and pop music history like Richard Avedon's famous 1967 lithographs of The Beatles, or the silk-screened celebrity portraiture of Andy Warhol (including, of course, Warhol's famous portrait of Jackson). A custom typeface was also created that drew inspiration from the pixelated letterforms then common on the cell phones that were rapidly becoming one of the key means of pop culture dissemination—a development that Jackson himself keenly understood and cannily embraced.

Uncharacteristically minimalistic and intimate by the standards of the famously larger-than-life showman, the cover of *Invincible* showed Jackson up close and personal, staring back at the world without any of the usual superstar trappings. The stripped-down image of Jackson stared out defiantly and invited audiences to stare back and try to see beneath the surface. And yet, even in this intimate and seemingly vulnerable pose, there was still something impenetrable, unknowable, and indeed invincible about Jackson.

THERE WAS A TIME WHEN THE WORLD ASKED ORDINARY MEN TO DO EXTRAORDINARY THINGS.
"...UNLIKE ANYTHING ELSE ON TV..."
– TV Guide
"...THE BEST MINI-SERIES..."
– Time.com
GRADE: "A" "...HBO IS BETTER THAN EVER."
– Entertainment Weekly
"...THE BEST MINI-SERIES..."
– Time.com
"...BAND OF BROTHERS IS FANTASTIC— RUTHLESSLY EMOTIONAL,."
– People Magazine
GRADE: "A" "...HBO IS BETTER THAN EVER."
– Entertainment Weekly
"...RIVETING TV..."
– NY Post
GRADE: "A"
– Entertainment Weekly
TOM HANKS AND STEVEN SPIELBERG PRESENT
BAND OF BROTHERS
AN EPIC 10-PART MINISERIES BASED ON THE TRUE STORY OF THE MEN OF EASY COMPANY
SPECIAL TWO-HOUR PREMIERE SUNDAY, SEPT. 9 AT 9PM
HBO
DREAMWORKS
PLAYTONE
Jeep
The Jeep brand is honored to celebrate the men of Easy Company and all those who served in WWII.

JEEP BAND OF BROTHERS

2001

The Willys Jeep became renowned for its indispensable role during World War II combat—a vehicle that proved indispensable to soldiers fighting overseas and captivated the imaginations of the people back home. With over 640,000 of these vehicles deployed in support of the war effort between 1941 and 1945, the Jeep became the primary light-wheeled transport vehicle of the United States military and its allies, with Supreme Allied Commander (and later president) Dwight D. Eisenhower even characterizing it as "one of three decisive weapons the U.S. had during WWII."

Decades later, this proud legacy had remained central to Jeep's brand character, so it was only natural that the automaker would play a central role in the epic World War II series *Band of Brothers* produced by Steven Spielberg and Tom Hanks.

Working under the close guidance and mentoring of Spielberg, and in collaboration with senior producers Anne Thomopoulos and Tony To, an ambitious shoot was arranged on Normandy Beach—the first time any commercial production had been allowed to film there. The result was a sweeping television ad that painstakingly recreated the landing of American and Allied forces at Normandy on June 6, 1944. The narration of the spot highlighted the tremendous sacrifice of these soldiers, undertaken in the name of freedom and liberty, and honored all of the men and women who served in World War II.

Released to coincide with the premiere of *Band of Brothers* on HBO in September, 2001, the final spot proudly celebrated Jeep's storied legacy and quietly reminded viewers of the unthinkable sacrifices that had come before them to secure the freedoms we all enjoy.

Jeep

The Jeep Brand is honored to celebrate the men of Easy Company and all those who served in WWII

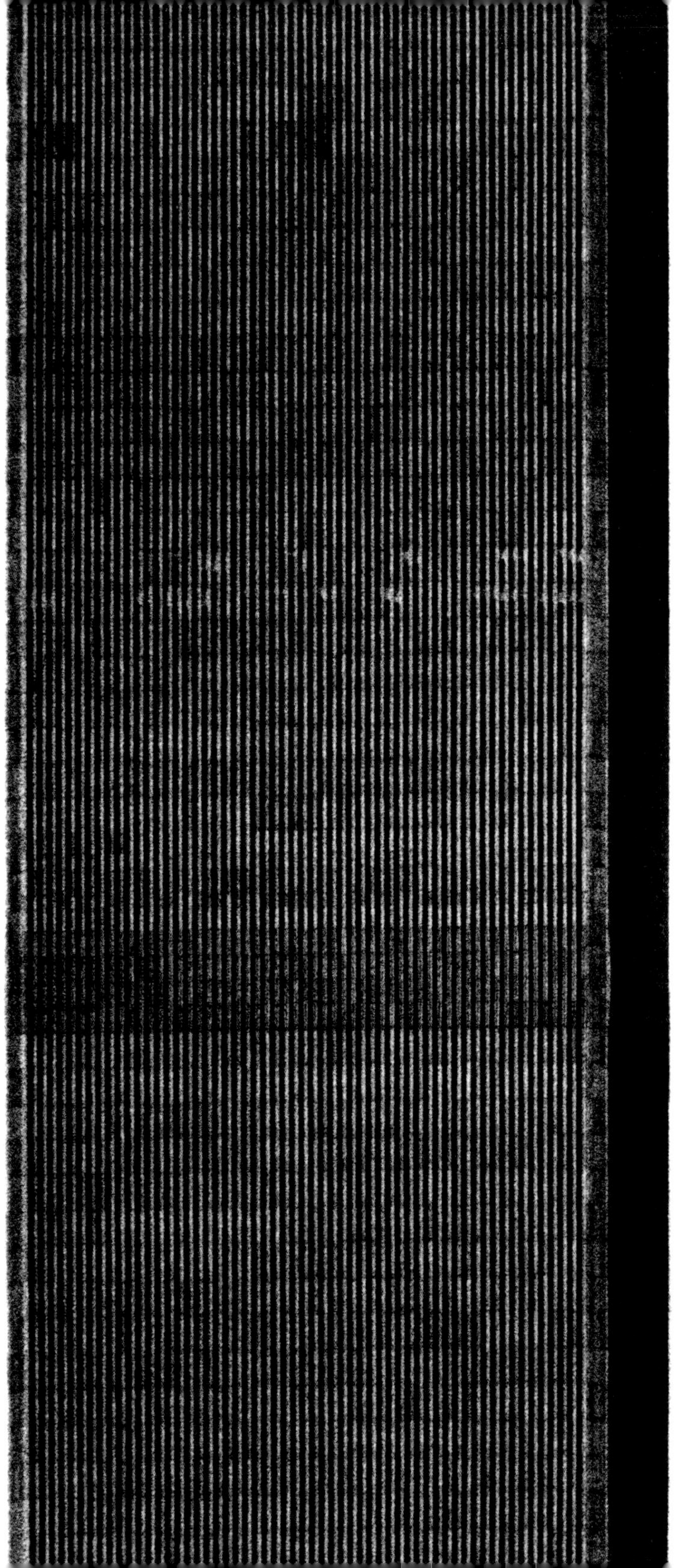

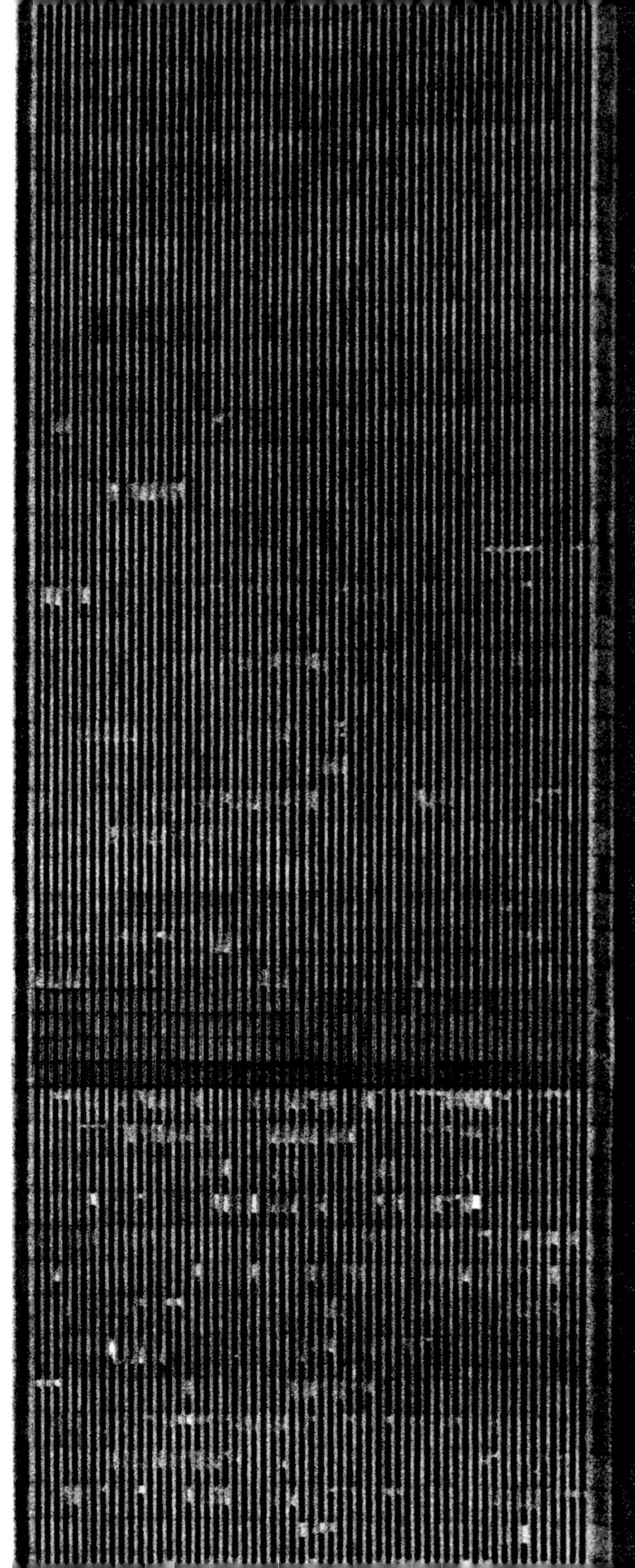

9/11 TRIBUTE MUSEUM

2001

In the aftermath of the unthinkable 9/11 tragedy, the 9/11 Tribute Museum was established to honor the memory of those whose lives were lost, and highlight the stories of family members, survivors, first responders, and residents who survived this immense tragedy and helped the city to rebuild. The 9/11 Tribute Museum highlights the tremendous spirit of resilience, compassion, and service that arose after the attacks and offers inspiration for visitors to honor the legacy of these actions through volunteering and acts of kindness in their own communities.

Working hand in hand with legendary graphic designer Lance Wyman, a new identity was created for the Center to coincide with the opening of its first permanent space in 2006. The design of the logo—two offset squares set within a circular form—represented the Twin Towers' absence: a void that will never be fully healed and a sense of loss that will never fully abate. This new logo was complemented by designs for communications, a complete merchandising curriculum, and a comprehensive visitor experience strategy, among other endeavors—an ongoing effort that continues to this day.

A similar meditation on absence also came to be echoed in the nearby 9/11 Memorial: comprised of two one-acre reflecting pools that each occupy the footprint of the towers that once stood there, this unfillable void in the heart of lower Manhattan gives the space over to the memory of those who perished there on September 11, 2001.

9/11
TRIBUTE
MUSEUM

VOGUE & CFDA

2001

To honor the heroes of 9/11, raise funds for the victims' families, and help the hard-hit fashion industry of New York get back on its feet, *Vogue* and the Council of Fashion Designers of America collaborated to create new charitable initiative called Fashion for America. *Vogue's* editor in chief Anna Wintour personally invited Peter to create a new symbol for the venture—a needle and thread stitching an American heart back together—that represented the healing of America: a nation bound together in grief, but also united by hope, faith, and common purpose.

The logo, at once a clever nod to the program's fashion industry roots and an optimistic symbol of collective healing in the aftermath of unfathomable tragedy, went on to adorn a wide range of limited-edition merchandise, including most famously the white T-shirt that supermodel Gisele Bündchen wore on the December 2001 cover of *Vogue*.

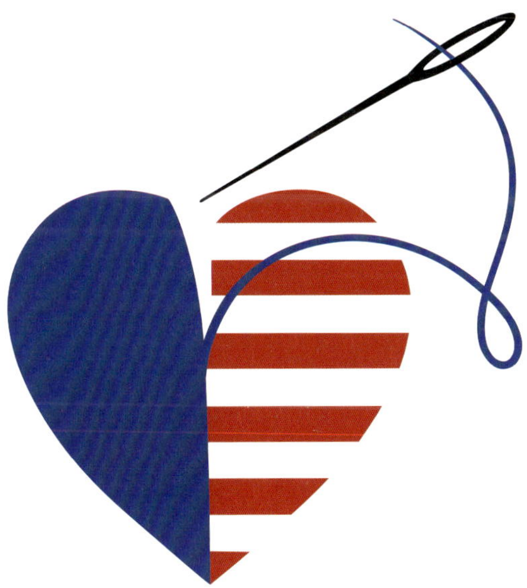

RAM 1500
DODGE
4X4

DODGE RAM

2002

For the introduction of the 2002 Dodge Ram, rock superstars Aerosmith were enlisted to helm a multi-platform promotional partnership that introduced the all-new Ram pickup truck to consumers everywhere.

Two thirty-second television spots were crafted, featuring front man Steven Tyler urging audiences to "Take Life by the Horns," while an accompanying print campaign hit newsstands at the same time, including poster-sized inserts in *USA Today.*

As part of the wide-ranging collaboration, Dodge served as a sponsor of Aerosmith's blockbuster *Just Push Play* tour, and a limited number of Aerosmith-branded Dodge Rams were also produced. Meanwhile, Aerosmith performed a series of impromptu private concerts for a select group of customers who had purchased the new Ram: a once-in-a-lifetime thrill for some of Dodge's most loyal fans.

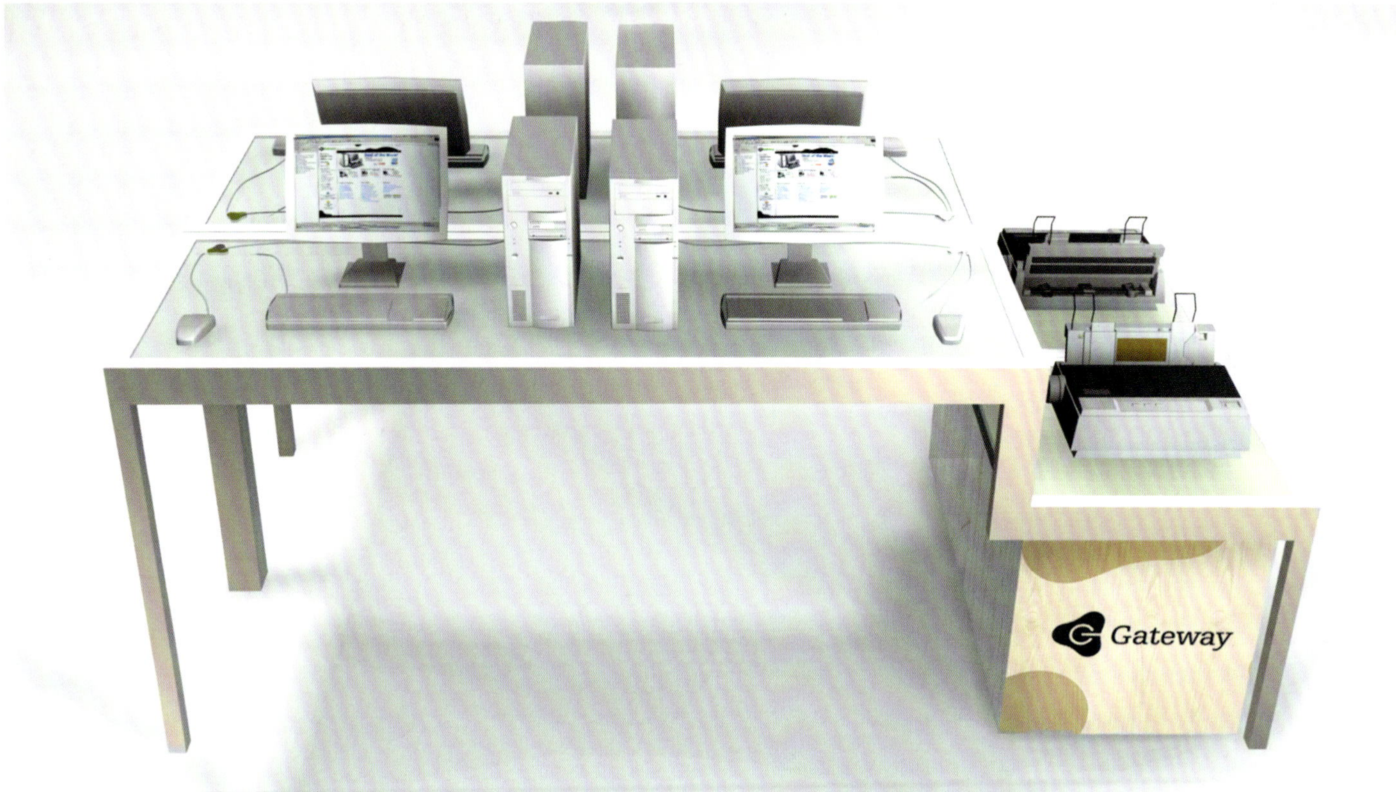
Gateway
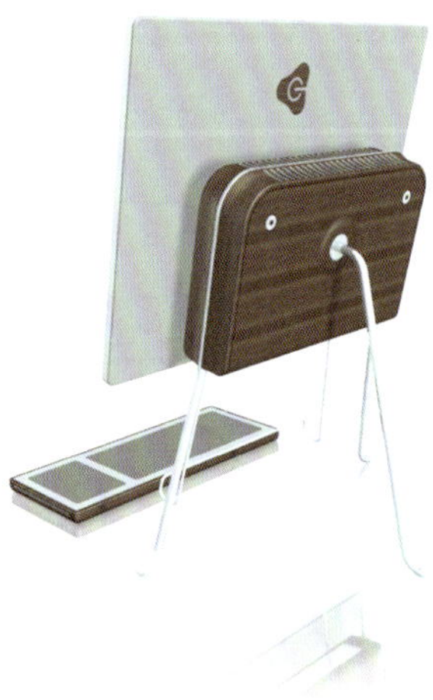
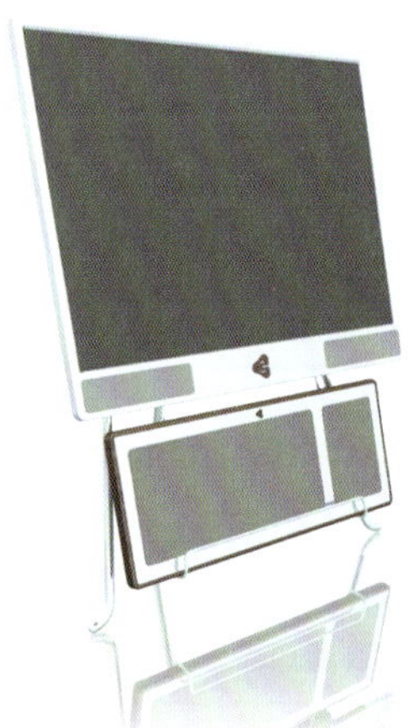

GATEWAY

2002

Founded in 1985 on a farm outside of Sioux City, Iowa, by entrepreneurs Ted Waitt and Mike Hammond, Gateway Inc. first came to widespread attention in 1991 when it started shipping its computer hardware in distinctive cow print-patterned boxes—an unexpected move for a technology company, but one that nonetheless became an instantly recognizable brand signature.

By the early 2000s however, recognizing the need to reimagine the company's role in a rapidly changing world, Waitt and the rest of Gateway's senior leadership sought out a fresh perspective to help them envision a new set of possibilities for the brand and engineer a comprehensive, company-wide refresh.

As part of this brand reimagining, a new brand mark cleverly rotated a power symbol from its usual vertical orientation to a horizontal one, and in doing so, formed a stylized "G"—itself nestled inside a sleek amoeboid form reminiscent of both the cow print patterns associated with the Gateway brand. Packaging featured a clean and sparse visual style punctuated by unexpected natural textures that lent a warmer, more approachable air to a product that was still, for many consumers, intimidating and foreign.

Speculative designs for a range of new desktop and portable devices were developed in an effort to bring Gateway's bold new design ethos to the hardware itself. Employing both bright and colorful plastics and natural materials, these concepts prefigured many of the form factors that are now commonplace, including all-in-one desktop units, razor-thin displays, ultracompact computer towers, and touch-screen tablets.

Concept designs for new retail experiences were also explored, with sleek, minimalist flagship stores proposed as aspirational destinations to complement Gateway's highly successful direct-sales, built-to-order business model.

Gateway

2
3
EXPERIENCE

3
4

need for speed
need for speed
700L
700SE

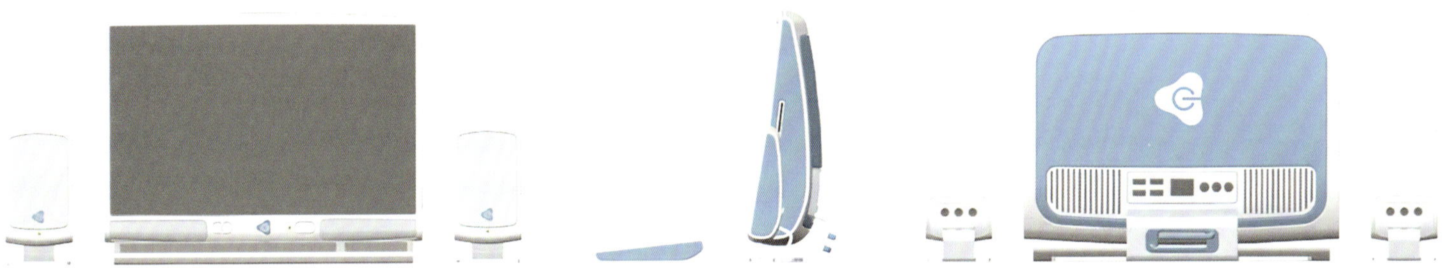

AIWA

2003

For consumer electronics giant Aiwa, a concept for a new brand identity system was developed and proposed. Rooted in the idea of "digital senses," the proposal sought to combat the ever-increasing complexity of technology by creating an entirely new, natively digital syntax purpose-built for the Digital Age.

The Aiwa logo, already bold and energetic in character, was animated with pulsing shockwaves and digital noise to convey a sense of energy, emotion, and dynamism that would help to position Aiwa as a powerhouse at the vanguard of innovation. A custom alphabet was also designed that evoked the different forms of digital communications—from Morse code to computer code—and gave Aiwa a distinctive and proprietary language for communicating with its customers.

Bank of America

BANK OF AMERICA

2003

To help Bank of America build trust and rapport with its customers, a completely reimagined approach to its brand standards, customer journey, and retail operations was developed to transform the experience of in-person banking.

This effort involved rolling out a new look and feel for retail environments, with all furniture, finishes, fixtures, and materials chosen to create an inviting, warm, and human-scaled experience in-branch. The idea of transparency was also applied as a design tool in their new bank environments, with redesigned branches featuring glass walls, open façades that broke down barriers between bank and customer, and design concepts for lighter, smaller, more user-friendly self-service kiosks.

ATMs were completely redesigned to improve and streamline the customer experience through every step of the process. The vivid red color of the new machines was designed to be attention grabbing and stand out against the bland backdrops that most bank ATMs find themselves in, while machine facings, keypads, and on-screen directions were simplified for increased ease-of-use and greater accessibility for a wider range of customers.

To encourage customers to browse and linger at branches, new customer paths were introduced to eliminate the traditional lineups that were an unfortunate feature at nearly every other bank. This radically redefined new approach to the customer's journey completely inverted the traditional rhythm of a visit to the bank: rather than waiting in a line to speak to a teller behind a counter, free-floating kiosks and mobile attendants offered customers assistance and guidance as soon as they walked in the door.

Now a commonplace feature of bank branches and retail environments more broadly, this free-roaming, customer-centric retail model helped to transform in-person banking from a frustrating and time-consuming ordeal to a pleasant, efficient, and stress-free experience.

Bank of America

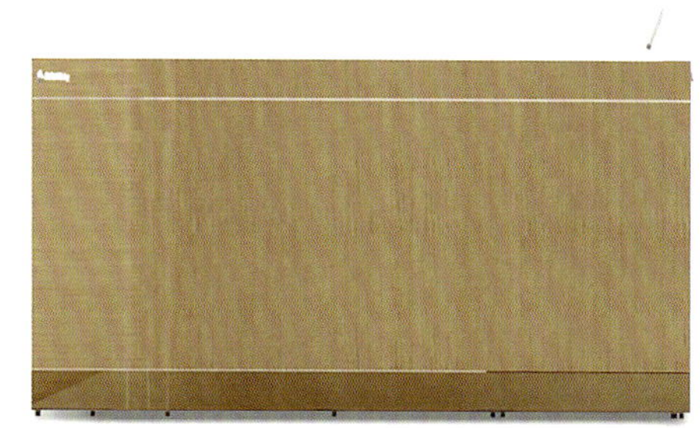

Bank of America

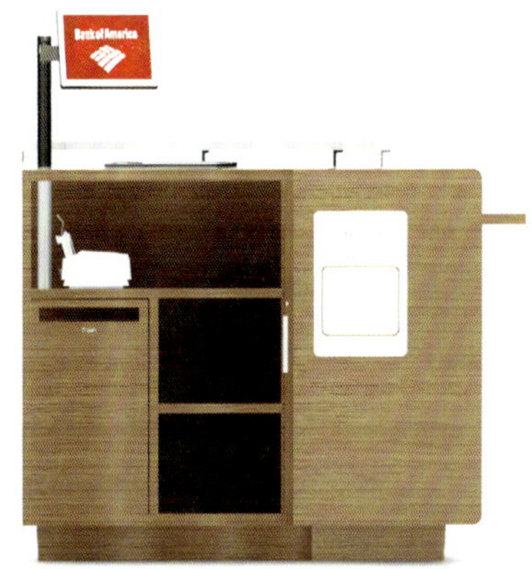

Bank of America

Bank of America

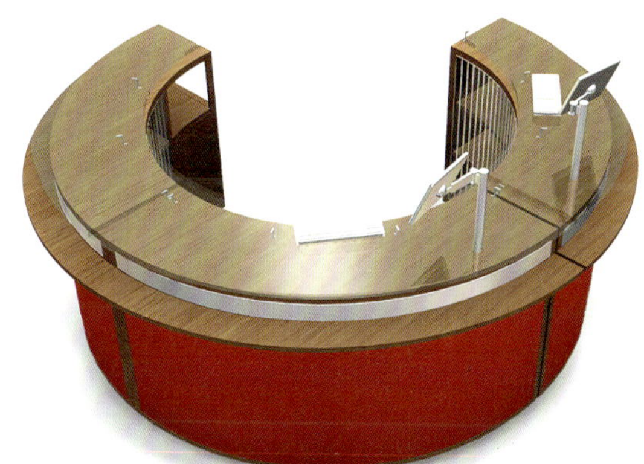

Bank of America
Bank of America
Bank of America
Bank of America

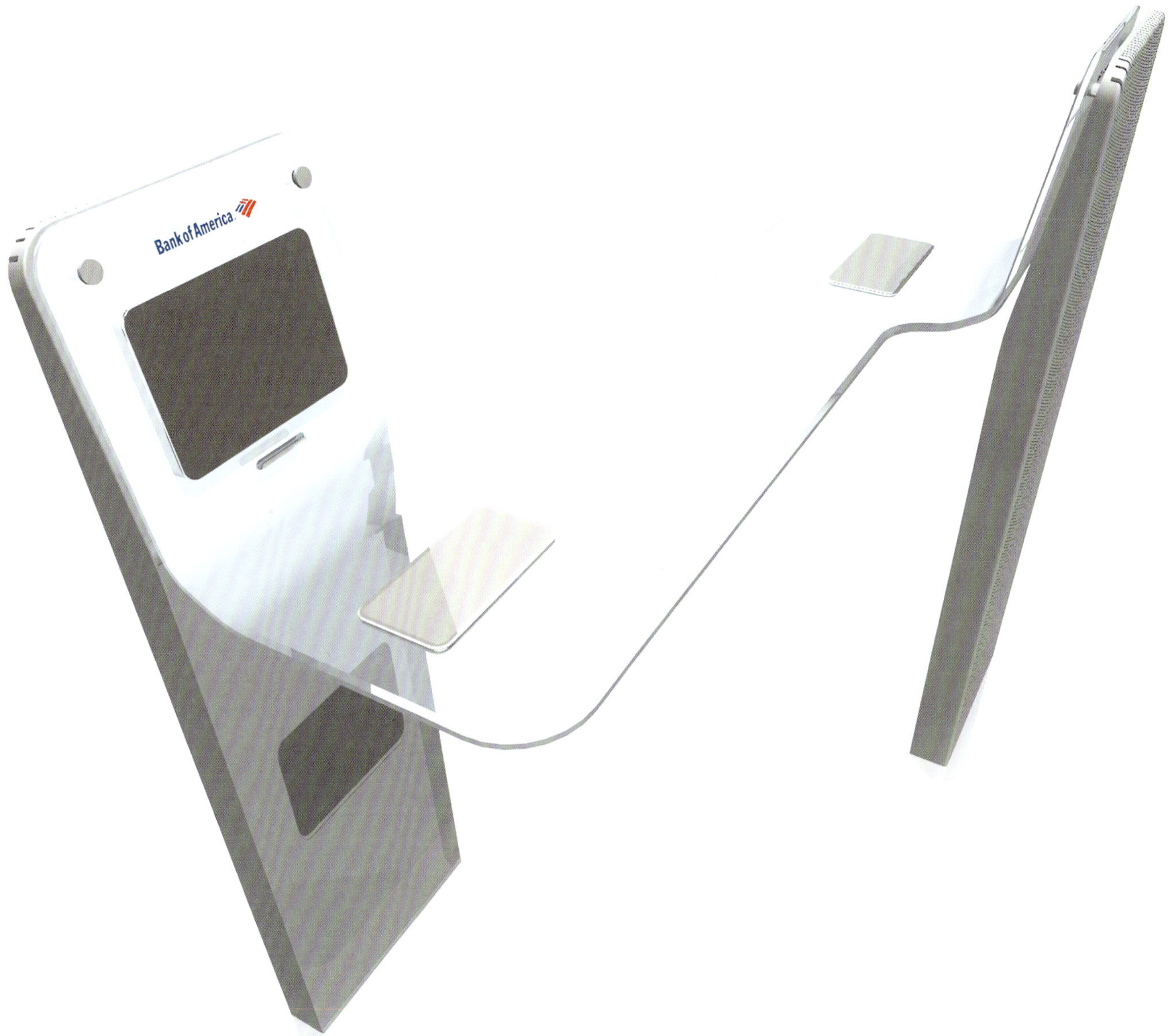
Bank of America

Wynn

WYNN RESORTS

2003

When Wynn Resorts was contemplating a new name for their new flagship Las Vegas property, their first choice was Le Rêve—so named for the Picasso masterpiece of the same name that was planned to be hung in the lobby of the luxury resort and casino. And while that name was undeniably beautiful and evocative, to name a soaring, nearly five thousand-room hotel after a diminutive painting discreetly presented in the lobby felt somehow out of step with the ambitious scale of the development.

But as is so often the case, the solution was ultimately found hiding in plain sight: in the name of the company, and its founder, Steve Wynn. Wynn Resorts … as in "win." It was a distinctive, authentic, and radically understated alternative to the themed and gimmicky resorts that Las Vegas was known for. The decision to name the resort after its founder also followed a long legacy of business titans and magnates "self-branding" the companies they ran: legendary figures like Conrad Hilton, Henry Ford, and J.P. Morgan. By lending his name to his flagship resort, Steve Wynn in essence personally underwrote a guarantee to every guest: Wynn had to be the best because it was his name on the line.

In developing the visual identity for this new destination, once again, the most obvious choice was the best one: the simple handwritten signature of the founder. In a sea of flashing lights and aesthetic excess, the simple, austere, and elegant Wynn signature stood out for all the right reasons.

Wynn Las Vegas represented the next generation of destination resorts and casinos in Las Vegas: less about spectacle and more about providing an elevated, refined, world-class experience.

CHUBBY
SHORTY
GRAND MASTER
GRAND MASTER
STRETCH
STOUT

DAVIDOFF ZINO PLATINUM

2003

In the early 2000s, a new generation of stylish, city-dwelling twenty-somethings had begun to embrace cigars as a new kind of status symbol. Among this younger set, cigars began appearing at high-end nightclubs and restaurants, but given that most cigar brands were explicitly marketed toward an older clientele and represented a staid and stodgy kind of sophistication, there was a huge market opportunity for a new brand of cigars to be launched that would speak directly to this new generation of cigar aficionados.

Building on this insight, a new brand concept called Zino Platinum was developed and presented to world's premier cigar manufacturer, Davidoff. A fusion of time-honored cigar tradition and current pop culture and lifestyle trends, Zino Platinum cigars were engineered as a large-format, hand-rolled product—Old World charm in a next generation wrapper. Davidoff immediately gave the green light to accelerate this new brand and product range to market.

A high price point was matched by ultra-luxurious branding befitting the prestige of Davidoff: everything from its name, to its logo band, to its packaging and custom accessories—and even its elegant, slender, tapered form—spoke to the elevated aspirational currency of this new generation. Immediately after launch, Zino Platinums could be seen being enjoyed by countless celebrities and other young power players, which led to a rapid adoption of this new brand of cigars among those who valued and embodied a new type of status symbol.

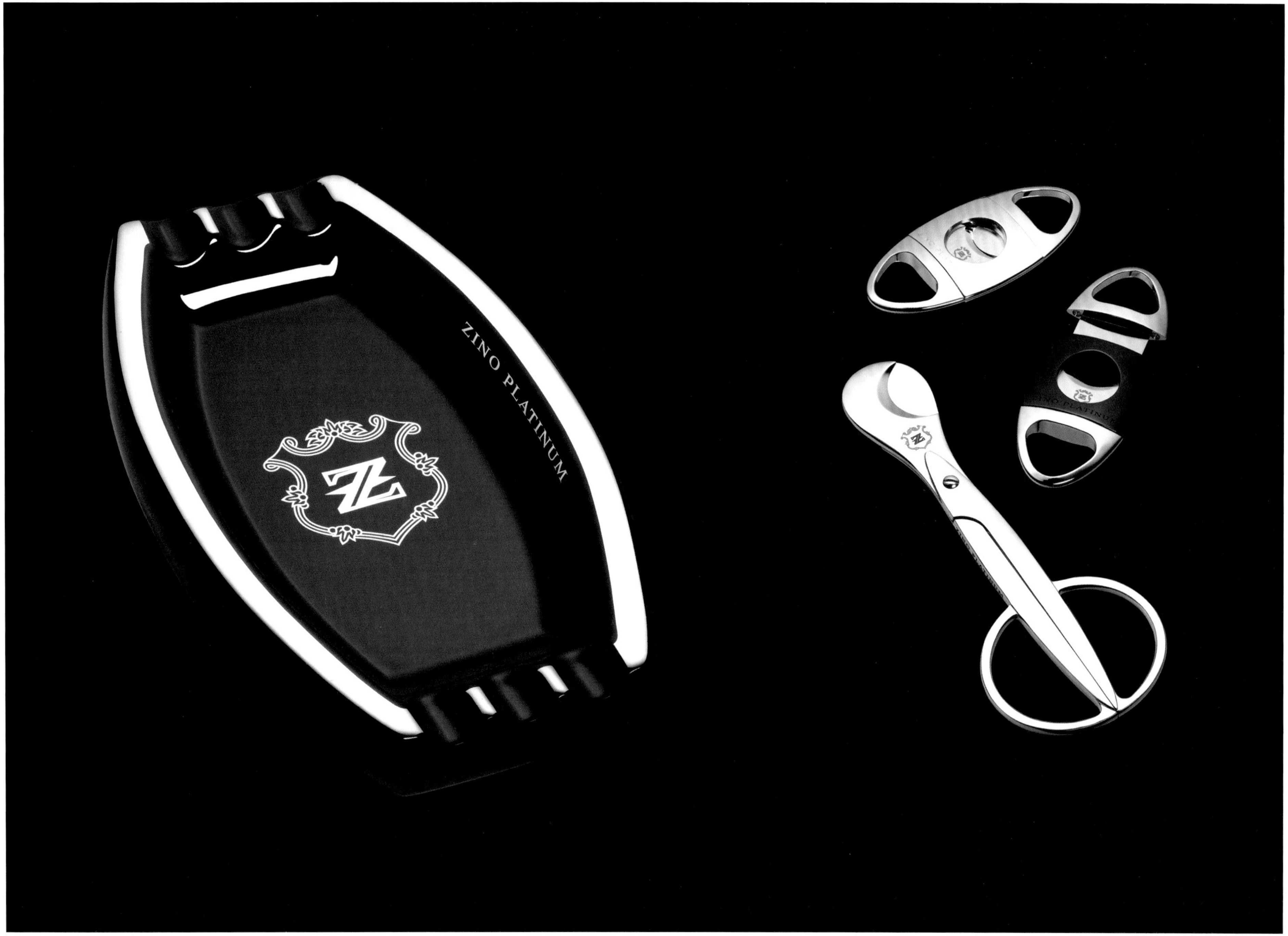
ZINO PLATINUM

Zino
ZINO PLATINUM CROWN SERIES
STRETCH
Zino
ZINO PLATINUM CROWN SERIES
CHUBBY ESPECIAL
Zino
ZINO PLATINUM CROWN SERIES
DOUBLE GRANDE

Zino
ZINO PLATINUM CROWN SERIES
BARREL

ZINO PLATINUM
HAND MADE
GRAND MASTER
Zino
ZINO PLATINUM
SCEPTER SERIES
GRAND MASTER
12 FINE CIGARS
HANDMADE IN THE DOMINICAN REPUBLIC

Zino
ZINO PLATINUM
SCEPTER SERIES
Shorty
Zino
ZINO PLATINUM
SCEPTER SERIES
Stout
Zino
ZINO PLATINUM
SCEPTER SERIES
Low Rider
Zino
ZINO PLATINUM
SCEPTER SERIES
Grand Master
Zino
ZINO PLATINUM
SCEPTER SERIES
Chubby

DAVIDOFF DAWGS

2003

Davidoff's Zino Platinum line proved so popular that it spawned a limited edition line of cigars called Davidoff Dawgs. This new line retained the dapper flair of the Zino Platinum range, while adding a youthful, whimsical twist. A memorable campaign was created to evoke the mood and ambiance of a Victorian-era gentlemen's club, albeit one with some decidedly unconventional canine members.

ROCKPORT

2003

Known for selling comfortable footwear that's rock solid, Rockport had always enjoyed a reputation for practicality and durability, yet the brand lacked an enduring emotional connection with its customers. They needed to make their brand stand for something more than just well-made, sensible footwear by expressing the bigger, more emotionally resonant idea at the core of their identity—namely, love and care for the foot.

The Rockport Comfort campaign featured evocative images that conveyed the brand's unrivaled care for the feet of their customers and spoke to the appeal of pure comfort. Featuring images of supported, soothed, and cared-for feet, the campaign told consumers everything they need to know about the brand without even showing the product.

A fresh and modern brand mark was developed to accompany the campaign: a sleek, lowercase "r" in a circle that announced Rockport as a vibrant, design-focused contemporary brand.

A new quarterly publication, *Feat,* was also launched, featuring inspirational content that spoke to the mantra of "adventures in living" and continued to elevate Rockport's reputation as a true lifestyle brand.

The unique combination of comfort and beauty honored Rockport's heritage of engineered comfort and fit while also helping to reposition the brand as a uniquely stylish and aspirational brand.

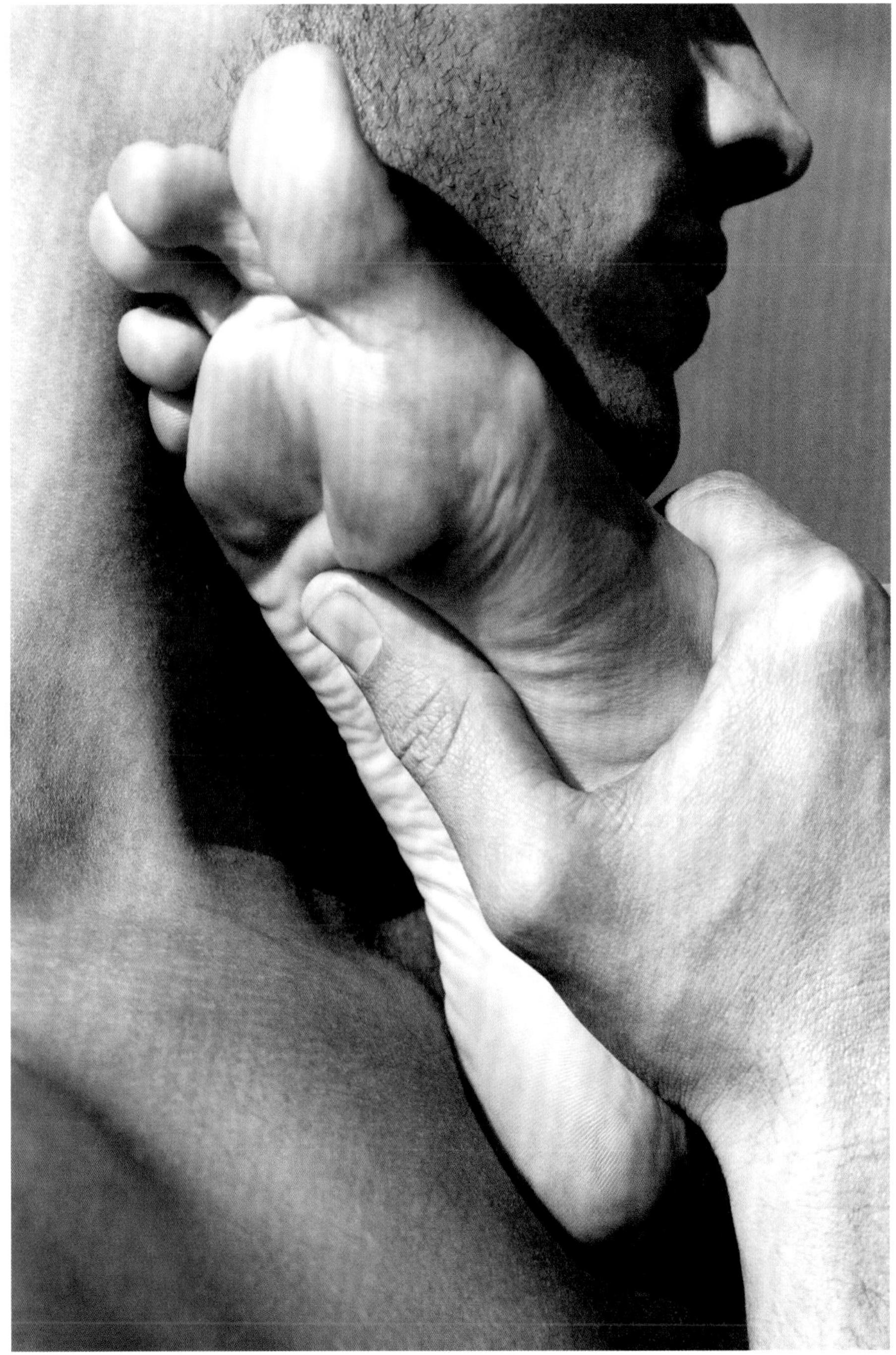

Gibson
Lucille
B.B. King

feat
ROCKPORT ADVENTURES IN LIVING
FALL/WINTER 2003 $4.95
blues master
B.B. King keeps the good times rollin'
secret hideaways
Three undiscovered islands
into the wild
One family's Alaskan adventure

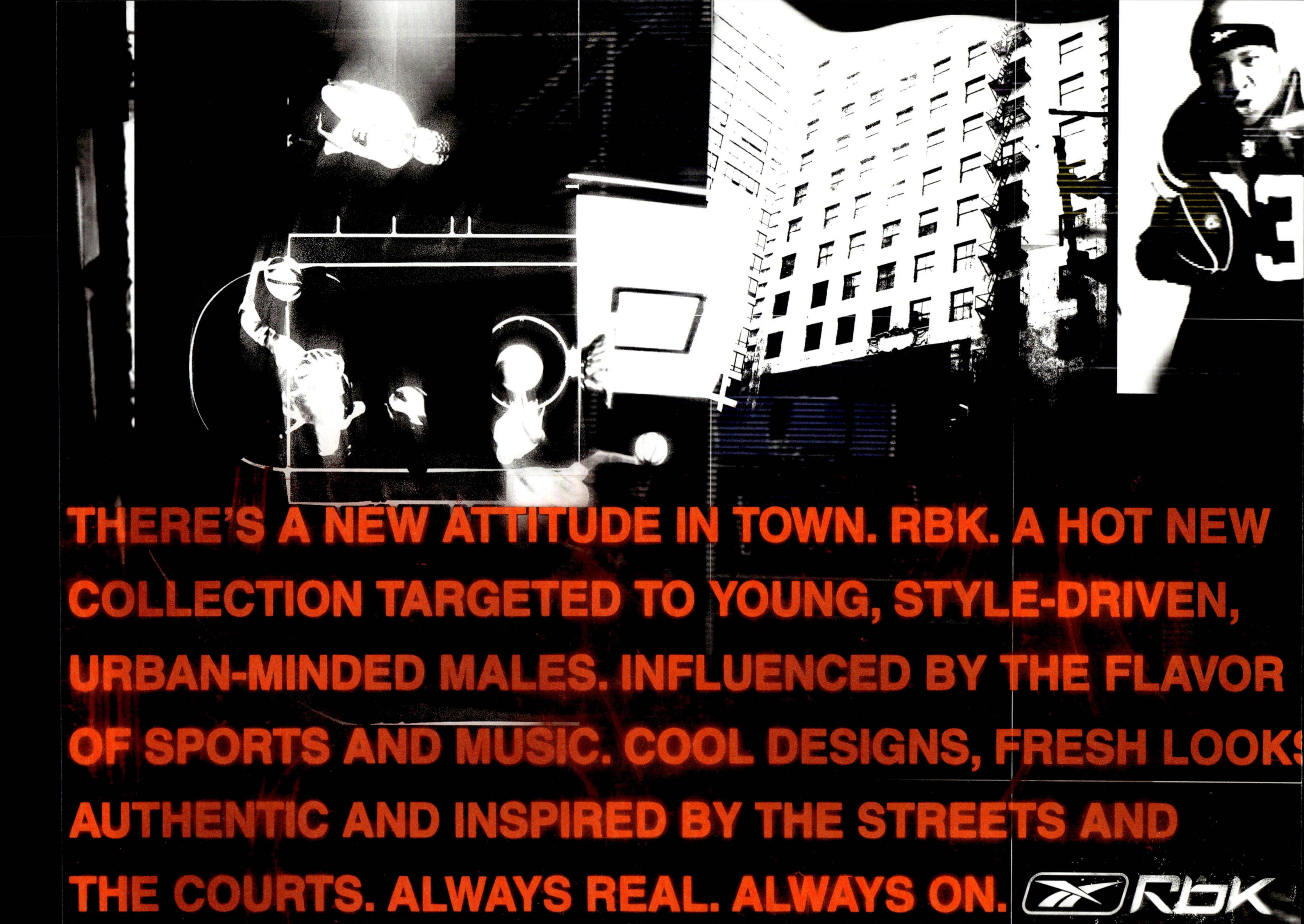
THERE'S A NEW ATTITUDE IN TOWN. RBK. A HOT NEW COLLECTION TARGETED TO YOUNG, STYLE-DRIVEN, URBAN-MINDED MALES. INFLUENCED BY THE FLAVOR OF SPORTS AND MUSIC. COOL DESIGNS, FRESH LOOKS AUTHENTIC AND INSPIRED BY THE STREETS AND THE COURTS. ALWAYS REAL. ALWAYS ON.
Rbk

RBK

2003

Though Reebok had always enjoyed significant success with its performance footwear, by the early 2000s the venerable sneaker brand had found itself sidelined by other players in the lifestyle segment. To maintain relevancy in this heavily competitive, trend-driven market segment, Reebok was in urgent need of a reboot.

The overarching ambition was to connect to younger consumers in a way that felt authentic and natural to them; to talk to them the way they talked to each other—which in the early 2000s was increasingly over text, using quick, condensed, short-form abbreviations. The quickest way to text "Reebok," for example? Simple. Just like the symbol on the NYSE stock ticker does: eliminate the vowels. R-B-K.

Rbk as a bold effacement of its own name positioned Reebok as a fearless, dynamic, confident brand that didn't take itself too seriously. To bring this new attitude to life, a completely refreshed brand identity was developed that drew from the visual language of street art and graffiti, and a new flagship retail experience was opened in downtown Philadelphia that served as a multimedia, multisensory immersion into the new energy of the brand.

As an energetic and youthful brand, Rbk needed to authentically reflect and engage the voices that were most relevant to the younger generation. To help accomplish this, music industry executives were brought in to help broker relationships with artists like Jay-Z and 50 Cent. Far from conventional sponsorship deals, these 360-degree partnerships leveraged the creative genius and business acumen of these icons to create signature products and new Rbk sub-brands that became genuine pop culture phenomenons.

Rbk's 2003 launch and subsequent breakout success helped to revitalize the Reebok brand, repositioning it as a youth-driven and style-conscious lifestyle brand, backed up by true performance engineering and Reebok's respected heritage of speed, power, and agility.

Rbk

RbK

RbK

ONE WAY
STOP
Rbk
ENTER
CHAMPS
VIACOM
Rbk

Rbk
Rbk

Rbk

RbK
S. Carter
COLLECTION

RBK S. CARTER COLLECTION

2003

By the early 2000s, musicians—and particularly hip-hop artists—had indisputably become the driving force in American popular culture. Artists like Jay-Z were not only breaking new ground with their prolific musical output; they were branching out into all facets of entertainment, working across mediums and sectors to create wildly successful lifestyle empires that brought their unique style to vast new audiences.

Musicians—from the Ramones to Run DMC—had always been known for their allegiances to certain brands of sneakers, but no sneaker company had truly capitalized on that connection before with a dedicated line of products personally designed by a music icon.

The Rbk S. Carter Collection capitalized on Jay-Z's massive appeal through a fully integrated partnership that was timed to feature an S. Carter Collection sneaker on Jay-Z's then-upcoming album cover, which helped the S. Carter become the fastest-selling shoe in Reebok's history and made Reebok newly relevant, accessible, and desirable to a vast new demographic.

G Unit
FOOTWEAR BY RbK
G Unit

RBK G-UNIT

2003

In partnering with 50 Cent to develop a signature shoe for his G-Unit brand, Reebok helped to bring the superstar rapper's signature style to a massive new audience, a move that opened the door for countless other artists and entertainers to leverage their personal brands into multi-million dollar global juggernauts. While other brands tended to broker conventional endorsement deals or issued limited edition items bearing a celebrity's name, Reebok opted to create a new and stand-alone sub-brand with 50 Cent, built from the ground up.

Communications developed for the initial product launch featured 50 in a series of striking, stoic poses: unexpectedly intimate snapshots of the superstar and his signature shoe that heralded the arrival of an entirely new kind of lifestyle brand.

REEBOK YAO MING

2003

Conceived of as a partnership between Reebok and Chinese-born basketball phenomenon Yao Ming, the Reebok Yao Ming brand was created to appeal to a truly worldwide audience. Recognizing the unprecedented international phenomenon that Yao had become during his time playing for the Houston Rockets—the first Chinese NBA superstar—this new brand was engineered to celebrate his proud heritage, while also embracing a truly global outlook.

A distinctive new brand mark was created that stacked the characters of Yao's English name into a vertical block—a gesture that subtly nodded to the graceful calligraphic style of Chinese characters and the conventions of the writing system.

A campaign for his signature shoe and line of athletic wear was accompanied by a publication, *Yao's Big Red Book*. Filled with striking collaged visuals, traditional Chinese aphorisms, and inspirational quotes from Yao himself, the publication was a clever remix of centuries-old cultural artifacts and contemporary pop culture, and a nod to Yao's global advocacy for Chinese contributions to sports and culture.

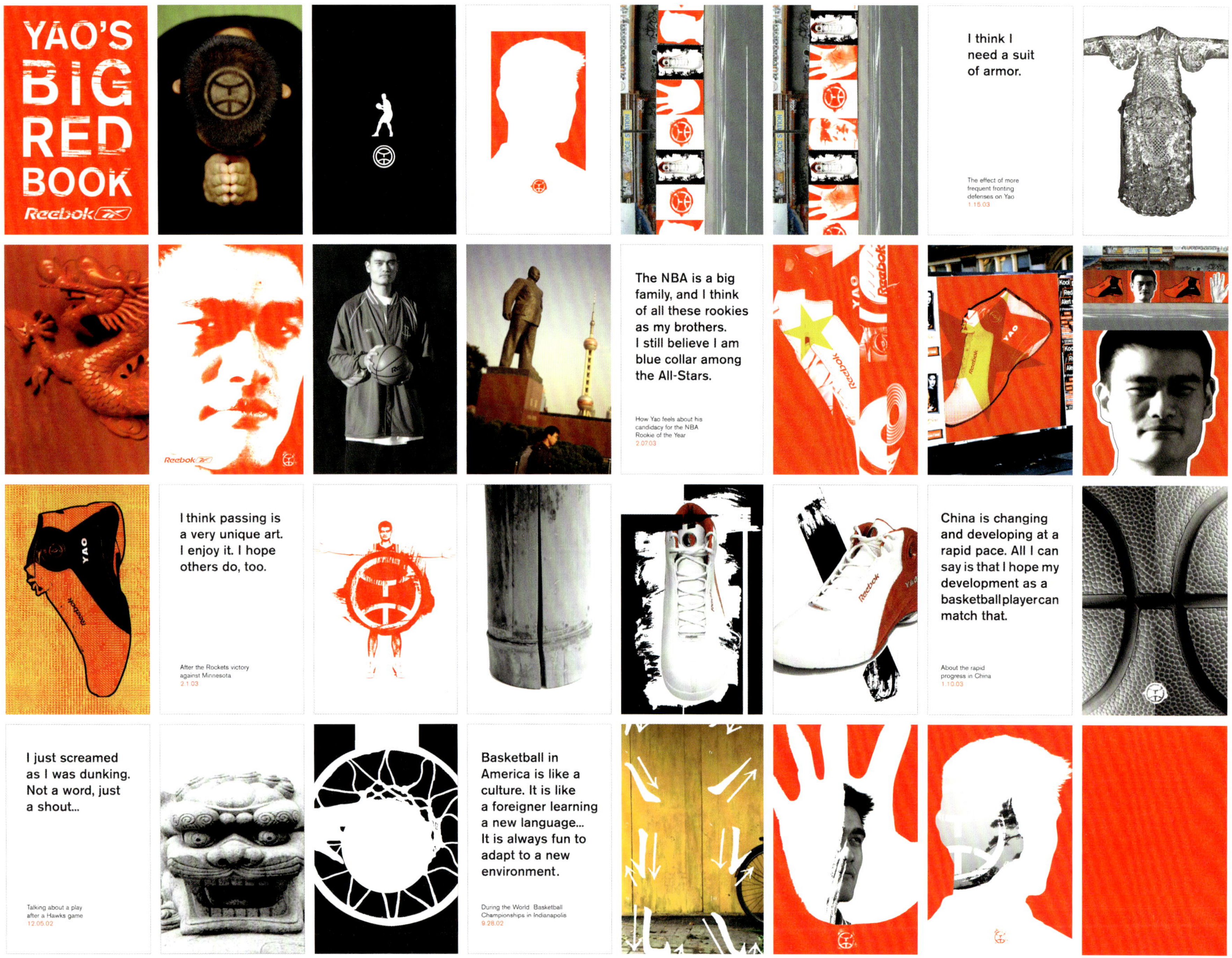
YAO'S
BIG
RED
BOOK
Reebok
I think I
need a suit
of armor.
The effect of more
frequent fronting
defenses on Yao
1.15.03
Reebok
The NBA is a big
family, and I think
of all these rookies
as my brothers.
I still believe I am
blue collar among
the All-Stars.
How Yao feels about his
candidacy for the NBA
Rookie of the Year
2.07.03
YAO
Reebok
YAO
I think passing is
a very unique art.
I enjoy it. I hope
others do, too.
After the Rockets victory
against Minnesota
2.1.03
Reebok
China is changing
and developing at a
rapid pace. All I can
say is that I hope my
development as a
basketball player can
match that.
About the rapid
progress in China
1.10.03
I just screamed
as I was dunking.
Not a word, just
a shout...
Talking about a play
after a Hawks game
12.05.02
Basketball in
America is like a
culture. It is like
a foreigner learning
a new language...
It is always fun to
adapt to a new
environment.
During the World Basketball
Championships in Indianapolis
9.28.02

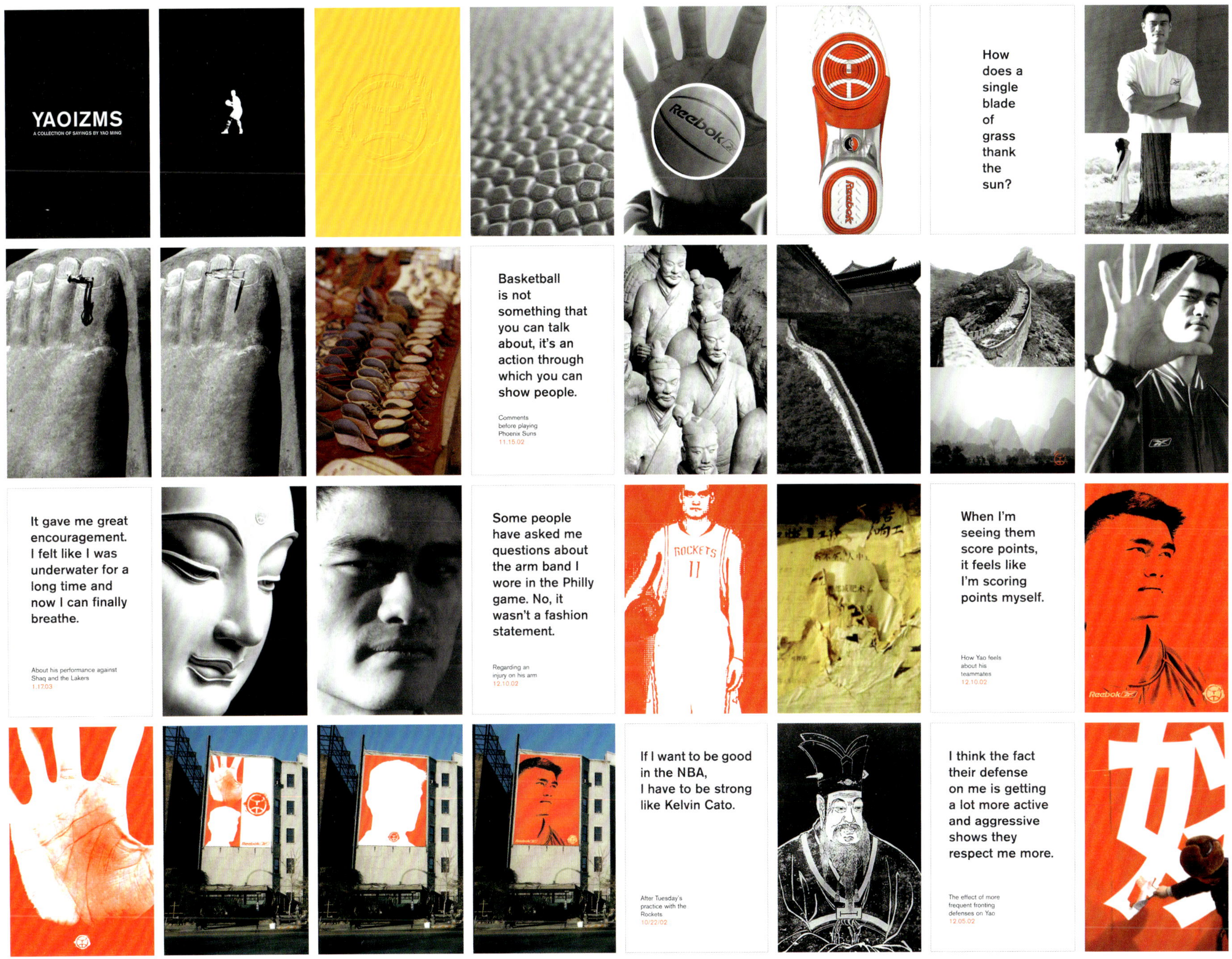
YAOIZMS
A COLLECTION OF SAYINGS BY YAO MING
How does a single blade of grass thank the sun?
Basketball is not something that you can talk about, it's an action through which you can show people.
Comments before playing Phoenix Suns
11.15.02
It gave me great encouragement. I felt like I was underwater for a long time and now I can finally breathe.
About his performance against Shaq and the Lakers
1.17.03
Some people have asked me questions about the arm band I wore in the Philly game. No, it wasn't a fashion statement.
Regarding an injury on his arm
12.10.02
ROCKETS
11
When I'm seeing them score points, it feels like I'm scoring points myself.
How Yao feels about his teammates
12.10.02
If I want to be good in the NBA, I have to be strong like Kelvin Cato.
After Tuesday's practice with the Rockets
10/22/02
I think the fact their defense on me is getting a lot more active and aggressive shows they respect me more.
The effect of more frequent fronting defenses on Yao
12.05.02

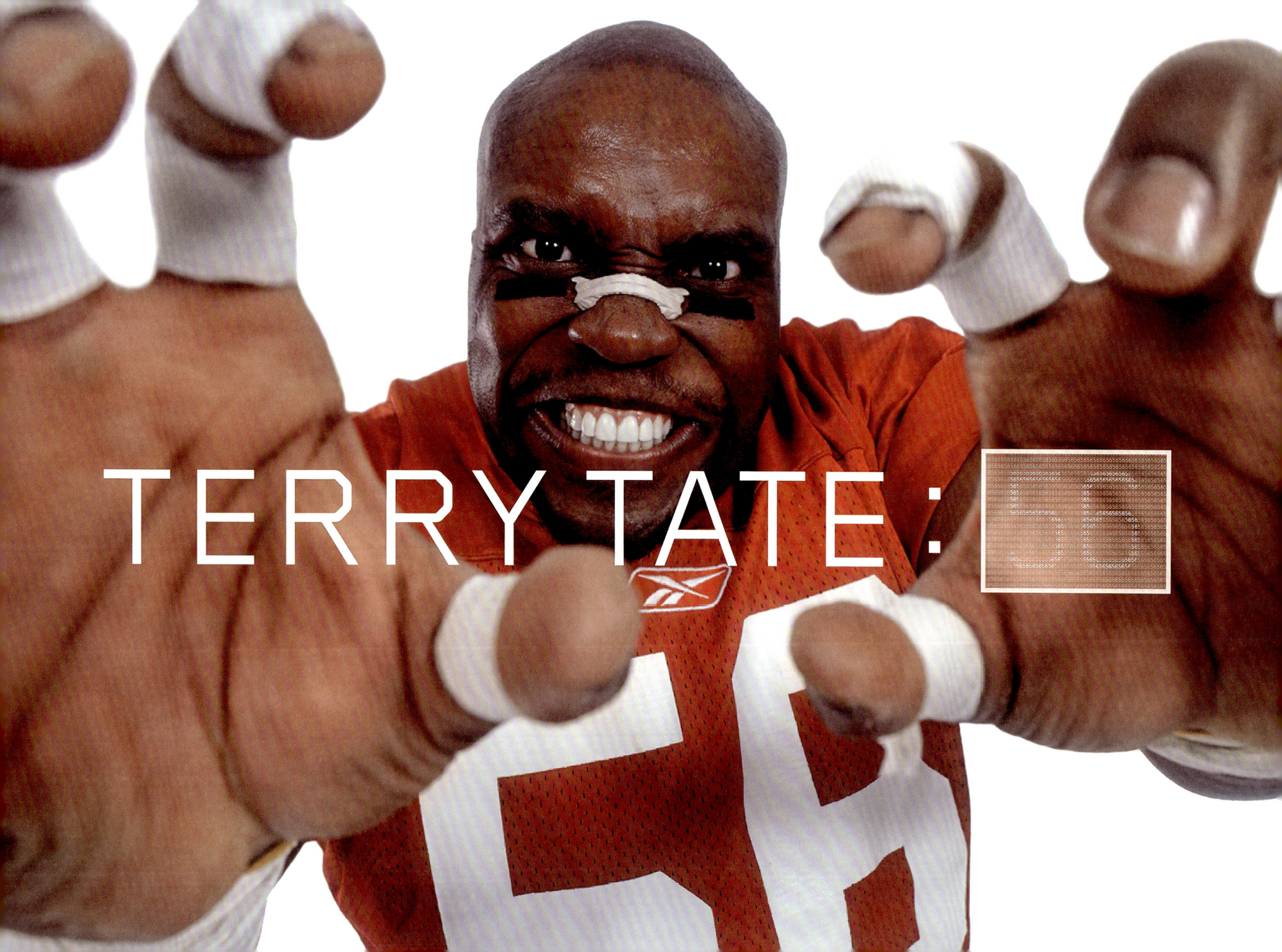
TERRY TATE:
56

TERRY TATE: OFFICE LINEBACKER

2003

As a comically confrontational linebacker who fiercely sacked office workers with minor workplace transgressions, Terry Tate was an instant sensation in Terry Tate: Office Linebacker, an extremely popular Super Bowl campaign crafted for Reebok.

Created in collaboration with director, producer, and screenwriter Rawson Marshall Thurber, the spot featured NFL linebacker Terry Tate as a hyped-up office linebacker who "gives out pain" to coworkers who aren't following standard office policies. An instant viral smash, the spot also birthed a number of iconic catchphrases including "when it's game time, it's pain time!" and "the pain train's comin'!"

Premiered as a sixty-second TV spot during Super Bowl XXXVII in 2003, a total of nine episodes were produced altogether, and were made available for download on Reebok's website following the Super Bowl broadcast.

Terry Tate: Office Linebacker is widely considered one of the most popular Super Bowl ads of all time, and also became one of the first (and most) viral Super Bowl ads ever, with more than thirty million views on YouTube in the early days of that platform, and over twelve million downloads from Reebok's website.

Repeat visits to Reebok's website created extraordinary traffic—so much so that the company hosting Reebok's website had to immediately purchase more server space just to keep up with the flood of visitors and downloads. The success of the spot in converting attention and word-of-mouth buzz to actual views and meaningful engagement helped to demonstrate the full potential of online engagement at a time when many companies were still skeptical of the value of digital media. In June 2003, the ad won the Golden Lion at Cannes for Best in Category, and every year since, when round-up lists of the most iconic Super Bowl spots of all time begin to circulate, Terry Tate: Office Linebacker invariably makes an appearance.

56
TERRY TATE
OFFICE LINEBACKER
Reebok

WATCH MORE
TERRY TATE
OFFICE LINEBACKER
Reebok
reebok.com

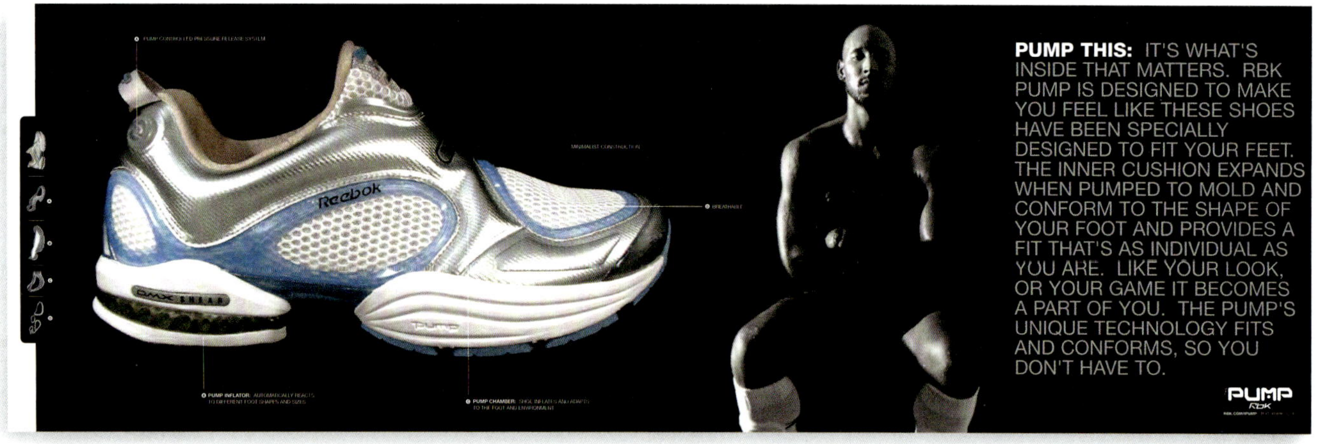
Reebok
PUMP THIS: IT'S WHAT'S INSIDE THAT MATTERS. RBK PUMP IS DESIGNED TO MAKE YOU FEEL LIKE THESE SHOES HAVE BEEN SPECIALLY DESIGNED TO FIT YOUR FEET. THE INNER CUSHION EXPANDS WHEN PUMPED TO MOLD AND CONFORM TO THE SHAPE OF YOUR FOOT AND PROVIDES A FIT THAT'S AS INDIVIDUAL AS YOU ARE. LIKE YOUR LOOK, OR YOUR GAME IT BECOMES A PART OF YOU. THE PUMP'S UNIQUE TECHNOLOGY FITS AND CONFORMS, SO YOU DON'T HAVE TO.
PUMP

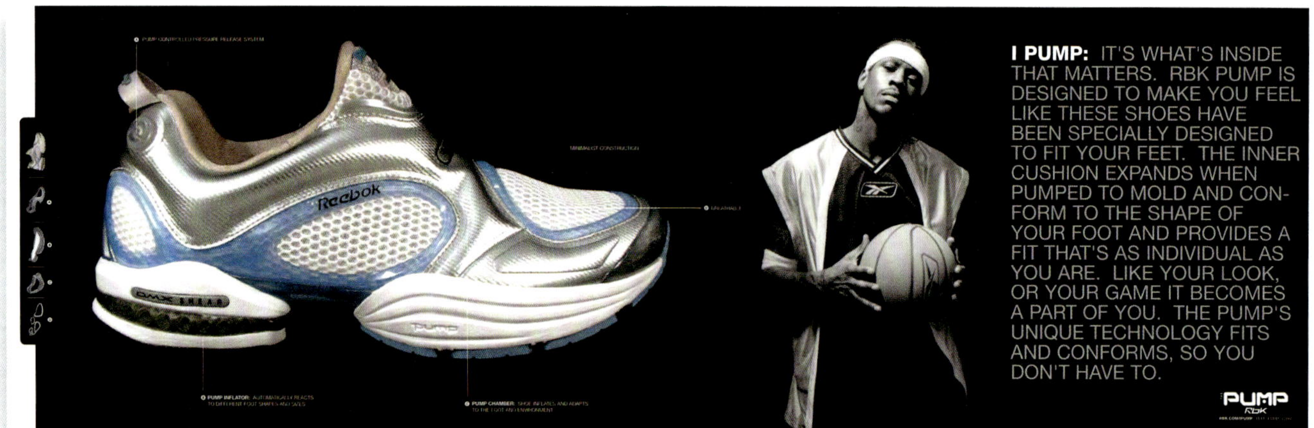
Reebok
I PUMP: IT'S WHAT'S INSIDE THAT MATTERS. RBK PUMP IS DESIGNED TO MAKE YOU FEEL LIKE THESE SHOES HAVE BEEN SPECIALLY DESIGNED TO FIT YOUR FEET. THE INNER CUSHION EXPANDS WHEN PUMPED TO MOLD AND CON-FORM TO THE SHAPE OF YOUR FOOT AND PROVIDES A FIT THAT'S AS INDIVIDUAL AS YOU ARE. LIKE YOUR LOOK, OR YOUR GAME IT BECOMES A PART OF YOU. THE PUMP'S UNIQUE TECHNOLOGY FITS AND CONFORMS, SO YOU DON'T HAVE TO.
PUMP

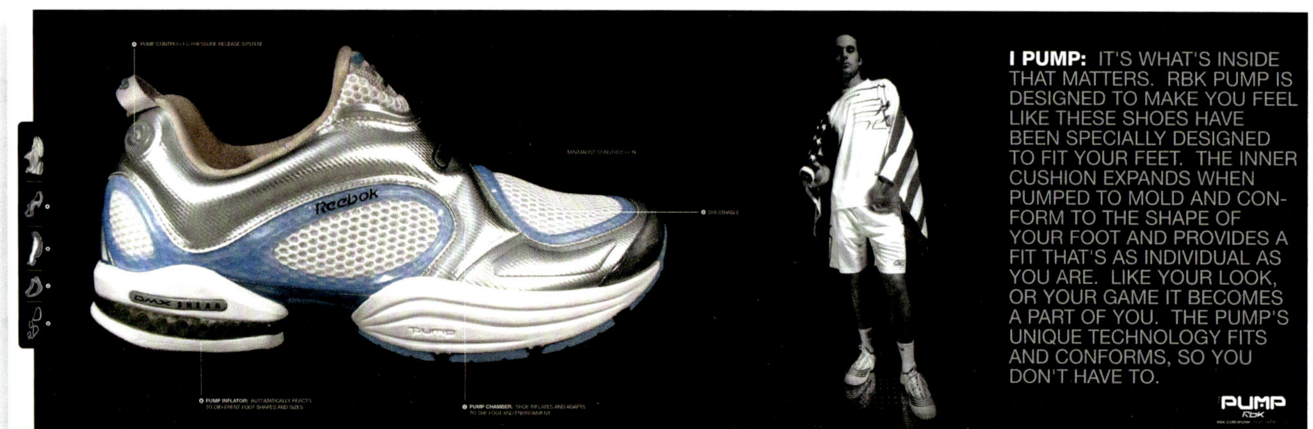
Reebok
I PUMP: IT'S WHAT'S INSIDE THAT MATTERS. RBK PUMP IS DESIGNED TO MAKE YOU FEEL LIKE THESE SHOES HAVE BEEN SPECIALLY DESIGNED TO FIT YOUR FEET. THE INNER CUSHION EXPANDS WHEN PUMPED TO MOLD AND CON-FORM TO THE SHAPE OF YOUR FOOT AND PROVIDES A FIT THAT'S AS INDIVIDUAL AS YOU ARE. LIKE YOUR LOOK, OR YOUR GAME IT BECOMES A PART OF YOU. THE PUMP'S UNIQUE TECHNOLOGY FITS AND CONFORMS, SO YOU DON'T HAVE TO.
PUMP

RBK IPUMP

2003

Since the early 1990s, Reebok's breakthrough IPUMP sneaker technology had allowed the wearer to adjust their cushioning with a proprietary pump that inflated and deflated for a custom fit.

To bring the technology firmly into the twenty-first century, a sleek new logo was designed: featuring puffed up lettering and an abstracted representation of the pump button, this new brand mark told the story of this unique technology while making it fresh, modern, and relevant.

A series of accompanying communications boldly proclaimed all the things that one could "pump"—your music, your game, your attitude—and in doing so, transformed what might have otherwise been seen as a gimmick into a distinctive sub-brand with an unforgettable personal mantra.

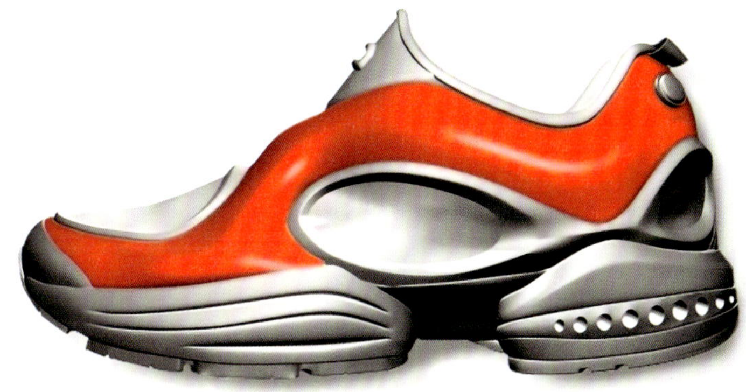

IPUMP

I PUMP MY FIST I PUMP MY HAIR I PUMP MY LOOK I PUMP MY CAR I PUMP MY MUSIC I PUMP MY GAME
I PUMP MY ATTITUDE I PUMP MY PLAY I PUMP MY SHOES I PUMP MY ICE
I PUMP
PUMP

I PUMP MY
MUSIC
GAME
FIST
HAIR
LOOK
CAR
ICE
ATTITUDE
PLAY
SHOES
I PUMP
THE PUMP
Rbk
RBK.COM/IPUMP TEXT "PUMP" 72587

I PUMP MY
MUSIC
GAME
FIST
HAIR
LOOK
CAR
ICE
ATTITUDE
PLAY
SHOES
THE PUMP
Rbk
IPUMP
RBK.COM/IPUMP TEXT "PUMP" 72587

REEBOK OUTPERFORM

2003

When Reebok set out to develop a new range of performance, the inspiration came from the company's long-standing classic brand mark, the symbol known internally as the "vector"—a twisting sweep of two lines intersected by another at a sharp angle. Representing speed, movement, dynamism, and relentless forward striving, "Vector" was the perfect name for this new line.

To herald the arrival of the Vector line, the Outperform campaign was created and launched with a series of bold print communications and an accompanying short film directed by legendary filmmaker Ridley Scott. The spot featured a truly diverse portrait of commitment, drive, and athleticism at all levels: from amateurs to professionals, each of these incredible athletes were showcased in their element, doing what they love, striving to outperform and attain their own definition of greatness.

WEAR THE VECTOR
OUTPERFORM
VECTOR

2003
REEBOK

REEBOK
WELCOME TO
THE MOVEMENT
VECTOR
A NEW COLLECTION OF
FOOTWEAR AND APPAREL BY
REEBOK
ROOTED IN INDIVIDUALITY
INSPIRED TO COMPETE
BUILT TO EXCEL
IN SPORT. IN LIFE
WEAR THE VECTOR
OUTPERFORM
IGNORE
ME?
WEAR THE VECTOR
OUTPERFORM
VECTOR
REEBOK BASKETBALL
REEBOK RUNNING

REEBOK OLYMPIANS

2003

To celebrate the 2004 Summer Olympics returning to their ceremonial home in Athens, Greece, Reebok honored the accomplishments of Olympic athletes with a striking campaign shot on location at the Agrigento ruins—a remarkably well-preserved Ancient Greek temple complex from the 5th century BCE that served as a stunning backdrop and for the men and women who were proudly carrying on the rich heritage of the Olympics, more than two thousand years later.

Working to capture the timeless spirit of power and resilience that these world-class athletes embodied, a striking campaign was created that hearkened back to the origins of the Olympic games. Dressed in authentic period costume, and posing among these monumental ruins, these Olympic hopefuls seemed to be transported to a timeless, almost mythical realm, and became a link to the heroes and legends that have captivated the world's imagination for more than two millennia.

CHRYSLER

2003

Following a decade of focusing on underwhelming mid-range family cars that were shared almost across their Chrysler, Dodge, and Plymouth brands, Chrysler decided to embark upon a total brand refresh that would drive the Chrysler marque upmarket and reclaim its legacy of design excellence.

The result of this dramatic reset was a new slate of innovative and high-style models—the family-friendly Pacifica Crossover, the retro-quirky PT Cruiser, and the sleek Crossfire Coupe. These striking new models needed to be introduced in an unforgettable way, so a multi-pronged campaign was created around Grammy-winning superstar Céline Dion and her hit song "I Drove All Night." Scenes of Céline driving different Chrysler vehicles down open desert highways were interspersed with imagery of the exceptional craftsmanship that went into the production of these new models, encouraging customers to see Chrysler and its vehicles in an entirely new light.

An accompanying print campaign featured elegant black-and-white photography taken on the assembly line, a move that was intended to restore customer faith in the quality of Chrysler's products, and to give value, pride, and dignity to the tradition of craftsmanship that underpinned the brand. These images—eschewing the usual glossy, dynamic, hyper-saturated photography style in favor of dignified, classically beautiful black-and-white imagery—helped to distinguish Chrysler from its competitors and heralded the beginning of a new chapter for this storied American automaker.

CHRYSLER

GABBANA
DOLCE

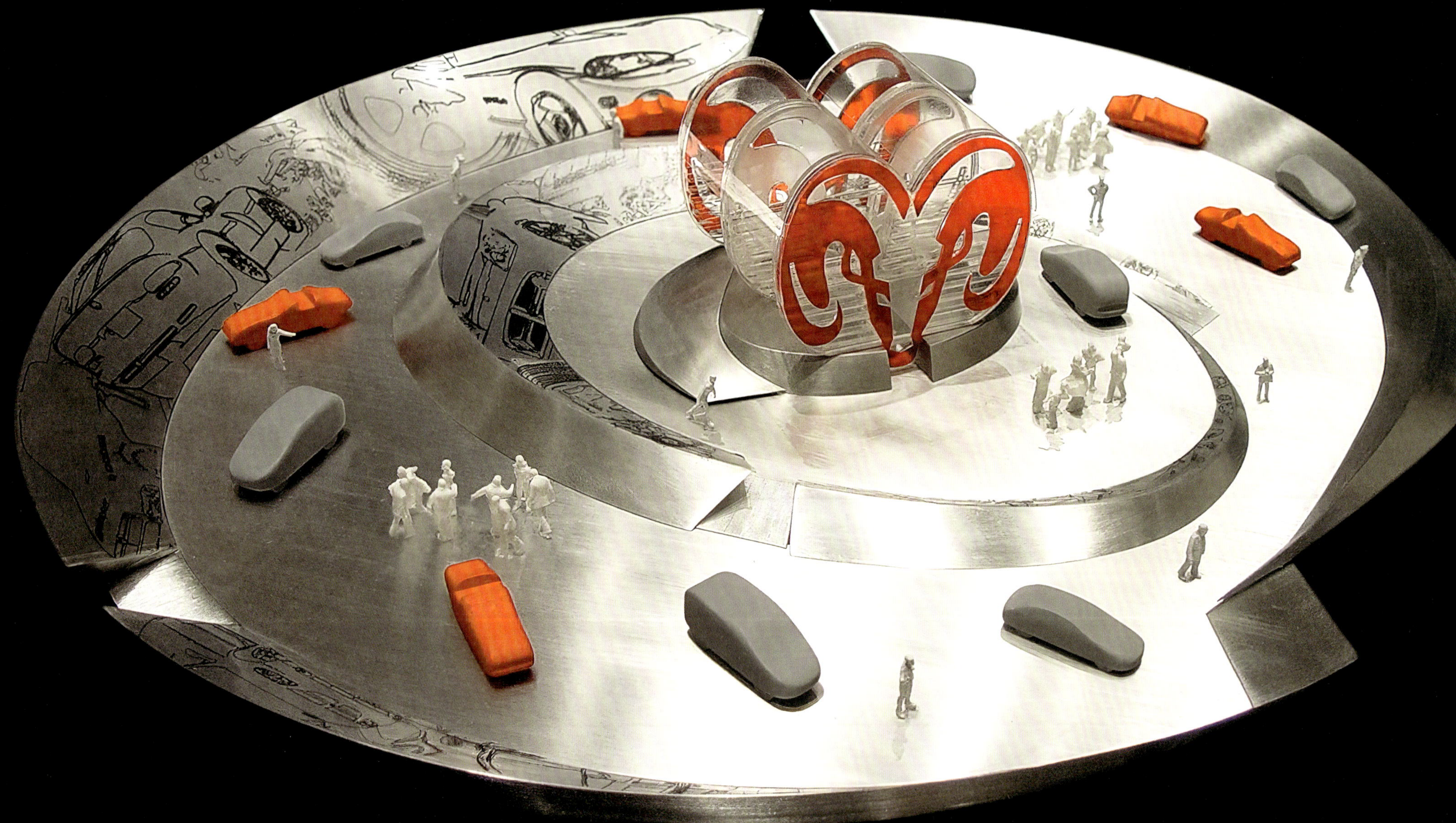

DODGE CAR SHOW

2003

For Dodge's presence at the North American International Auto Show in Detroit, an immersive brand environment was designed that would allow the many facets of Dodge's identity, history, and brand DNA to be brilliantly featured.

Taking the form of an ovoid racetrack—a configuration that was itself influenced by the dramatically curving form of Dodge's Ram brand mark—the installation allowed for a dynamic presentation of Dodge's entire product range. Passenger cars and trucks were displayed as though they were speeding alongside performance supercars and NASCAR race cars, while monster trucks and off-road vehicles seemed to crash over and through the dramatic terrain of the display.

At the center of the installation was a massive three-dimensional Dodge Ram logo, complete with multimedia presentations and interactive displays that allowed visitors to experience the energy and power of the brand firsthand.

RUBICON
Jeep

JEEP TOMB RAIDER

2003

When the blockbuster film *Lara Croft: Tomb Raider–The Cradle of Life* was released in 2003, audiences everywhere noted that the heroine, played by Angelina Jolie, had traded in her customary Land Rover for a Jeep Wrangler Rubicon. As part of a holistic, script-based, integrated marketing campaign, a limited-edition Wrangler Rubicon Tomb Raider model—identical to the model featured in the film—was debuted to coincide with the theatrical release of the new film.

Unique exterior features on the Tomb Raider model included sixteen-inch Alcoa-forged aluminum wheels, riveted fender flares in graphite, tubular grille guards, and diamond-plated bumper guards, while the interior featured dark slate fabric seats with red stitching, a customized silver surround instrument cluster, red seatbelts, and a Tomb Raider badge with serial number.

This limited-edition collector's vehicle highlighted the unbeatable capability and durability of Jeeps in even the most rugged extreme conditions: both on-screen and off.

i’m lovin’ it i’m lovin’ it
i’m lovin’ it i’m lovin’ it
i’m lovin’ it i’m lovin’ it
i’m lovin’ it i’m lovin’ it
i’m lovin’ it i’m lovin’ it
i’m lovin’ it i’m lovin’ it
i’m lovin’ it i’m lovin’ it

MCDONALD'S I'M LOVIN' IT

2003

By 2003, McDonald's found itself in need of jolt of fresh energy: sales were softening, and they needed to shift perceptions around their brand, and find inventive new ways to build authority and relevancy with a younger demographic.

Earlier that year, one of McDonald's international agencies—the Munich-based Heye & Partner—had come up with the new brand line "I'm Lovin' It." McDonald's president Charlie Bell and CMO Larry Light both loved the line and saw "I'm Lovin' It" as a great way to give the McDonald's brand a friendly and approachable consumer voice that would build credibility and cachet with younger audiences.

To help McDonald's define and deliver on the "I'm Lovin' It" attitude, a series of new brand activations was developed that tapped into the energy and spirit of the younger generation. "I'm Lovin' It" was a brilliant idea that had the potential to become a genuine pop culture phenomenon, but it needed to be presented in just the right way in order to truly make an impact and break through. Working closely with the creative leadership of Heye & Partner, a new multi-pronged strategy was devised to deliver on this marketing masterstroke.

Justin Timberlake, at the time just beginning his successful solo career, was invited to be the voice of the brand, and he quickly wrote and recorded a new song built around "I'm Lovin' It": a simple but ingenious tune that indelibly linked the new slogan to the McDonald's brand. To coincide with the roll out of this new campaign, Timberlake embarked upon his new McDonald's-sponsored "I'm Lovin' It" Tour: a highly effective cross-promotional exercise that would be endlessly imitated by other brands and A-list pop stars in the years to come.

To this day, thanks to Timberlake's performance, the "ba-da-ba-ba-ba" audio logo remains one of the most iconic and recognizable brand expressions ever created—in just five simple notes.

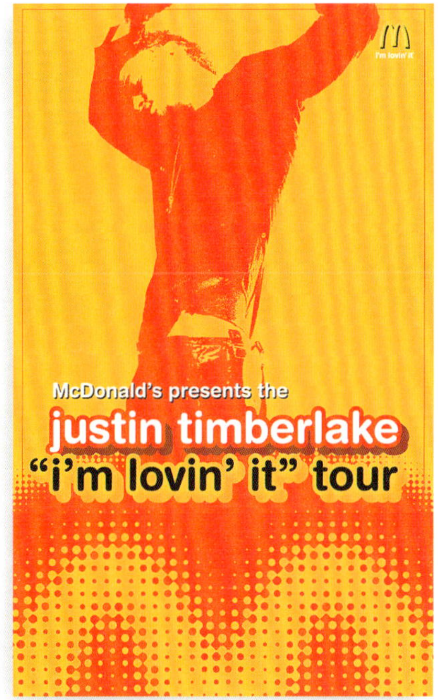

MCDONALD'S FOREVER YOUNG

2004

In the early 2000s, facing slumping sales and waning relevancy, a new management team under the leadership of CMO Larry Light embarked upon a brand revitalization effort to return McDonald's to its foundational promise of providing its customers happiness in a happy place. Inspired by the Bob Dylan song of the same name, the "Forever Young" strategy was focused on making McDonald's a place that provided a youthful sense of optimism, freedom, and sense of joy for all.

To help deliver on the main tenets of "Forever Young," a next-generation, modernized Ronald McDonald was developed to provide fresh appeal to children, teens, and adults alike. The longtime mascot and pop culture icon was reimagined to be hipper and more mature—a relatable ambassador of an emerging lifestyle brand, rather than just an old-fashioned kids' cartoon clown. The new Ronald was also given a whole new crew of friends, known as the M-Bunch, each of whom had their own captivating backstories, distinctive personalities, and unique points of view.

In tandem with this refreshed identity, the concept of being "Ronaldized!" was developed: a special flare or flourish that can make any ordinary moment extraordinary.

1964 2004

1 hr photo
Pharmacy
Garden

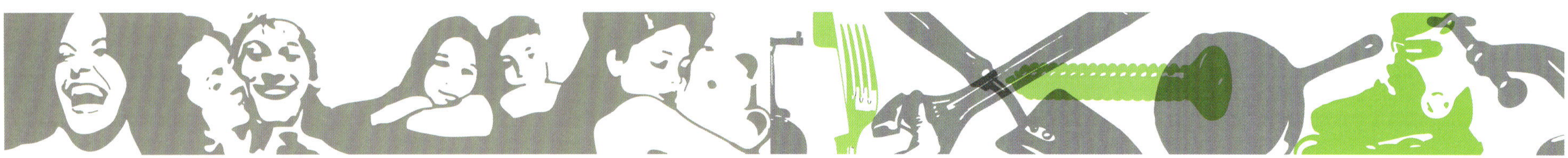

KMART

2004

To help reinvigorate Kmart's retail operations, a refreshed and revitalized brand identity was developed along with a proposed new retail design strategy that reimagined its stores as multifaceted "brand theaters"—where every consumer touchpoint was designed to engage, inform, and delight.

The basic form of the iconic brand mark was retained—the bold "K" symbol and the cursive "mart"—but it was given a bold new color scheme that lent a more upmarket feel and also spoke to Kmart's renewed emphasis on quality design. A series of vibrant, stylized murals was also developed to adorn the frequently blank exterior expanses of Kmart stores, signaling a new sense of energy and excitement.

Inside stores, a signature wayfinding system was implemented: simple icons and bold, readable text that quickly oriented customers within the store and allowed them to find what they were looking for quickly and easily. This simple gesture, along with improved lighting, broader aisles, and a more curated and intentional approach to merchandising made store environments much more inviting to customers and transformed Kmart into a welcoming and engaging retail environment.

Careful studies of customer behavior and traffic patterns followed the implementation of this system, which led to further strategic merchandising efforts to continually elevate the efficiency and enjoyment of the shopping experience.

M&M'S

2004

For decades, Mars, Inc.'s M&M's has been one of the world's favorite treats: bright, colorful, and delicious. But the venerable treat was also so ubiquitous that it was in some senses taken for granted, and sales had begun to soften. To help recreate a sense of excitement around the classic candies, a bold and counter-intuitive campaign was engineered: without any warning, M&Ms rolled out more than thirty-seven million packages of the chocolate candies completely stripped of their usual vivid colors—both the packaging and the candies themselves suddenly appeared in monochrome. Paradoxically, the bright, eye-catching colors of M&M's had become so commonplace that they were all but invisible to the average consumer, and it took removing those signature hues altogether to get their attention once again.

As confused customers everywhere started to take notice of the candies' absent colors, a choreographed series of ads were rolled out that placed the usually vibrant M&Ms characters into a black-and-white scene from *The Wizard of Oz* searching for their lost colors.

The subsequent Great Color Quest vote allowed consumers to vote on their favorite classic M&M's colors, and also cast a ballot for the addition of an entirely new color to the classic M&M's rainbow of red, orange, yellow, green, blue, and brown. The eventual "re-introduction" of the once-again colorful candy featured a brighter-than-ever mix of the six classic colors, along with the new, limited-edition purple shade selected by the public vote.

M&M'S MINIS

2004

M&M's Minis were created specifically for kids—a smaller, cuter, more playful version of the classic candy. To help drive appeal among this young audience, a series of new mascot characters was also developed to appeal to a younger generation of M&M's lovers, giving M&Ms Minis their own unique cool factor, and a brand life that was completely distinct from classic M&M's.

Revitalized packaging concepts were also developed based on the insight that the durable, colorful, resealable plastic tube that the candies were housed in had become one of the main selling points: kids saw the tubes as collectable and loved to hang on to them after the M&M's were finished, using them for storage, play, or simply adding them to their collections. These inventive, ad hoc usages of the packaging signaled a massive opportunity for the brand: rather than focusing on the candies themselves, the focus shifted to what they facilitated, which in this case was socializing, collecting, and active, imaginative play.

A number of schemes were designed to make the tubes more functional, more collectable, and offer a greater potential for play: adding a small light to the cap, turning the tube into a fully functional flashlight; tweaking the shape of lid to make each tube function as individual snap-together components in a Lego-like interlocking construction system; embedding a watercolor paint palette in the flip-up lid; and transforming the entire tube into a musical instrument with interchangeable, snap-on mouth pieces. Along with all of these new tube designs, a wide array of puzzles, games, and online interactive tie-ins were also developed to extend the playability of every tube of M&M's Minis.

MILK CHOCOLATE CANDIES

3M POST-IT

2004

For 3M, a series of strategic brand extension opportunities and innovative next-generation product designs was developed for their iconic Post-it brand of note pads. The secret of Post-it's success was its proprietary adhesive: a formula made up of microscopic spheres that form a dynamic structure, allowing Post-it products to be stuck and re-stuck multiple times to almost any surface imaginable.

These core brand tenets of flexibility and adaptiveness inspired a new design language which was then applied to the core product range. The result was a series of spectacular, sculptural form factors for Post-it note pads that transformed a mundane office supply into whimsical, gravity-defying works of art.

MR. POST-IT

2004

As part of a comprehensive brand growth plan for 3M's Post-it, a range of new strategies and brand activations were developed to encourage customers to see the ubiquitous paper pads as something more than just a purely functional office item.

A new brand mascot, Mr. Post-it was created as an ambassador to guide consumers through all of Post-it's new offerings and open their eyes to the infinite creative potential offered by the simple, colorful square sheets of paper. Mr. Post-it and his cast of paper friends offered a much-needed injection of imagination, whimsy, and delight into what was otherwise a staid office staple, and helped to position Post-its as a must-have item in every creative toolkit.

SEEDS of CHANGE
better flavor from better seeds
greek feta with corno di toro sweet peppers, fresh cucumbers, kalamata olives and feta cheese
crisp and bright salad dressing
SEEDS of CHANGE
better flavour from better ingredients
tomato & basil
rich and zesty pasta sauce
SEEDS of CHANGE
better flavour from better ingredients
sundried tomato
bold & rich stir in sauce

SEEDS OF CHANGE

2005

Offering a wide range of both prepared foods and fresh produce, the essence of the Seeds of Change brand was "truth in flavor"—that is, awakening people to how good food truly tastes when it is carefully and sustainably produced through organic farming. The overarching aim was to promote a more thoughtful production and distribution model for high-quality, organic food that still fit within the framework of a traditional supermarket shopping experience.

Beautiful, ergonomic packaging was developed to help differentiate Seeds of Change products from the other products on the supermarket shelf and highlight the exceptionally high quality of the food within. Packaging showcased selected growers, including details on farmer profiles, other regional crops, and unique recipes that could be made with these ingredients. Fresh produce was labeled to clearly indicate the date of harvest as well as the name and location of the farm where it was grown.

Additional items outside the range of fresh and prepared food offerings were also developed, including grow-it-yourself seedling kits, proprietary produce shipping and display crates, and custom lids for prepared food jars with seed inserts to allow customers the chance to start growing food for themselves, encouraging greater mindfulness and deeper engagement around food production and consumption.

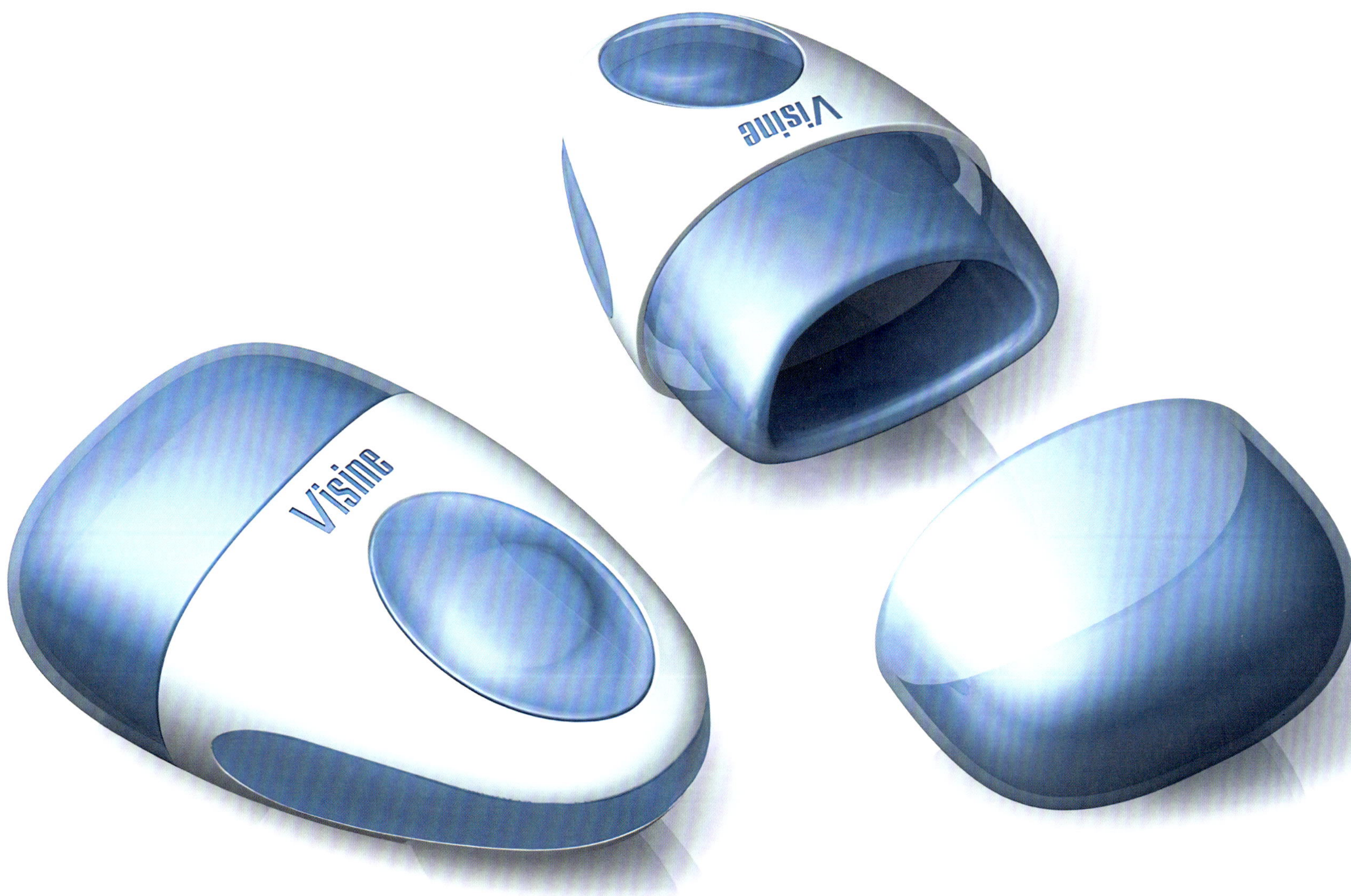
Visine
Visine

VISINE GENTLE MIST

2006

The Gentle Mist eye drop dispenser was an innovative new product concept conceived and designed for Johnson & Johnson's Visine brand to help make eye drop application quick, simple, and comfortable.

In observing users as they applied standard eye drops from a conventional squeeze bottle, it became clear that the typical position assumed in this scenario—tilting the head to the back or to the side—was not only uncomfortable and awkward, it didn't even ensure that the eyedrops hit their intended target. Instead of the drippy mess that typically accompanied eye drop application, the Gentle Mist pocket mister delivered a gentle and soothing fine cloud of mist to the eye.

Easy and gentle to apply, the user only needed to place the soft and ergonomic eye-shaped charger over the eye and press with a single squeeze. An advanced atomizing nozzle generated an ultrafine micro mist of droplets twenty microns in diameter. This fine particle spray significantly reduced the blinking reflex action that typically undermined the efficacy of eye drops, making the application process simple, comfortable, and completely mess-free.

PFIZER WOBBALL

2006

The WOBBALL concept was developed as an innovative, on-the-go oral care system as part of Pfizer's ImagineLab initiative that had been established to study next-generation product innovation around at-home oral care. Envisioned as a first-of-its-kind chewable toothbrush, WOBBALL's silicon shell had an internal cavity that could be filled with a proprietary oral hygiene gel to provide a convenient, reusable delivery mechanism for unparalleled breath freshness.

A small cylindrical silicone piece the size of an average toothbrush head, the WOBBALL could simply be popped into the mouth and chewed. The unique form factor and material technology cleaned teeth while the soft center released a concentrated dose of refreshing mouthwash. WOBBALL pods could also be attached to a handheld electric unit for enhanced oral hygiene at home.

A breakthrough experience in oral care, WOBBALL offered the efficacy of a conventional manual toothbrush in a compact, no-mess, on-the-go solution.

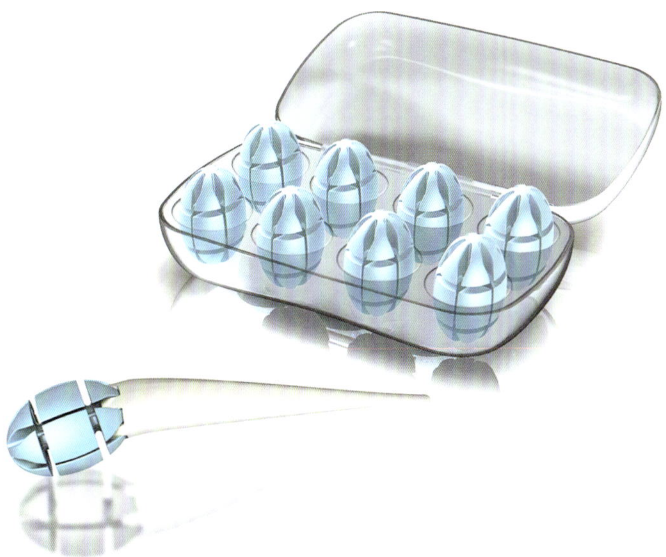

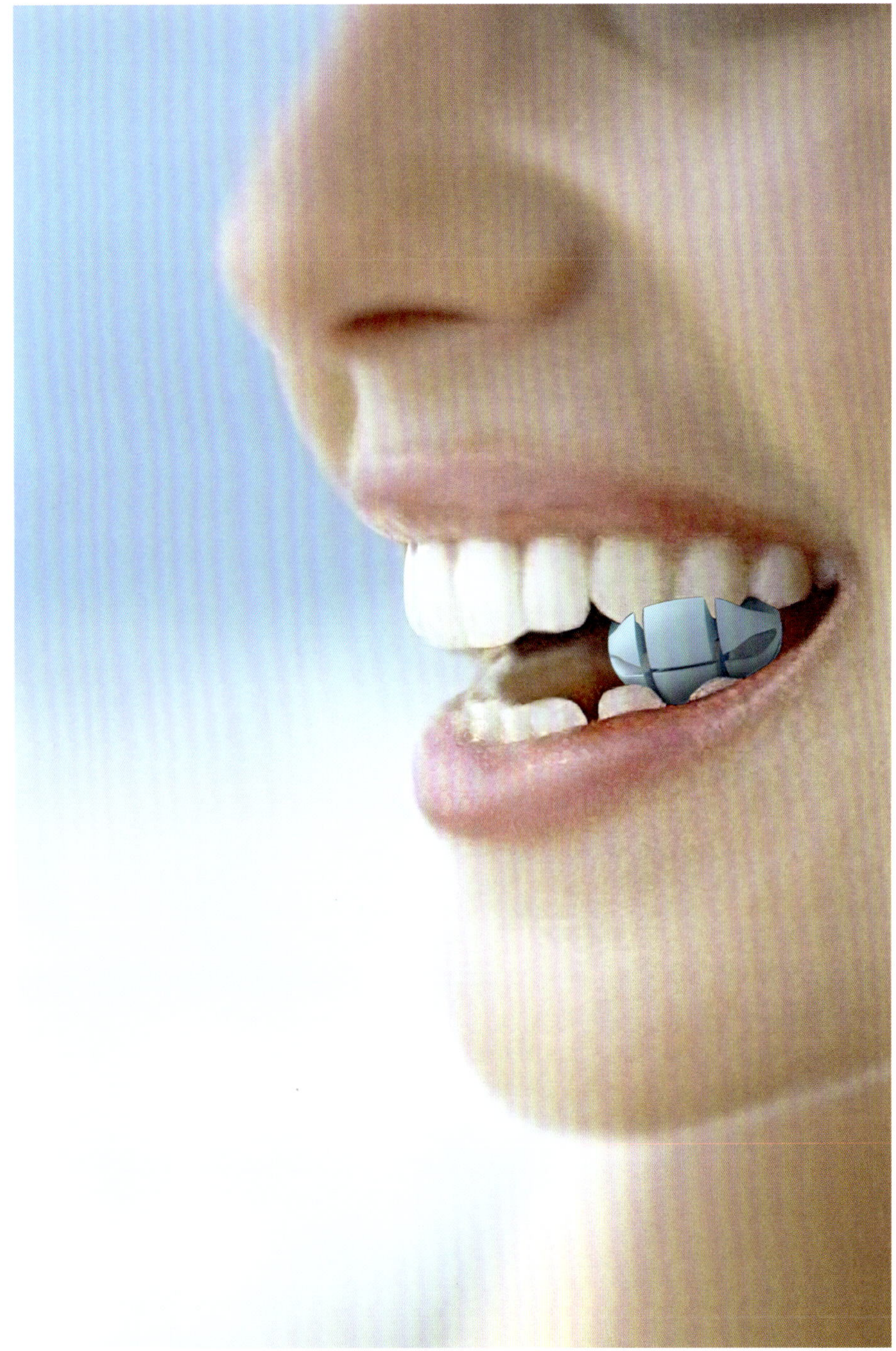

UNILEVER OMO SPLATS

2006

Developed by Unilever's in-house R&D team for their OMO detergent brand, the five-armed Splats detergent capsules were conceived of as a new all-in-one laundry solution. Each Splat contained detergent powder, detergent booster, whitening and brightening agents, fabric softener, fragrance, and a rinse aid, with each element queued to dissolve at the appropriate time during the wash cycle, releasing the different formulas sequentially to achieve perfectly clean laundry. Splats were designed to inject a little bit of color and friendliness into what was otherwise a mundane and onerous chore, while the targeted release system engineered into every Splats capsule took the guesswork out of laundry and guaranteed perfect results every time.

To help launch this innovative new product, new packaging, brand identity, and launch communications were developed that celebrated the Splats' all-in-one ease and superlative cleaning properties, and showcased their playful, endearing, and totally distinctive form factor.

LIPTON TEASPOON

2006

Since its founding in 1890, Lipton Tea's signature black tea has long been the gold standard among tea drinkers worldwide. And while Lipton's venerable tea blends didn't require any updating, there was an enormous opportunity for innovation in the way that the tea itself was brewed—a process that had remained unchanged for over a century since the introduction of disposable paper tea bags, even as people's habits and rituals around tea consumption had shifted dramatically.

The Lipton Teaspoon product concept was designed to remove the mess from the process of making a cup of tea. A sturdy cardboard frame neatly held the tea bag in place and served as a stirring device, which eliminated the need to dirty a spoon. The design also featured heat-sensing material which provided a precise visual measure of when the tea was ready, guaranteeing a perfect cup of tea, every time.

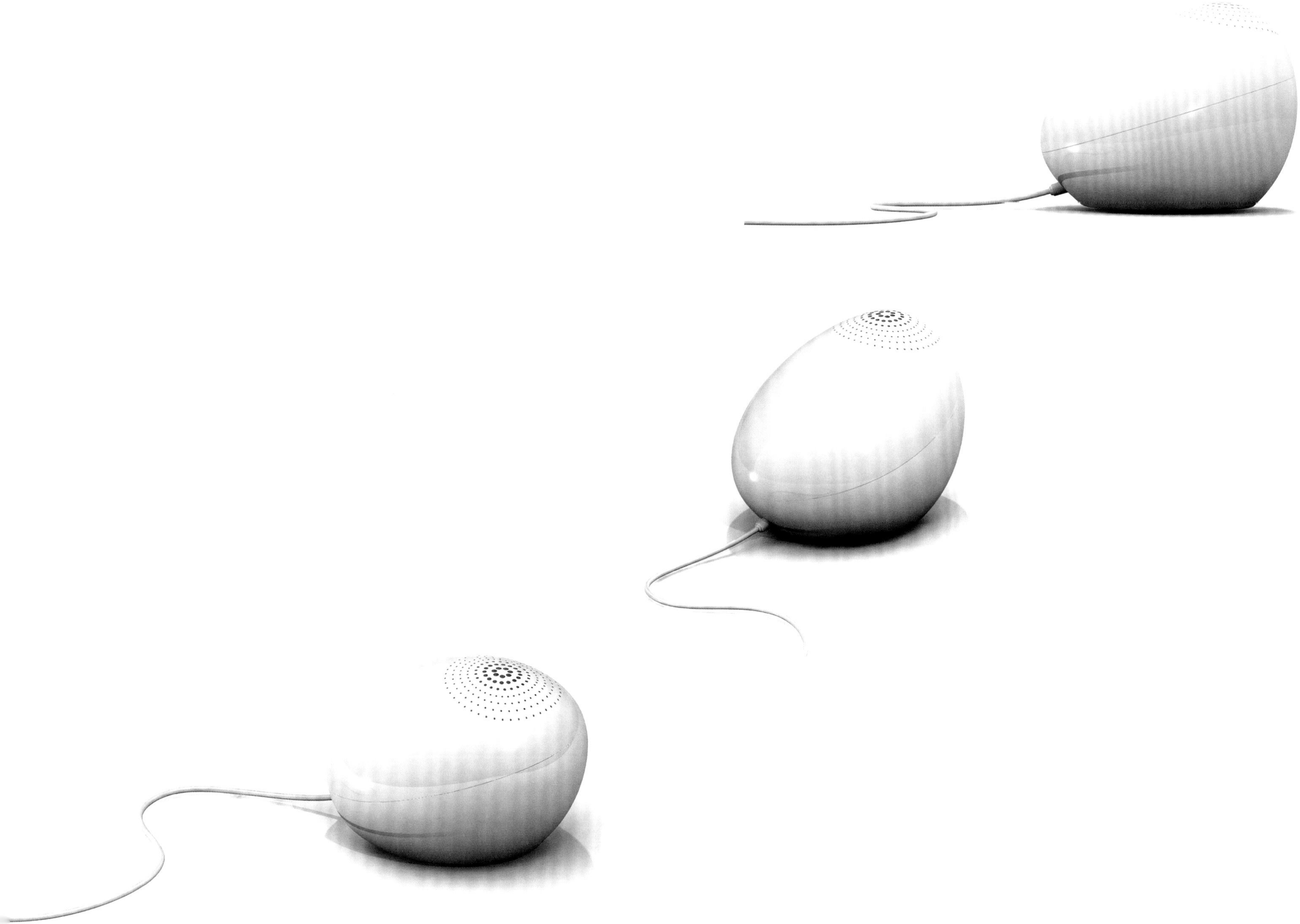

SUDAFED SUDACARE

2006

Johnson & Johnson's Sudafed sinus relief tablets have long been the decongestant that countless people reach for when they're experiencing cold symptoms and sinus congestion. In an effort to find new ways to bring Sudafed's renowned relief to more people, every day (not just when a cold or flu strikes), a series of design proposals for an innovative range of networked home health devices was developed.

The Sudacare aromatherapy pod product concept conveniently dispensed soothing aromatherapy at work, at home, or anywhere in between. Designed to plug into a standard USB port to charge, the self-contained, portable unit provided a simple, everyday health solution to help customers live better, healthier, more balanced lives.

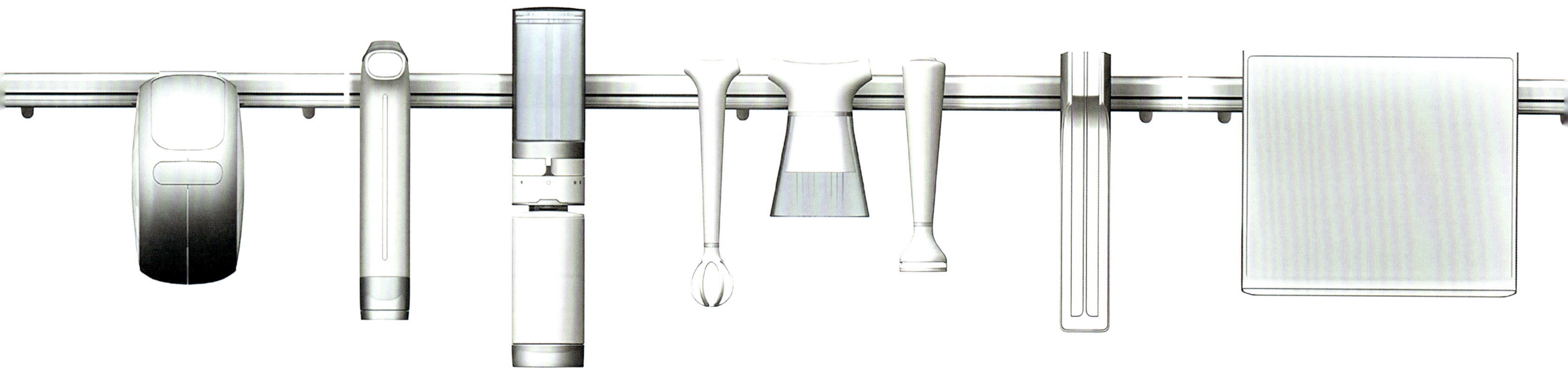

ELEMENTS

2006

Building off of insights that had been developed while working on a project with Electrolux that studied and mapped out the future landscape of kitchen and home appliances, Elements was developed as a design concept proposal for an innovative new small appliance system with the potential to redefine kitchen ergonomics. The Elements line of cordless, portable small appliances was designed to transform the home by bringing a radical level of mobility, versatility, and ease-of-use to daily life.

At the core of the system was the Power Bar: a powered rail that the Elements small appliances docked on for both charging and storage. With counter space at a premium in most kitchens, the power bar allowed for dramatically more free, flexible workspace.

Compact, lightweight, beautifully designed, the proposed Elements product family comprised a full range of small household appliances, including a toaster, kettle, microwave, fridge, hand mixer, single-cup coffee brewing system, music player and radio, and a Smart TV system called Digiframe.

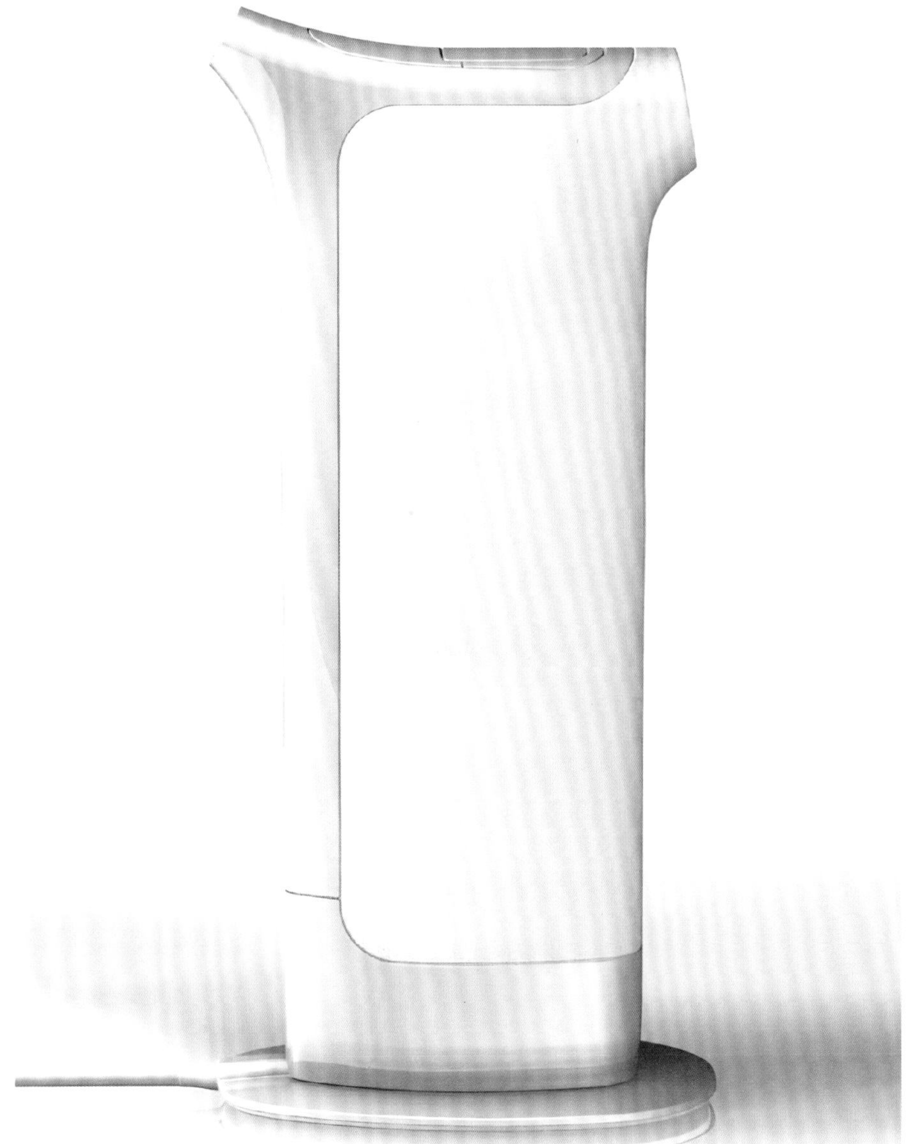

SONIA
SCULPTURE
SCULPTURE
GABRIEL OROZCO

7 WTC

2006

7 World Trade Center was one of the first new buildings to open on the site of the original World Trade Center in the aftermath of the devastating attacks of September 11, 2001. The elegant, understated, and technologically advanced fifty-two-story skyscraper was designed by global architecture firm SOM and built by legendary developer Larry Silverstein's firm Silverstein Properties. Because 7 WTC was separate from the main one-acre World Trade Center site, Silverstein was able to proceed quickly with the rebuilding effort, with construction beginning on May 7, 2002, and the official opening of the building occurring just over four years later on May 23, 2006.

The relocation of the Arnell Group to this site in lower Manhattan was, in effect, a vote of faith in the resilience of New York City and a direct contribution to the rebuilding of the area: part of a larger wave of businesses and civic leaders throwing their support behind the reinvigoration of the area around Ground Zero.

The open, airy, minimalistic space featured raw concrete floors, sleek metal fixtures, expansive planes of glass, and a library of over eighty thousand books on design, art, architecture, photography, industrial design, marketing, and countless other disciplines. The design of the space also included, for the first time, a gallery space which housed the complete history of the company's design work, and a full fabrication shop staffed by designers, architects, model makers, and other specialists who had the space, tools, and resources to bring the studio's designs to life. Thousands of other artworks and design objects from the company archives were also finally given a permanent and proper home, at hand for easy reference, thorough study, and spur-of-the-moment inspiration.

The overall effect was more of an experimental lab than a conventional office, which spoke to the unified embrace of environment, experience, product, and interactive design—a dynamic, multidisciplinary approach that remains the backbone of the practice to this day.

LIBERTY NATIONAL GOLF COURSE

2006

Liberty National is one of the premier country clubs in the country. Located in Jersey City, New Jersey, the ultra-luxurious golf course was developed and conceived by entrepreneur Paul Fireman and featured such amenities as an on-site helipad, yacht services, spas, and multiple fine dining restaurants. But by far the most remarkable feature of all is its location: located across the Hudson River from lower Manhattan and the iconic Statue of Liberty.

In devising a visual identity for this landmark facility, there was no need to look any further than the panoramic views of the world's most famous beacon of hope just across the river. Designed to completion in less than one day, the Liberty National brand mark is a beautifully stylized rendition of Lady Liberty: optimistic, bold, and dynamic, just like New York City.

LIBERTY
NATIONAL GC

DR. KRAXTON

TRIX

DR. PEABIX

ALIX THE ANT

WARRIOR

MINISTER VAAJAR

PIX

QUAX THE DUCK

MOBOT

ROX THE ROBOT

EMPEROR REX XX

PIX & TRIX XPLORER

2006

Pix & Trix Xplorer was developed as a proposal for as an interactive gaming adventure that followed the escapades of Pix, a gifted child prodigy with skills far beyond his years, and his sister Trix, a clever trickster with an unrivaled ability to set traps. Together, Pix and Trix embark on a series of wild adventures, befriending a host of eclectic characters along the way.

The passion project of longtime Arnell Group design director Mike Doyle, Pix & Trix Xplorer was initially developed as a video game concept, with plans for subsequent extensions into other forms of media, merchandise, and immersive experiences. Doyle was given the support and the resources to build out his vision for Pix & Trix Xplorer as an original educational entertainment property that would ignite children's curiosity and encourage them to adopt an "explorer" mentality as they learned about the vastness of the world around them.

WHISKAS

2006

The Mars, Inc.-owned Whiskas is one of the oldest and most popular brands of cat food, having served the health of cats since 1936. As the pet care market started to explode in the early 2000s, Whiskas sought new offerings for its customers. Faced with the challenge of revitalizing the famed pet care brand within a saturated and crowded market, Arnell designed irresistible cat milk bottles and food bins to entice shoppers, lending a playful aspect to mealtime. Using the form of the cat's body, the innovative Whiskas packaging design enhanced the presentation of the brand and ensured that Whiskas would maintain a presence in the consumer's home long after its contents had been devoured.

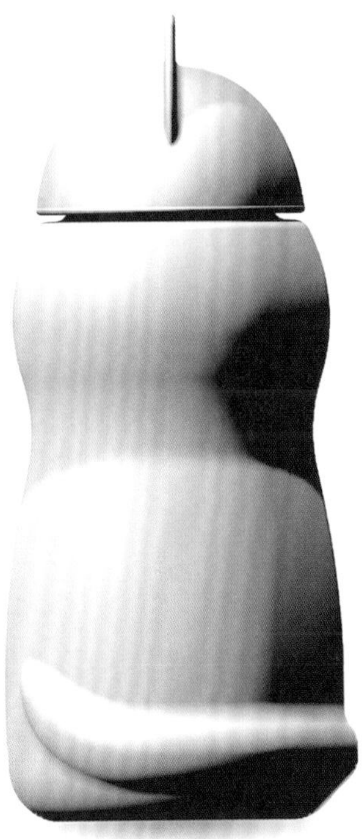

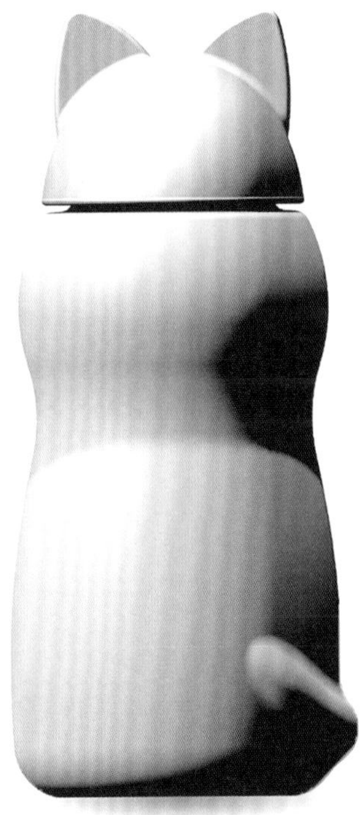

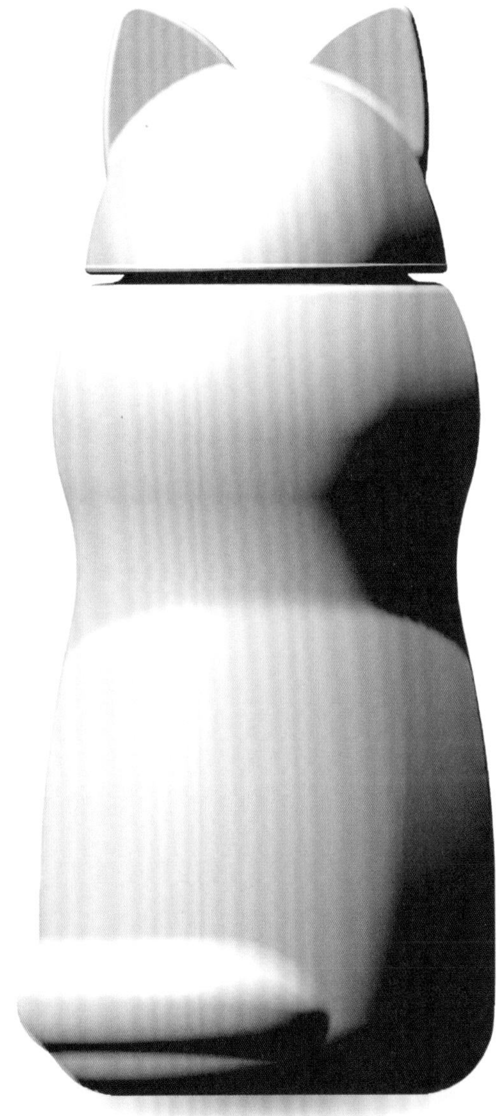

E 72 ST
TAXI

DISNEY MOUSEUM

2006

The Disney Mouseum concept was created as a new landmark cultural entertainment destination for Disney fans around the world.

Developed as part of the Altitude project—a sweeping initiative developed in collaboration with Disney CEO Robert Iger to encourage a culture of next generation design-led thinking at the company—the Mouseum plans included an open archive that allowed visitors to explore thousands of never-before-seen items from the company's history, along with extensive multimedia exhibitions that brought Disney's decades of unrivaled artistic and creative contributions to life.

The concept for the Mouseum also included an educational program that promoted a diversity of cultural viewpoints and encouraged young people to develop their imaginations through the power of storytelling. The vast archives, free public learning programs, and research activities at the Mouseum were all designed to serve a diverse public of varied ages, backgrounds, and interests, and to reinforce Disney's core mission of entertaining, informing, and inspiring people everywhere.

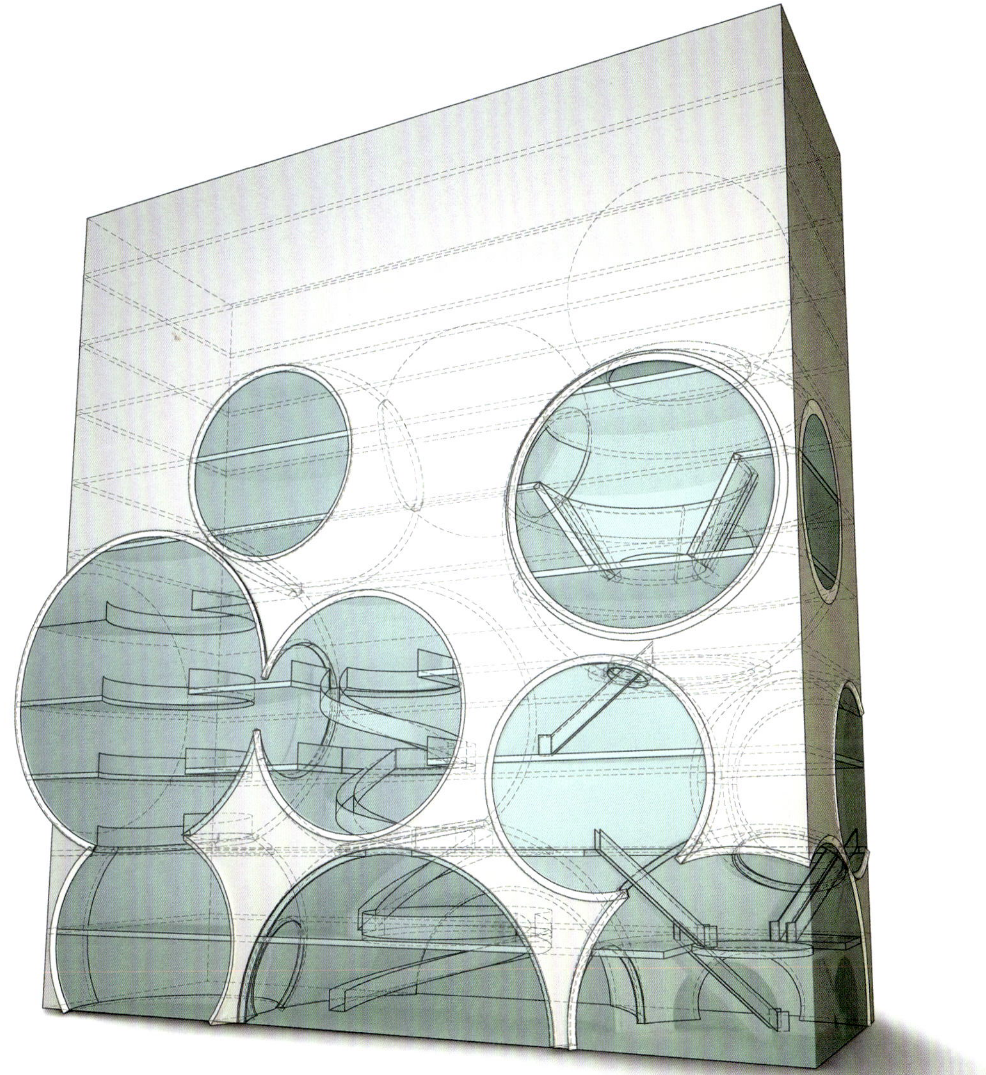

30
60
20
10
40
20
Disney
31
30
60
20
10
40
20
Disney
31
30
60
20
10
40
20
Disney
31
30
60
20
10
40
20
Disney
31

MICKEY MOUSE WATCHES

2006

Disney's beloved mascot Mickey Mouse first appeared on a wristwatch in 1933, five years after his debut in the animated short film *Steamboat Willie*. The watch, created and manufactured by Ingersoll-Waterbury, was an instant hit: more than eleven thousand were sold at Macy's in New York on its release day, where it retailed for just under three dollars, making it an affordable must-have item for every type of customer. Over the years, countless variations on this original watch have been released—from cute, kid-friendly novelties to ultra-luxe diamond-encrusted models.

This proposal for a reimagined version of the Mickey Mouse Watch was designed to appeal to a new generation of design-conscious Disney fans who wanted a sleek, modern take on a charming classic. The elevated and minimalist designs cleverly integrated multiple watch faces into the three-circle Mickey logo—a subtle but playful retooling of one of the world's most iconic symbols.

DISNEY PETS

2006

For Disney, a proposal was developed for a new line of pet products that playfully mined the equity of classic Disney animal characters. Given that Disney had a long history of developing unforgettable cartoon animal characters, it made perfect sense to leverage that legacy into a new series of beautifully designed bowls, leashes, collars, treats, and other pet accessories. A related design proposal for a national network of Disney-branded dog runs called Pluto Park was also explored: a magical everyday experience for pets and their owners alike.

Disney Pets cleverly leveraged Disney's organic brand and its vast existing trove of intellectual property to create beautiful, functional products that made perfect sense for the Disney brand, while still being surprising, delightful, and full of Disney magic.

G.O.A.T

GOAT

G.O.A.T.

2006

After sharing a close friendship with the Alis for years, Arnell was honored to be asked to develop a cohesive brand to represent Muhammad Ali, widely regarded as one of the most significant and celebrated sports and cultural figures of the 20th century. Utilizing the word most associated with Muhammad Ali, Arnell named the brand G.O.A.T., "Greatest of All Time," and designed a powerful, bold, and instantly identifiable identity architecture in the shape of a stylized boxing glove that formed the first letter of the brand's name, G.

Ali was known across the world as a champion, humanitarian, and a hero, but research showed that, with his retirement from boxing and the decline of his health, public perception of Ali's ability to continue to wield influence had diminished. However, Ali was a fighter both in and outside the ring, and his greatest fight had not yet been fought. In response, Arnell created a strategy that would tap into Ali's formattable spirit and convey Ali's undiminished determination, humanity, soul, and intelligence to continue fighting for the advancement of all people. Empowered by Ali, Arnell was charged with identifying strategic partners for the implementation of the brand's programs. The program commenced with Ali and Arnell joining forces as partners with Mars Inc. under the leadership of Paul Michaels to create a range of delicious, healthy snacks: good-for-you nutrition built to be the next generation of healthy foods and backed by the authority and legacy of the Greatest of All Time.

G.O.A.T.'s product line offered both sweet and savory snacks based on some of Ali's all-time favorite foods. Product names like "Jabs," "Shuffle," and "Rumble" recalled Ali's iconic career highs. G.O.A.T. snacks were made to be serious on-the-go fuel to help high-performing individuals achieve greatness and, per the brand's mantra, truly "Rise Above." Working closely with Ali, his wife Loni, and his longtime manager Bernie Yuman, a complete brand culture and 360-degree brand and marketing strategy was executed, including the name, identity, packaging design, product form factors, and a multi-touchpoint global campaign to introduce the brand to the world. To accompany the launch of G.O.A.T., a series of pop-up environments were designed and deployed throughout the country where one could sample these delicious new snacks and become immersed in the incredible life and legacy of a true icon.

★ G.O.A.T. ★

Research Findings

- A Hero
- A Great Memory
- People Don't Understand who he is today
- The Perception of Him Is Wrong
- People Think He is Diminished
- They Have Trouble Accepting His Physical Change
- They Feel Sorry About the Image They Have of Him

The Keys

- He is Still A Fighter
- We Need to Tap into the Person Behind the Fighter
- Show He is Still the Best Regardless of the Situation
- Show The Spirit is the Same
- Express Who He Is
- Give A Better Image of Ali to the People
- Represent the Same Mental Strength
- Represent Taking Yourself Somewhere
- Go to the Man, Not The Disease

Muhammad Ali

September 6, 2005

Peter Arnell
130 Prince Street, 5th Floor
New York, NY 10012

Dear Peter,

We admire your vision for the worldwide development of the GOAT brand. Consider this our authorization for you to explore with your strategic partners the framework for such an arrangement.

Best regards,

Muhammad Ali   Lonnie Ali

LA/ds

G.O.A.T.
GREATEST OF ALL TIME
ENERGY SNACKS
EAT LIKE CHAMPIONS. WALK LIKE KINGS.
Shuffle
RUMBLE & SHUFFLE
G.O.A.T.
GOATFOOD.COM

Jabs
Shuffle

VITAMIN-POWERED ENERGY SNACKS
IT'S SOMETHING THAT WILL GO DOWN IN HISTORY, LIKE THE LEFT HOOK.
CAN'T STAND A BLOW HARD, BUT THEY WILL ALWAYS LISTEN TO HIM. EAT YOUR WORDS! EAT YOUR WORDS! I AM THE GREATEST! I KNOW WHERE I'M GOING AND I KNOW THE TRUTH, AND I DON'T HAVE TO BE WHAT YOU WANT ME TO BE. I'M FREE TO BE WHAT I WANT. I'M THE PRETTIEST THING THAT EVER LIVED. I LOVE TO SEE MY NAME WHERE EVERYONE CAN READ IT. SOMEDAY I'M GONNA SEE IT IN BRIGHT, BRIGHT LIGHTS."
G.O.A.T.
FOOD & BEVERAGE COMPANY
EAT LIKE CHAMPIONS. WALK LIKE KINGS.
G.O.A.T.

G.O.A.T.
Rumble
VITAMIN-POWERED FRUIT CRUMBLE
nuts 'n' nana
Jabs
VITAMIN-POWERED ENERGY CRISPS
thrill-a dill-a
I'M ON EVERYBODY'S LIPS.
Shuffle
VITAMIN-POWERED ENERGY CRUNCH
slammin' salsa
NET WT 1.24 OZ (35.2g)
Shuffle
thrill-a dill-a
Shuffle
VITAMIN-POWERED ENERGY CRUNCH
hot picnic chik'
Jabs
VITAMIN-POWERED ENERGY CRISPS
holy guacamole
NET WT 3.0 OZ (85.1g)
EAT LIKE CHAMPIONS, WALK LIKE KINGS.
Rumble
VITAMIN-POWERED FRUIT CRUMBLE
apple punched
Rumble
VITAMIN-POWERED FRUIT CRUMBLE
fruit fight
Shuffle
VITAMIN-POWERED ENERGY CRUNCH
big bad bbq

G.O.A.T.

RISE ABOVE

parmalat
S. Carter
solo
SUNCLOUD
chesterfield
DE TOMASO
CE.ICEBERG
simply samsung.
SAMSUNG
OneRedCube
DEVON
SoLeMia
MIAMI
ANNE KLEIN
AUTOMOTIVE
blade
GUCCIMUSEO
GUGGENHEIM
MUG
ROOT BEER
GNC
WELLBEING
Trop5
one foundation
crush
360
icejeans
ICEBERG
sweetpea
surge
HENR
BENDEL
DORRIAN
breadblok
SAMSUNG DIGITall
RED AND WHITE
PIZZA
LIFETIME
TELEVISION
VISARUNO
diet
peps
sweetstuff
Vita
SAKE

PRIV
Dch*
theory
GOOP
Rbk
STY HARBOR
DONNAKARAN
ROCKP
NEWYORK
PHENOM
CHEZ BON BON
KFC
a.m
peapod
G.O.A.T.
WORKS
HomeHero
PUMP
PHD
G Unit
FOOTWEAR BY Rbk
DKNY
Tribute WTC 9/11
Person to Person
ALLURE
CHANEL
micro
naked
FDNY
FIRE ZONE
Zino
A Line
Anne Klein

PETER ARNELL

PORTFOLIO

1980–2020

HATJE
CANTZ

Editor
Nadine Barth

Managing Editor
Dan Zuzunaga

Project Management
Sonja Altmeppen
Claire Cichy, Hatje Cantz

Text
Tristan Thom

Copyediting
Aaron Bogart

Design Director
Liyuan Tong

Designers
Joanne Min, Yuko Yamazaki

Typeface
Adobe Garamond Pro
and Helvetica Neue

Production
Stefanie Kruszyk, Hatje Cantz

Image Preparation
Vincenzo Piscitelli

Reproductions
LVD Gesellschaft für
Datenverarbeitung mbH, Berlin

Paper
GardaMatt Ultra, 150 g/m²

Printing and Binding
Printer Trento S.r.l

Published by
Hatje Cantz Verlag GmbH
Mommsenstraße 27
10629 Berlin
www.hatjecantz.com
A Ganske Publishing Group Company

ISBN 978-3-7757-4667-0
Printed in Italy

Frontispiece
My grandfather Nathan Hutt (right)
and his brothers, 1922

Dedication Page Happy Birthday letter by Lucy Arnell

PETER ARNELL

PORTFOLIO

1980 – 2020

VOLUME II

Dear Papa:

HAPPY BIRTHDAY! I love you so much and really hope you have an amazing day!! You are so dedicated to everything you attempt/do, supportive, kind, understanding, and the best poppy anyone could wish for. You have supported me 100% through every thing I do, and I would just like to take the time to really, sincerely thank you for every thing. It means more than you would ever know. You touch the lives and hearts of everyone you touch in such positive and influencial ways; I would know first-hand. You are so gifted at what you do and If someone asked me what I would change about you, I wouldn't change a hair on your head. I love you soooo much. In the wise, poetic words of Elton John: "I hope you don't mind that I put down in words, how wonderful life is now you're in the world."

Happy Birthday, you're the best!

with all the love in the world, Lucy

For Anita, Lily, Lucy, and Thomas

In a life filled with extraordinary blessings and good fortune, being your father has been the greatest privilege of all.

He who seeks beauty will find it.

Peter Arnell's projects of 40 years of Big Idea thinking are stunning, thoughtful, and hubris at its most powerful. Like his impactful project for Muhammad Ali, Peter Arnell's humanistic spirit is the greatest of all time.

George Lois

VOLUME II

Selected Works

BKLYN Arena
Sentosa Resort
Pizza Hut Delivered
KFC AM
KFCity
The Home Depot
iCet & eCet
Paini
Kyoto Knives
HomeHero
One Foundation
Special Olympics
Dodge Challenger
Fontainebleau
LIV
GNC
Pepsi
Mug Root Beer
Mountain Dew
Sierra Mist
Tropicana
Trop50
SoBe Lifewater
Naked Juice

Phenom
Muji
ITO-YA
Uniqlo
SA Sake
Oribe
Goop
EKS
The North Face
Peapod
GEM
Dodge ATV
DeTomaso Pantera
Guilio Micro
High-Speed Trains
Devon
Samsung LED 9000
Entertainment Optics
Gucci
Gucci Museo
Russian Standard Vodka
Belvedere Vodka
Don Julio Tequila
Theory

Theory 38
Dubai
Milk Gallery
ThaiWORKS
ThaiMoon Sake
TukTuk
Chez Bon Bon
Denford Park Honey
Al-Ula
Camelus
Equestrian Center
DreamPlace
Sphere Arabia
Visconti
Unopiu
Martha Graham Dance Company
DVF
Booes
Sharecare
UFC
NYCFC
CityCage
BlackBerry
Aman

Solemia
Veon
Lifecell
TDLI
The Legends of Creativity & Design
BiP
1440 Font
YaaniMail
FDNY Firezone
Digicel
SportsMax
Digicel+
Digicel Business
Qatar America Institute
TB12
Allen + Roth
Kobalt

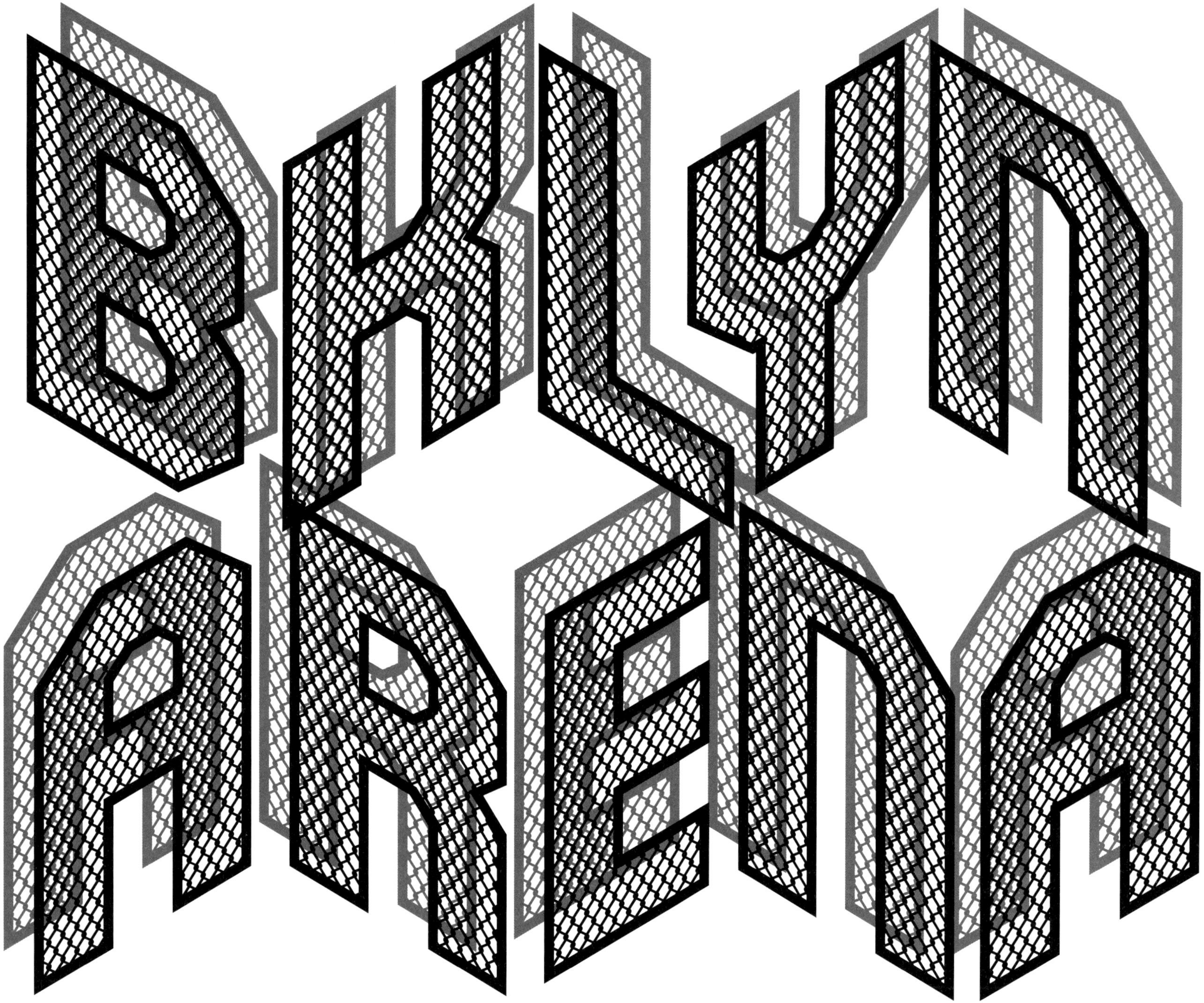
BKLYN
ARENA

BKLYN ARENA

2006

When Forest City Ratner Company and the Brooklyn Nets Arena Development Corporation set out to build a new home for the Brooklyn Nets in the heart of downtown Brooklyn, developer (and Nets co-owner) Bruce Ratner invited Frank Gehry to design a new arena and mixed-use development that would be an instant landmark befitting the long-awaited return of professional sports to Brooklyn.

Collaborating closely with Gehry, and working alongside legendary graphic designer Lance Wyman, a total brand identity architecture system and wayfinding program was developed for this new landmark destination.

In devising a name for the new arena, the main requirement was something that felt organically a part of Brooklyn, while still being new and novel. After working through hundreds of potential options, the name BKLYN Arena was settled on. The subtle transformation of "Brooklyn" to the simplified, all-caps spelling of "BKLYN" was a nod to the way in which shorthand abbreviations, popularized by texting and instant messaging platforms, were creeping into mainstream usage. This treatment also elevated the otherwise generic name "Brooklyn Arena" into something bold, proprietary, and totally unique. Naming it BKLYN Arena was a clear way of giving something back to Brooklyn: not just to be an arena in Brooklyn, but an arena for Brooklyn, of Brooklyn, by Brooklyn.

Borrowing from the blocky typefaces long used for jersey numbers in the NBA, a bold new typeface was developed. Inspired by the folding, undulating character of the arena that Gehry had designed, FRANKTYPE was created for use throughout the facility, including in signage and wayfinding applications—and even became the basis of the new brand identity system for the arena itself. Designed as a dynamic typeface with no one fixed expression, FRANKTYPE's alphanumeric characters were meant to be shifted, folded, dimensionalized, faceted, colored, and patterned in an infinite number of ways. This flexible typographic and identity architecture system ensured that there would be no one correct expression of BKLYN Arena, but rather a multitude of vibrant, dynamic, and flexible expressions—much like the endlessly diverse character of Brooklyn itself.

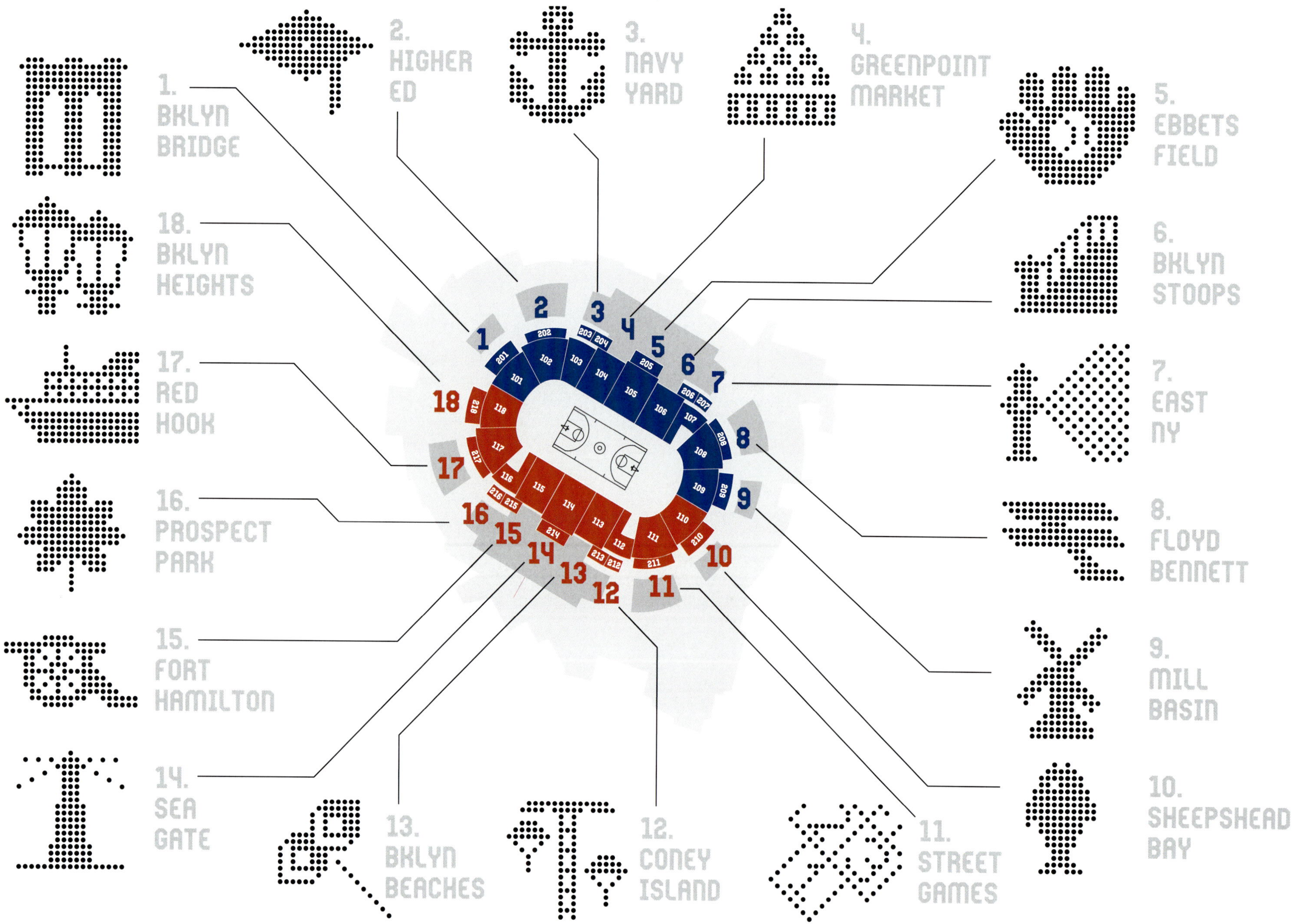

1. BKLYN BRIDGE
2. HIGHER ED
3. NAVY YARD
4. GREENPOINT MARKET
5. EBBETS FIELD
6. BKLYN STOOPS
7. EAST NY
8. FLOYD BENNETT
9. MILL BASIN
10. SHEEPSHEAD BAY
11. STREET GAMES
12. CONEY ISLAND
13. BKLYN BEACHES
14. SEA GATE
15. FORT HAMILTON
16. PROSPECT PARK
17. RED HOOK
18. BKLYN HEIGHTS

BKLYN
ARENA

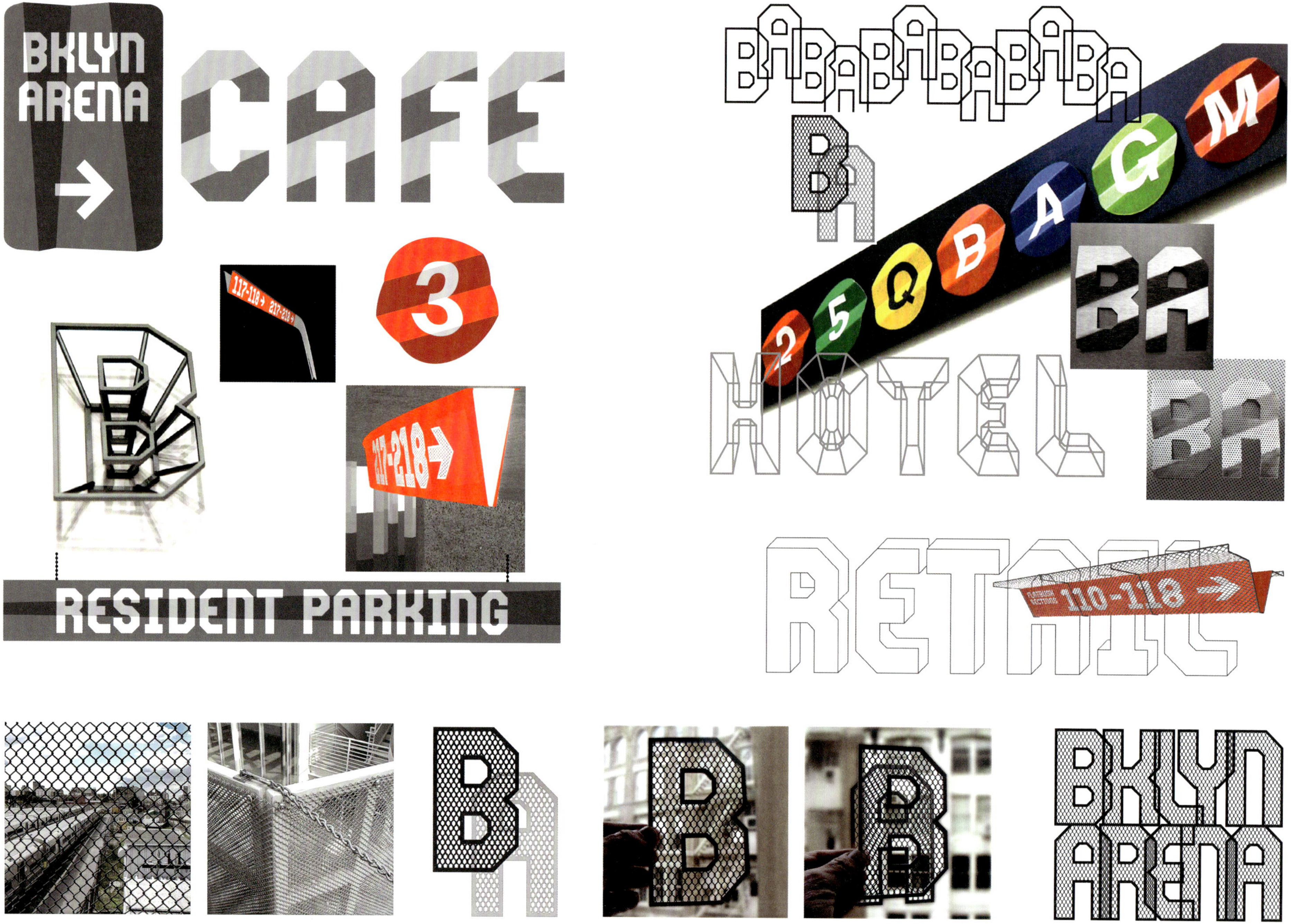
BKLYN ARENA
CAFE
3
117-118 → 217-218 →
217-218 →
RESIDENT PARKING
BABABABABABABABA
BA
2 5 Q B A G M
BA
HOTEL
BA
RETAIL
110-118 →
BA
BKLYN ARENA

FRANKTYPE
1234567
FRANK 1
FRANK 2
FRANK 3
FRANK 4
FRANK 5
FRANK 6
FRANK 7
FRANK 8
FRANK 9
FRANK 10
FRANK 11

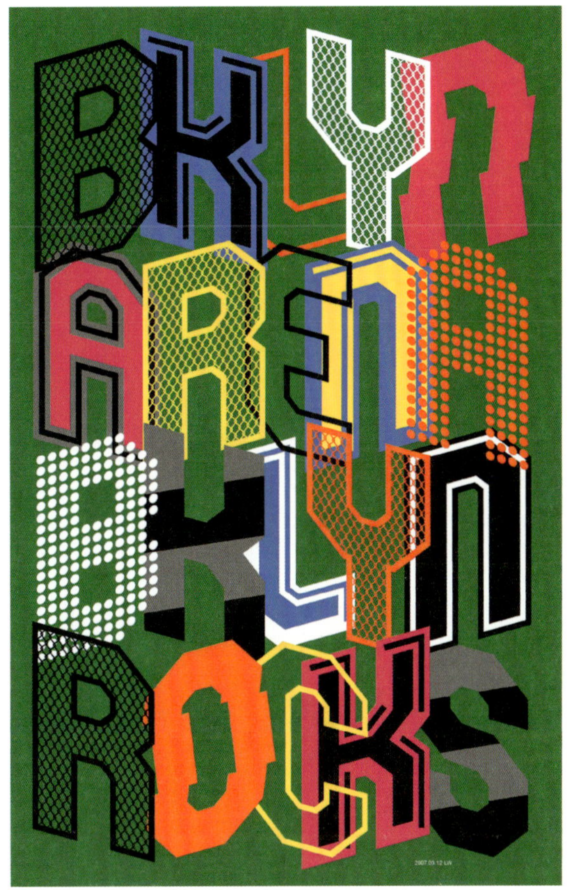
BKLYN
ARENA
BKLYN
ROCKS

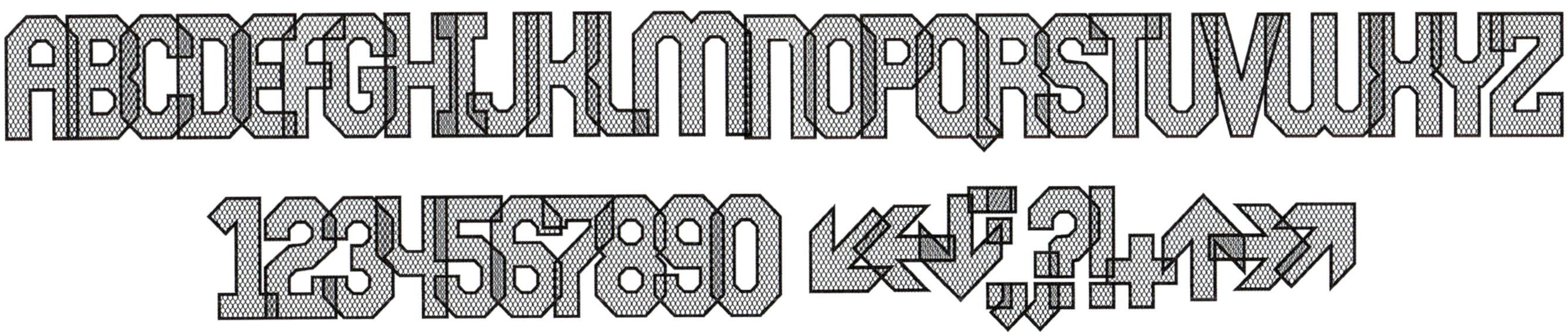
ABCDEFGHIJKLMNOPQRSTUVWXYZ
1234567890

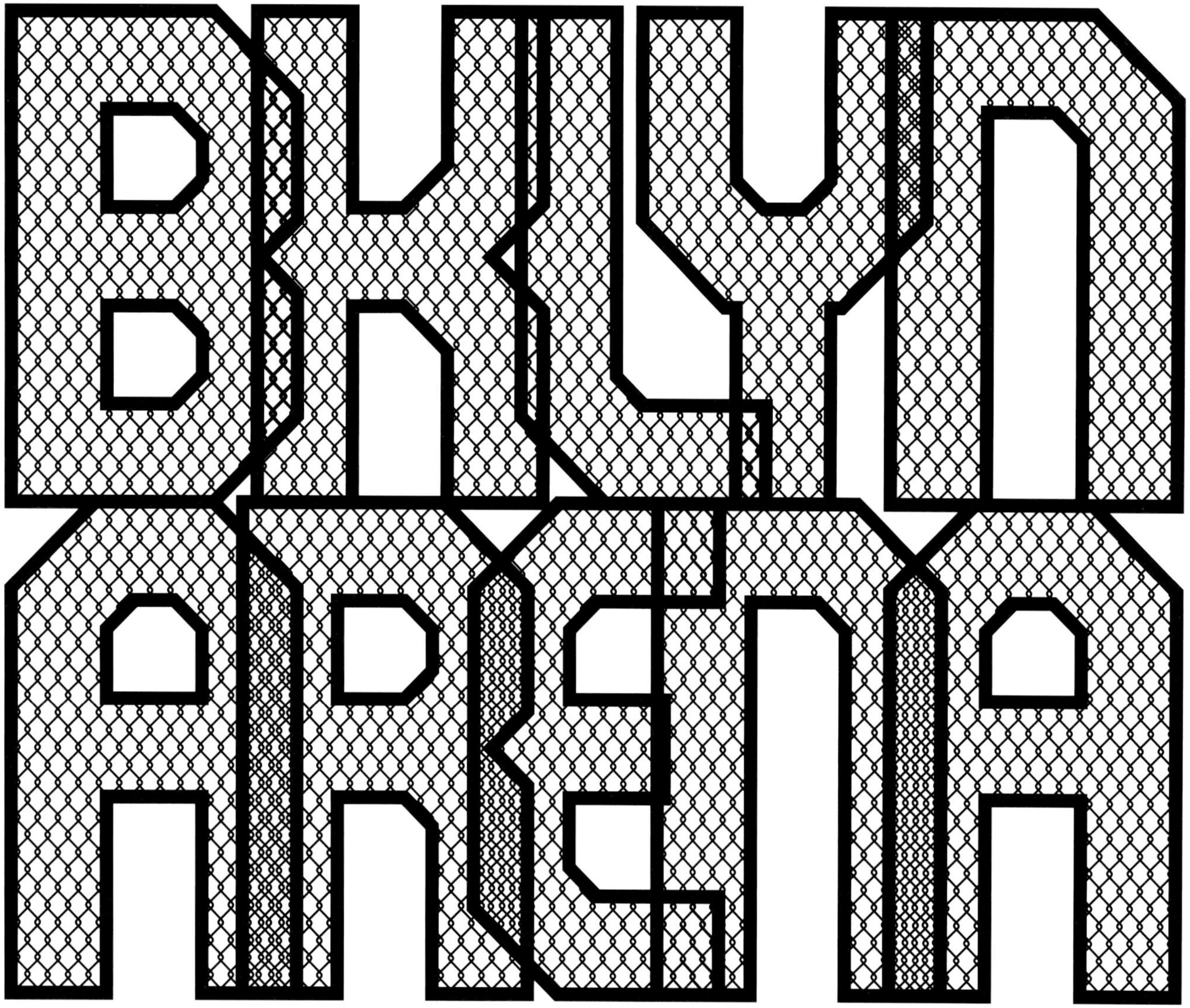
BKLYN
ARENA

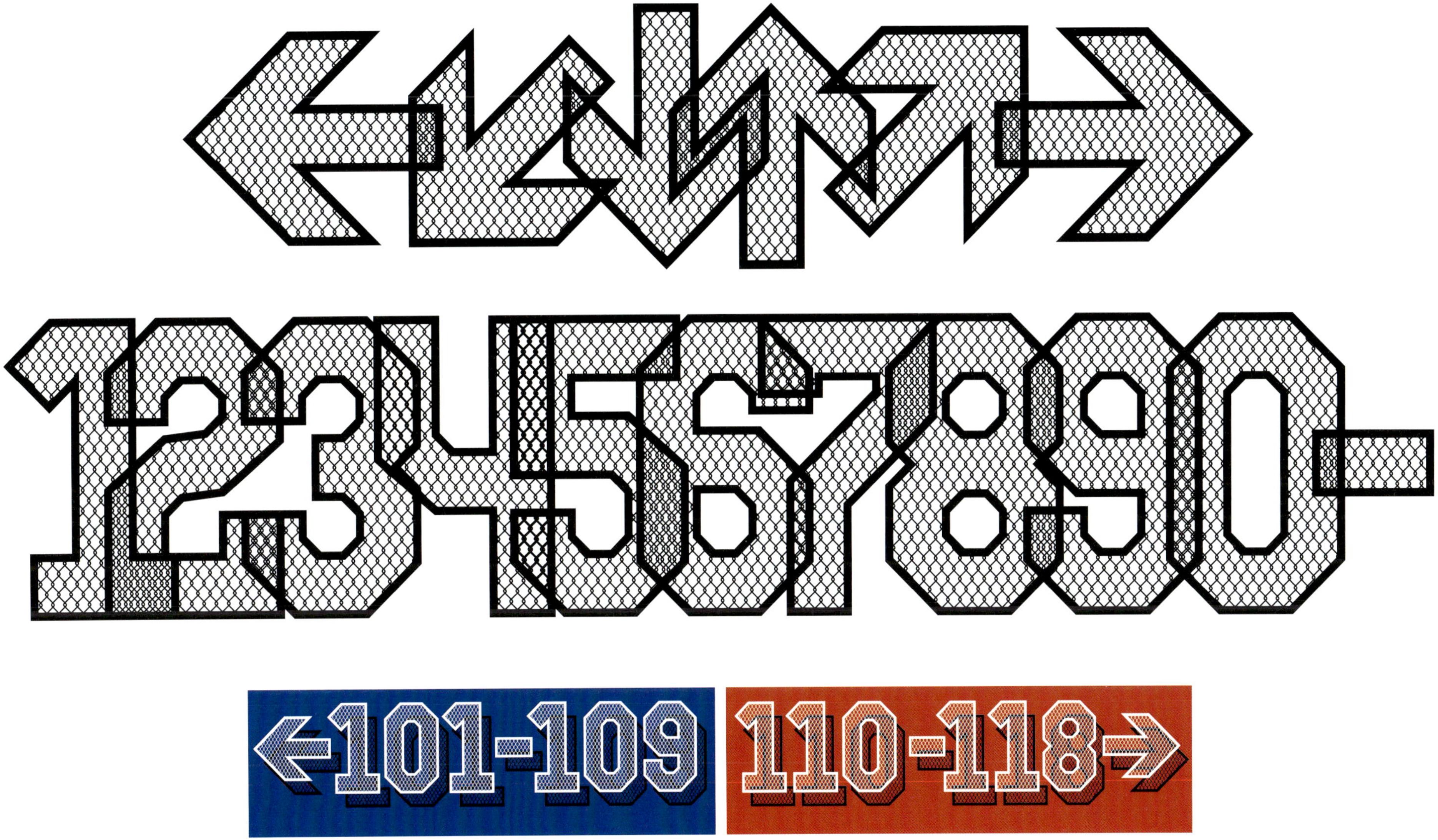
1234567890-
←101-109
110-118→

ABCDEFGHIJKLM
NOPQRSTUVWXYZ
1234567890
↙ ← ↓ : . , ; ? ! - + ↑ → ↗

BKLYN
ARENA

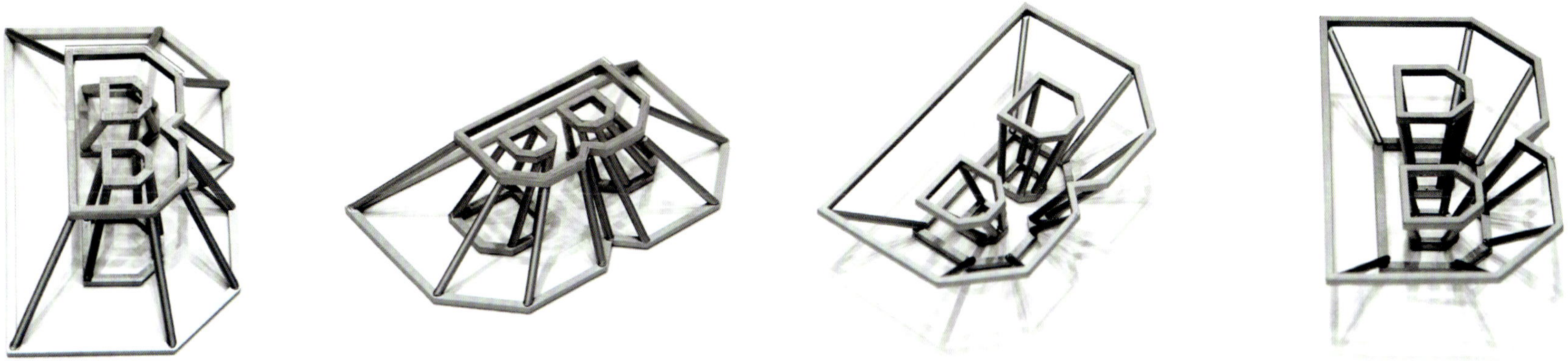

BA
HOME
BA
BA
BA
BKLYN
ARENA
BKLYN
ARENA

BKLYN ARENA MEDIA SKY

2006

The BKLYN Arena Media Sky concept was developed as a dramatic, sculptural, high-tech reimagining of the typical LED and dot matrix billboards found in stadiums above center court. The sweeping, organic character of the Frank Gehry-designed stadium was echoed in the undulating surfaces of the Media Sky, with the overall effect being an immersive, dynamic, multimedia experience that blurred the lines between the physical and digital realms.

Innovative seating design concepts were also developed to provide greater comfort, audience engagement, and increased revenue opportunities for the arena. Ergonomic mesh seat backs and an integrated lumbar support system were a vast improvement over the usual hard plastic shell arena seating, while armrest-embedded touchscreen LCD screens provided event-specific information throughout games, concerts, and other events, and an integrated ordering system that allowed every audience member to order food and refreshments directly from their seat with a just a few taps.

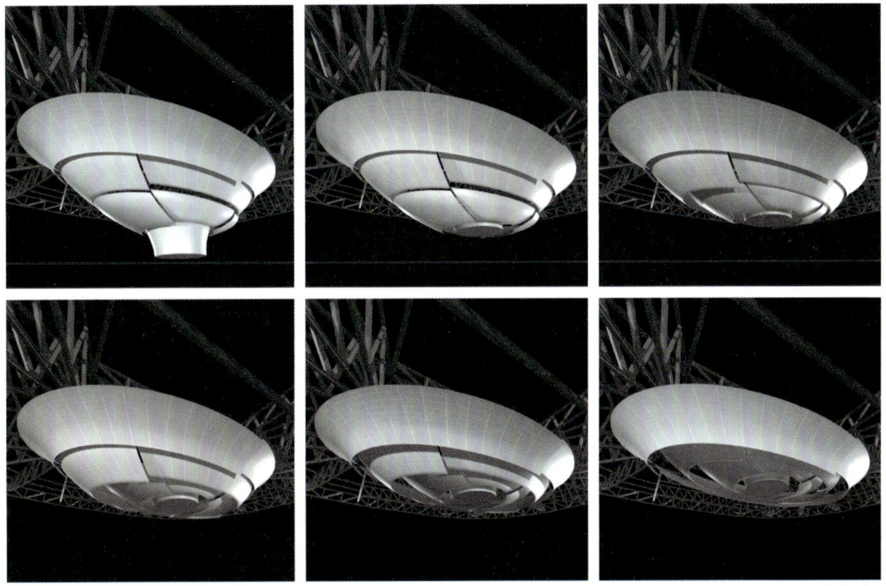

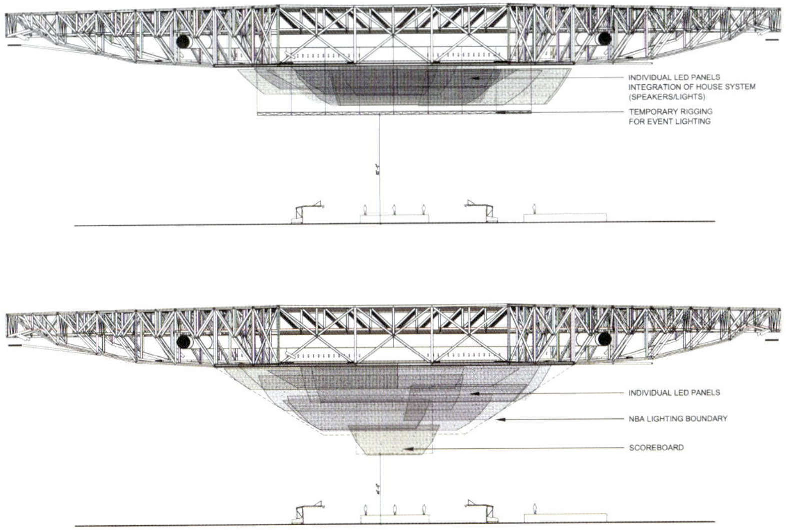

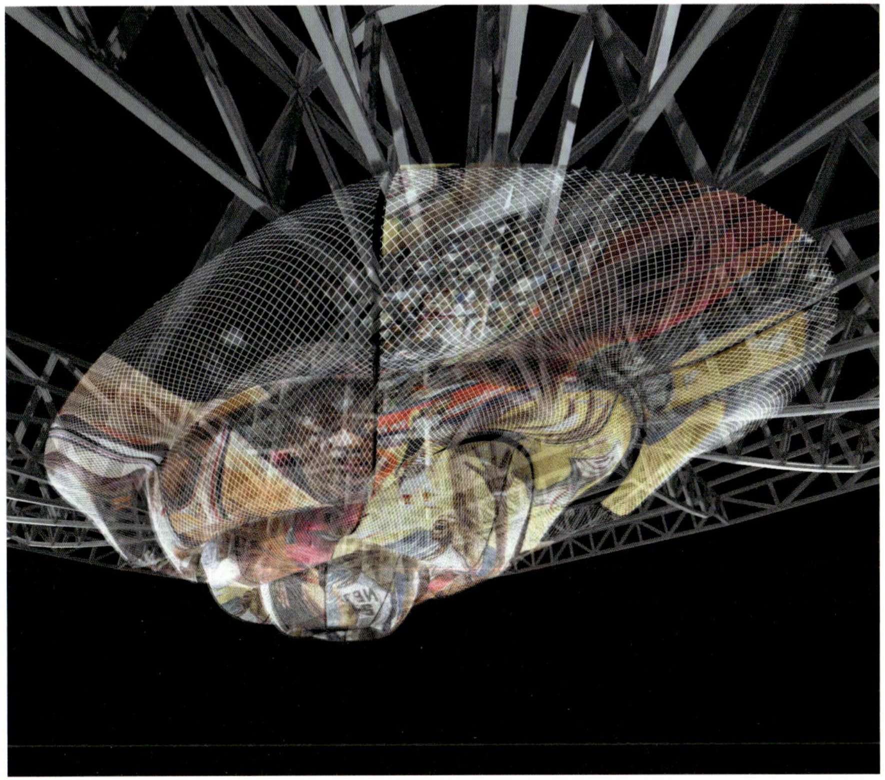

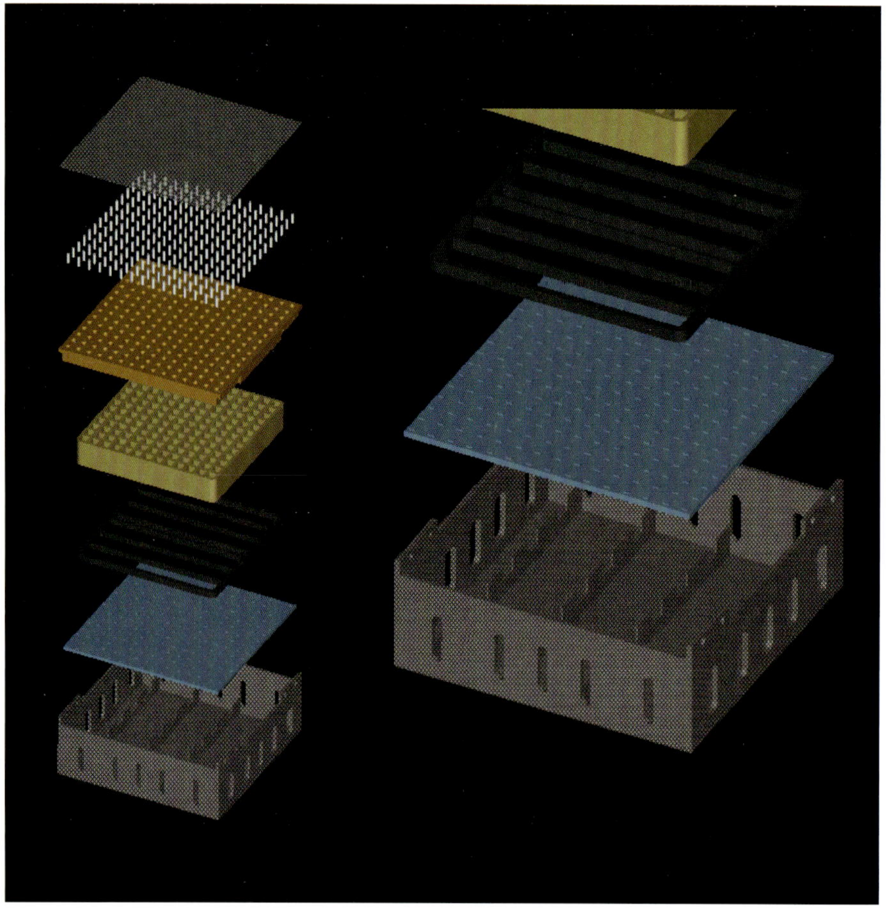

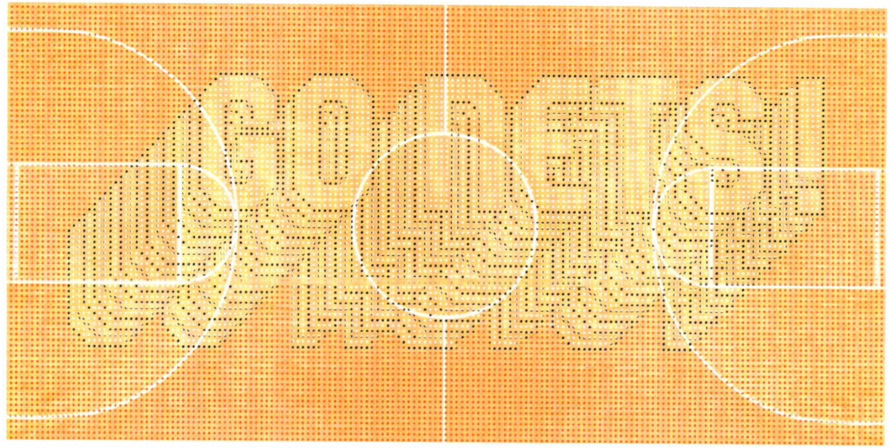
GO NETS!

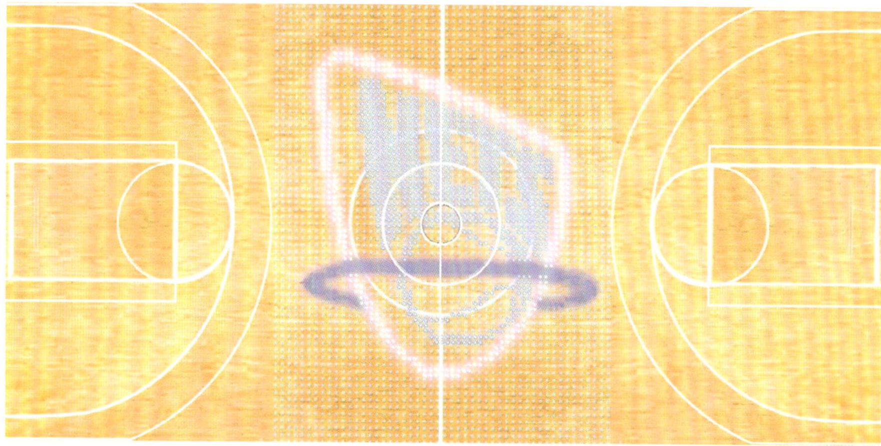

NETS
35

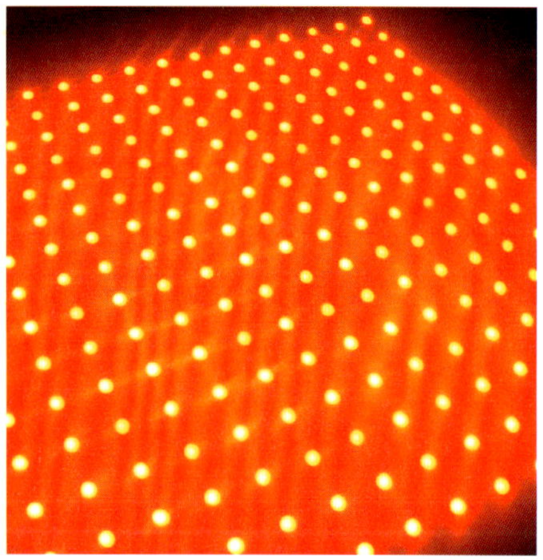

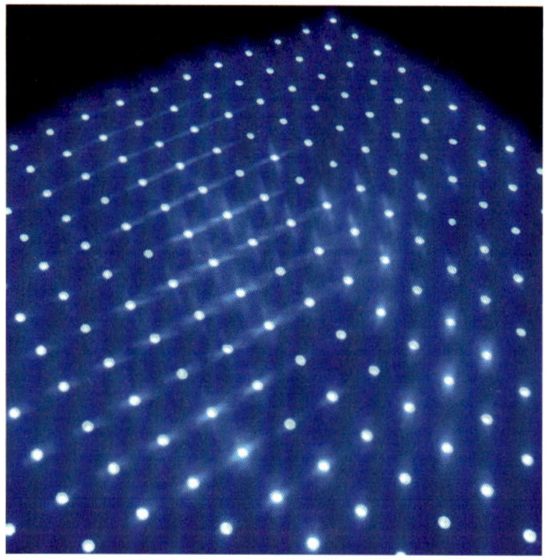

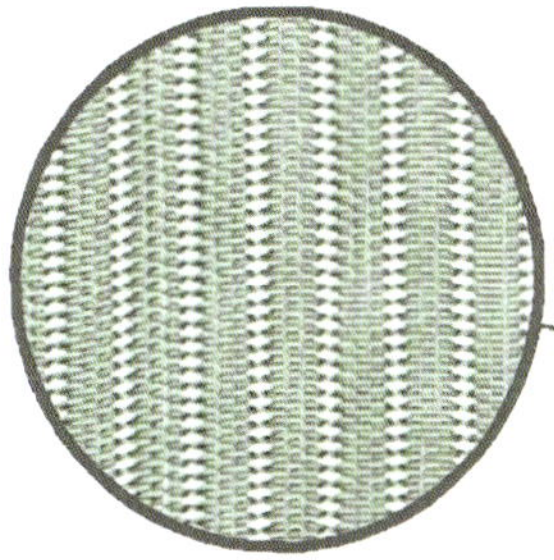

Metal mesh seat back ensures maximum back support and contouring to multiple the body forms

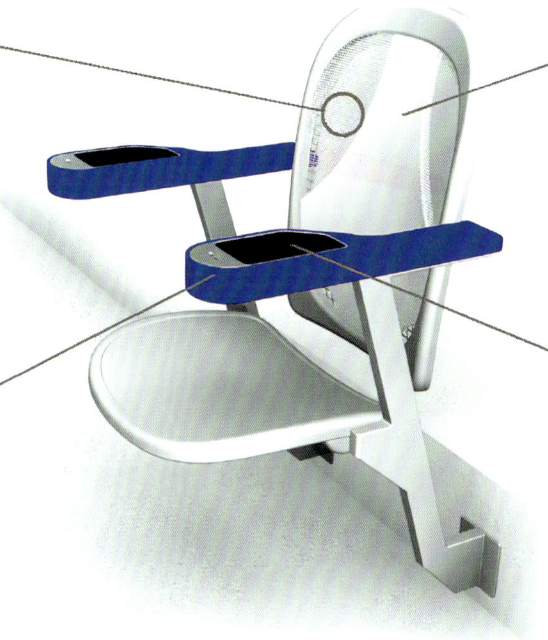

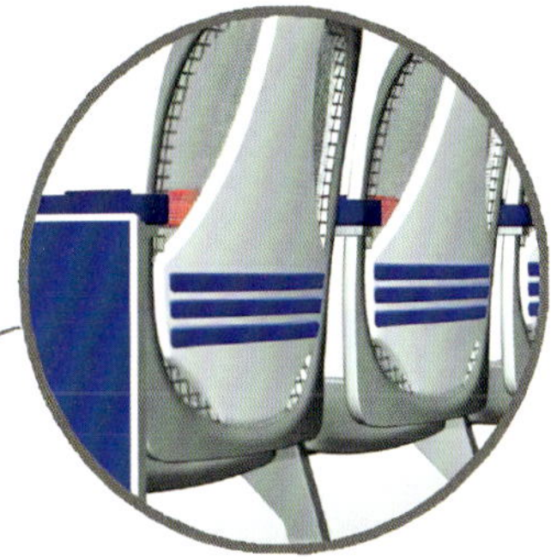

Lumbar support system with integrated storage bands that pull out to hold and store clothing etc.

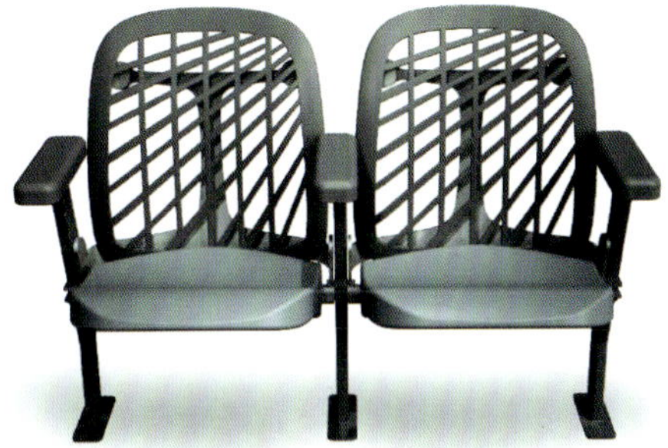

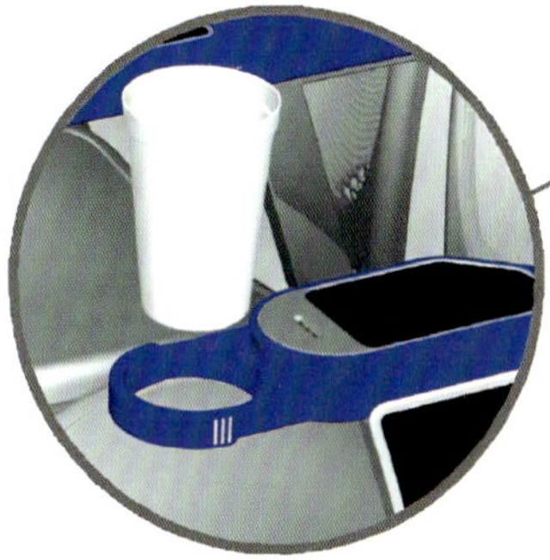

Integrated into the arm is a pull out cup holder. The holder is stored away inside the arm when not in use.

Embedded seat LCD welcomes you to the game/event and provide instant information and general connectivity

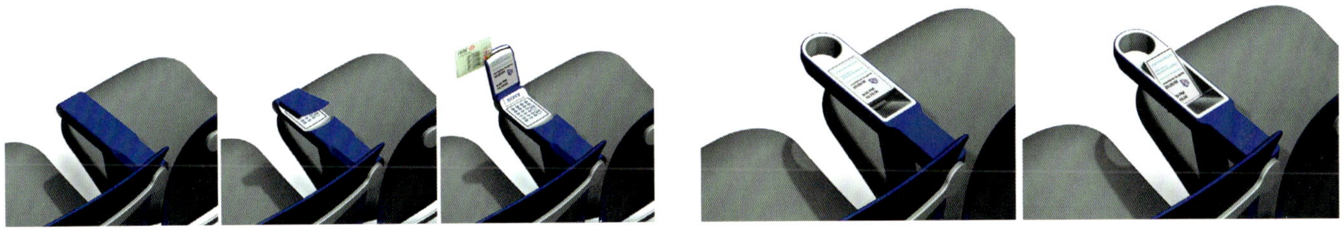

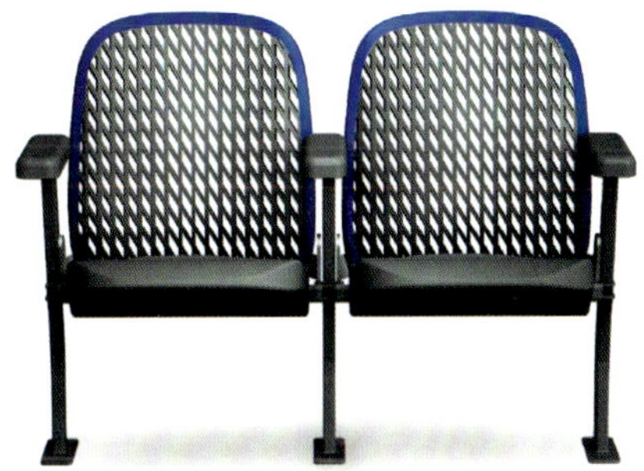

BKLYN ARENA MEDIA WALL

2006

Spanning multiple city blocks in a highly trafficked area of downtown Brooklyn, the scale and high degree of visibility of the BKLYN Arena development proposal offered numerous opportunities to make the arena into a new and different kind of landmark—one that was both responsive to, and reflective of its surrounding community.

As part of a broader community engagement program, a scheme was developed for an innovative digital façade that transformed the arena's exterior façade into large-scale, multimedia billboards that would be given over to content for and by the local residents who lived in the surrounding neighborhood. Thirty percent of the building's surface was embedded with small LED lights that formed a vast electronic canvas: community-specific information, news, art, and more would float across the surface of the building according to a dynamic messaging strategy built around real-time, real-world input.

This massive digital billboard was designed to disappear when not in use—the thousands of small lights in the façade adjusted in brightness depending on the time of day, dimming completely late at night out of respect for the many residential neighbors whose homes overlooked the arena.

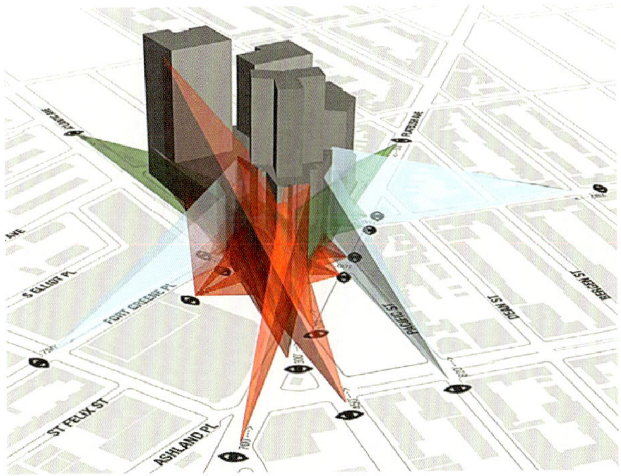

SENTOSA ISLAND RESORT

2007

Developed as a conceptual collaboration with architects Frank Gehry and Greg Lynn and developer Kerzner International in response to a competition call from the Singaporean Government, the Sentosa Island Resort scheme was designed to be an unprecedented destination experience that thrust visitors into a spectacular new world of discovery, imagination, and otherworldly beauty.

At the heart of the resort was an aquarium unlike any other: the facility featured a vast array of exotic marine life from all over the world, housed in massive, sculptural tanks made of specially engineered glass that was molded to mimic the fluid movement of water. The mesmerizing effect gave visitors the feeling of being in and among the creatures, as if swimming alongside them in the sea.

In addition to these exotic marine creatures from around the world, a concept was also developed for a new virtual lifeform: robotic-organic hybrid creatures called "Aquabrids." Designed alongside famed animatronics creator Walt Conti and roboticist Rod Brooks, a select number of robotic Aquabrids would "live" in the Aquarium among the actual sea life, further blurring the line between reality and fantasy.

Visitors were also invited to create their own original virtual Aquabrids using their RFID wristbands: as they traversed the aquarium, they could select specific attributes of the real-world sea life all around them, and this information would then be input into a platform that generated a new "species" according to their selections. At the end of their visit, they would be presented with detailed, lifelike 3-D visualizations of their unique Aquabrid creations: an unprecedented merging of physical and virtual worlds that prefigured much of the virtual and augmented-reality experiences that have since become commonplace in the worlds of entertainment, design, and beyond.

4.35 Million

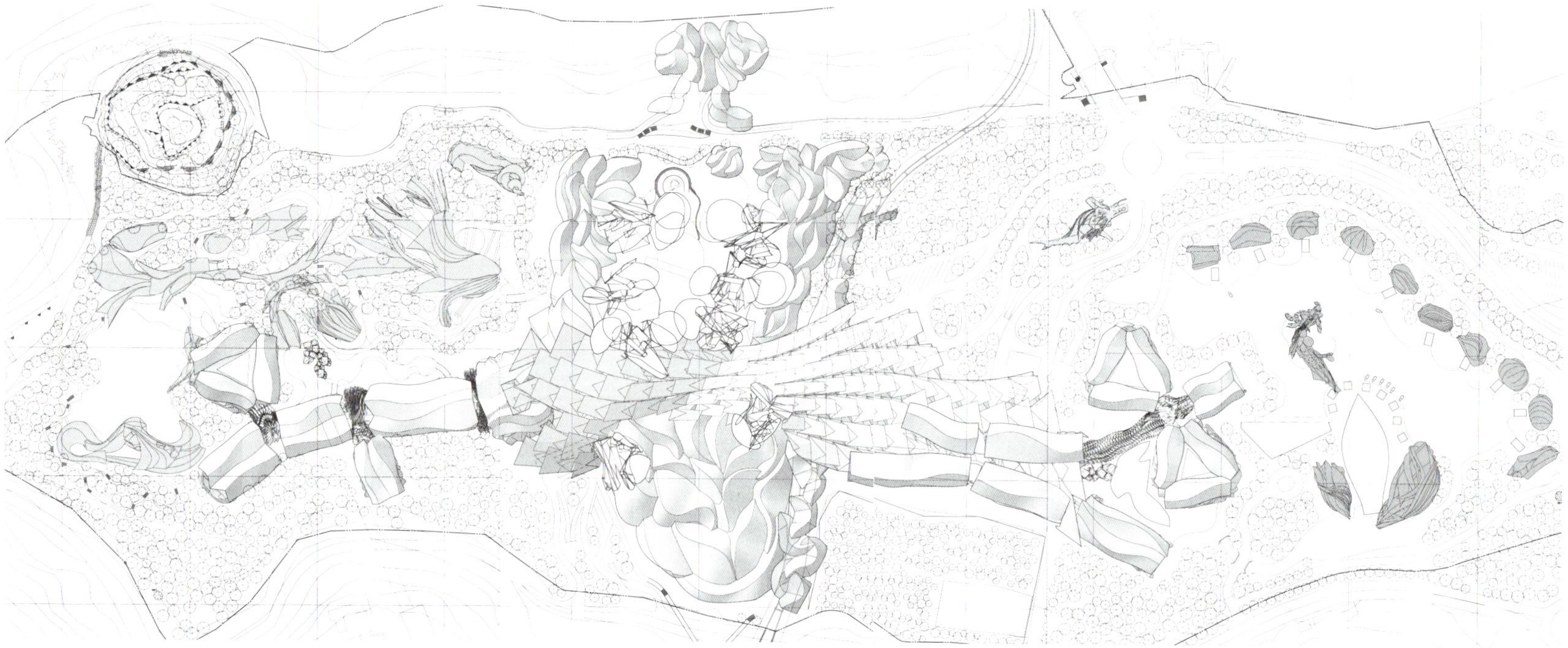

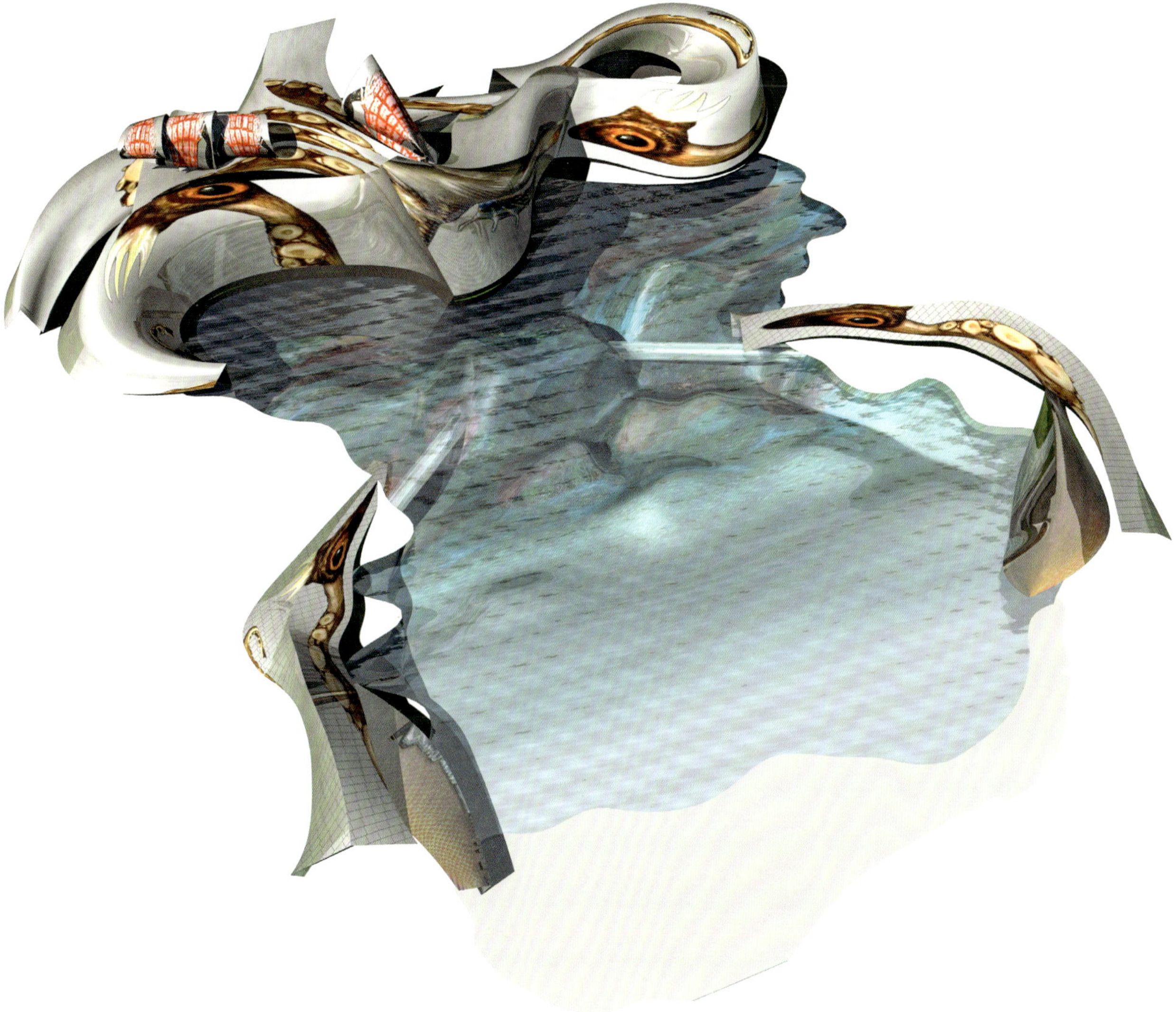

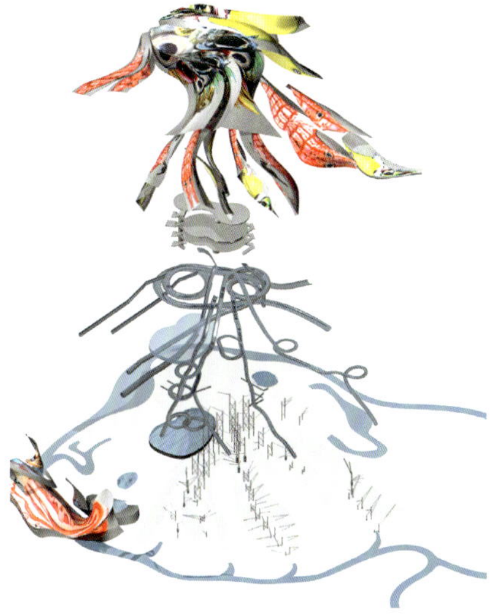

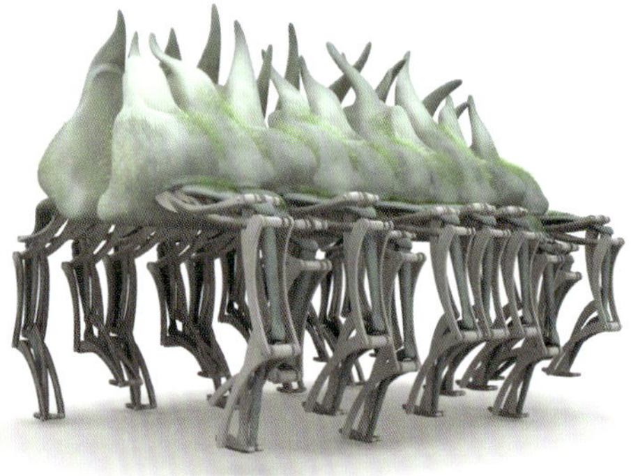

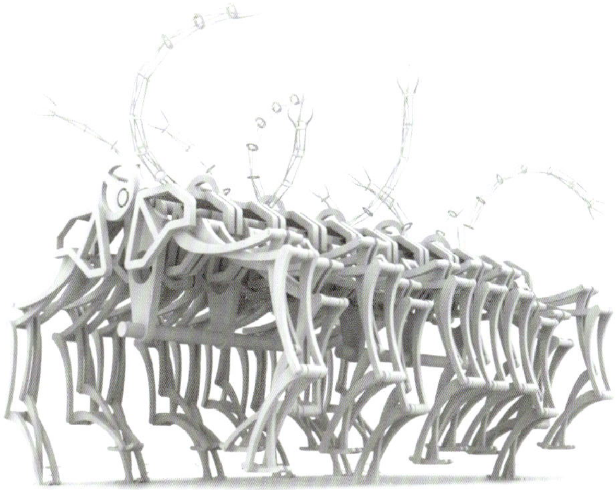

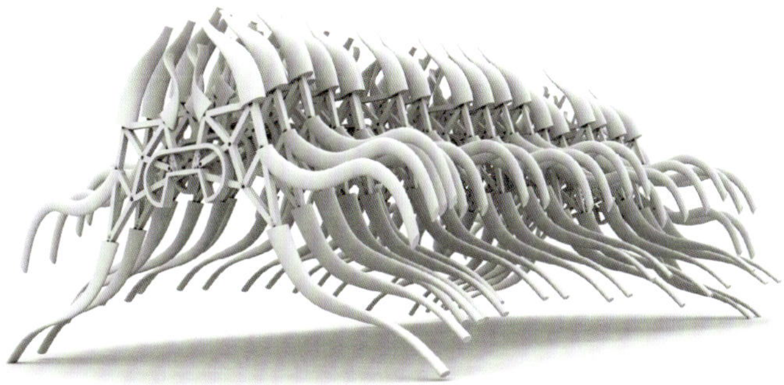

PHD

PIZZA HUT DELIVERED

2007

To help Pizza Hut reach more customers, a proposal was developed for standalone retail outlets that focused solely on Pizza Hut's fast, fresh home delivery service: the same great pizza, delivered faster than ever. Dubbed PHD, Pizza Hut Delivered was a comprehensive new program and extension of the core Pizza Hut brand that emphasized the speed, efficiency, and convenience of home delivery.

A new brand mark was developed for the venture—a speeding car inspired by the form of Pizza Hut's iconic red roofs—that implicitly promised customers the same classic dine-in experience that Pizza Hut had become renowned for, delivered to the comfort of their own home.

Design concepts were also created for innovative heat-preserving packaging designs, new delivery uniforms, and a pioneering online ordering platform, all of which helped to deliver on PHD's core promise of blazing hot pizza, delivered fast and fresh.

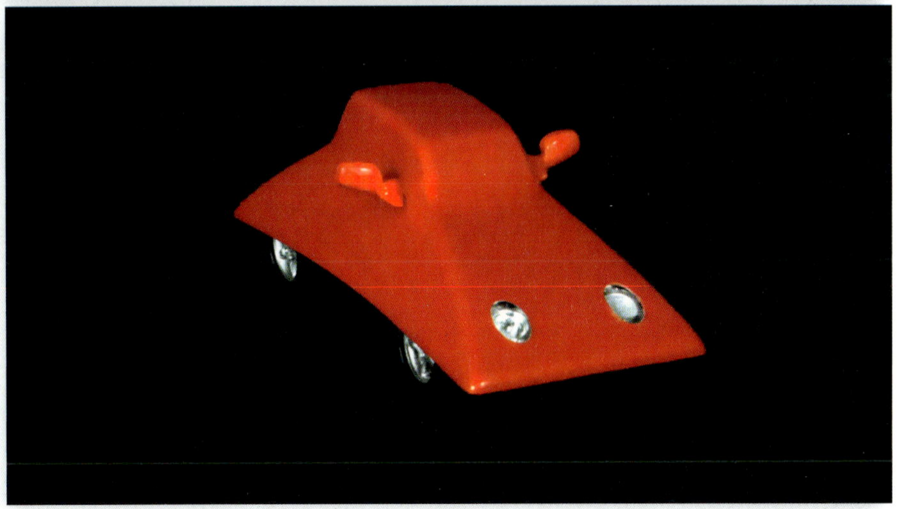

KFC AM

2007

After decades of exponential global growth, KFC was looking for ways to expand its menu offerings that still felt like an organic extension of its distinctive brand. It was important that any new menu items had an authentic connection to the core DNA of the KFC brand. If KFC had been built around chicken, what came next was obvious: eggs. And by extension, breakfast.

The KFC AM pilot program represented a natural evolution of KFC's existing brand equity into a lucrative new space. The expansion into the breakfast market gave existing KFC customers a new reason to come in, while also enticing those who might not have otherwise been drawn in by the restaurant's more traditional lunch and dinner menu items.

KFC AM was given numerous expressions across multiple customer touchpoints,including innovative new egg-inspired meal packaging, a family of eggy mascots decked out in Colonel Sanders's signature glasses and bow tie, and a cheerful, bright, modern design scheme for dedicated KFC AM restaurant locations.

KFC a.m.

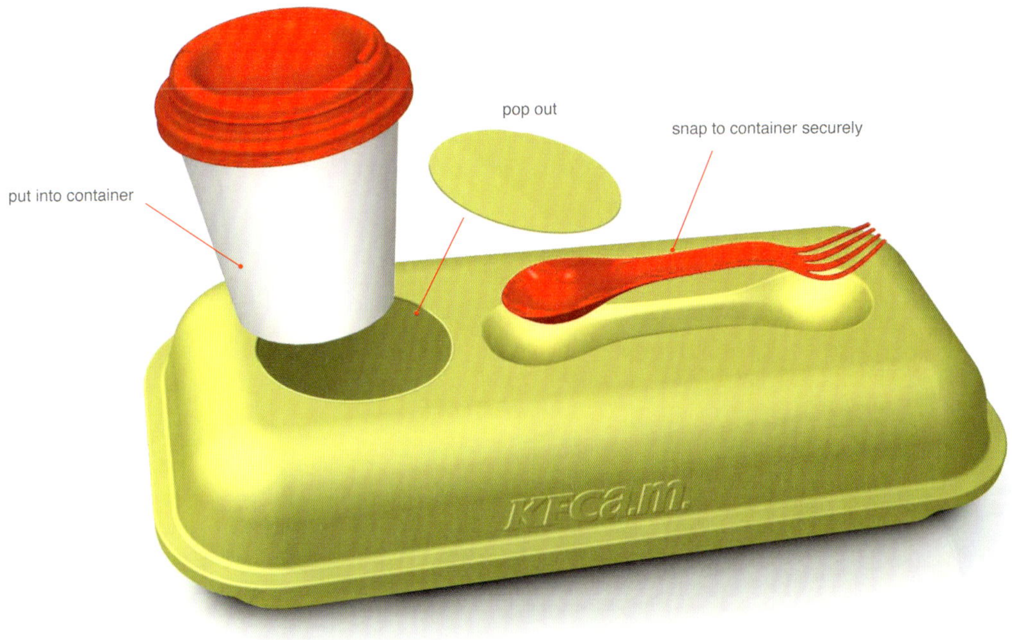
put into container
pop out
snap to container securely
KFCa.m.

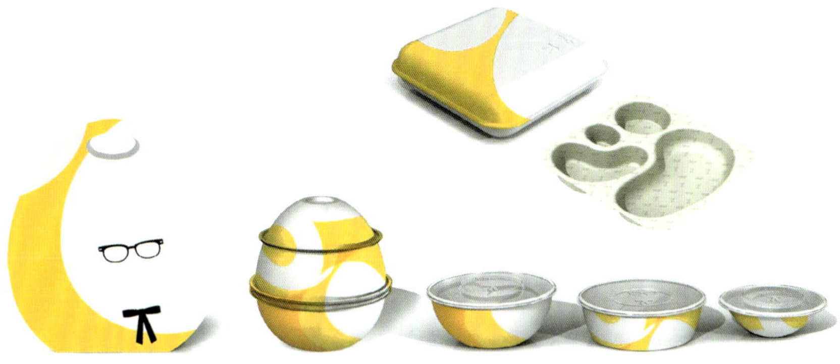

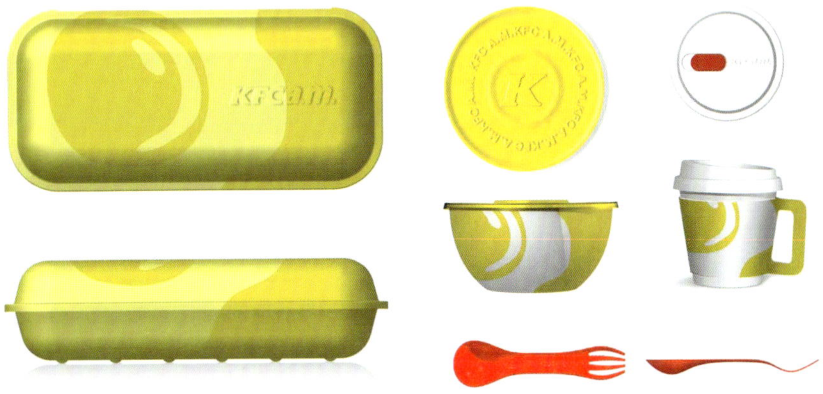
KFCa.m.

KFC

KFCITY

2007

With the specific needs of the city-dwelling customer in mind, a new streamlined, modernized, and elevated retail concept, KFCity, was developed for KFC and launched in select test markets across England.

A clean, simple, sophisticated design language was applied to product packaging, wayfinding, and store design, while carefully engineered, continuous customer paths sought to help customers avoid long queues, and ideally eliminate wait time altogether. The whole experience, while still distinctively KFC, was refined to appeal to a customer who wanted to enjoy an elevated take on the classic KFC menu.

The sleek, minimalist designs of KFCity locations coupled with the high-quality menu offerings took KFC's food and service to the next level, and anticipated the more recent proliferation of premium fast-casual dining options that balance convenience and efficiency with exceptional quality.

THE HOME DEPOT®

ORANGE WORKS™

HOME DEPOT ORANGE WORKS

2007

As part of a coordinated push to help The Home Depot cement its status as the world's preeminent home improvement retailer, under the leadership of CEO Robert Nardelli a new in-house innovation and design lab was launched to accelerate the development of industry-leading products for their private label brands.

Dubbed Orange Works, this group was charged with creating innovative products that embodied high levels of functionality, next-generation technology, consumer appeal, and design excellence across "mega trend" categories, including outdoor living, home safety, and smart home technology, among others.

Orange Works designed and developed house-branded products to replace national brands on Home Depot shelves, working closely with its manufacturing partners to develop proprietary products at an attractive price point for consumers. This process rapidly accelerated the speed to market of best-in-class products to meet the needs of consumers in new and emerging product segments.

Orange Works quickly proved its value, providing extraordinary margins through a series of innovative and high-quality product lines across categories, ranging from tools to furniture to appliances and more.

In addition to developing numerous high-margin consumer products, the many internal efficiencies and optimizations that Orange Works pioneered also provided the template for the next generation of supply chain management strategies for The Home Depot as a whole.

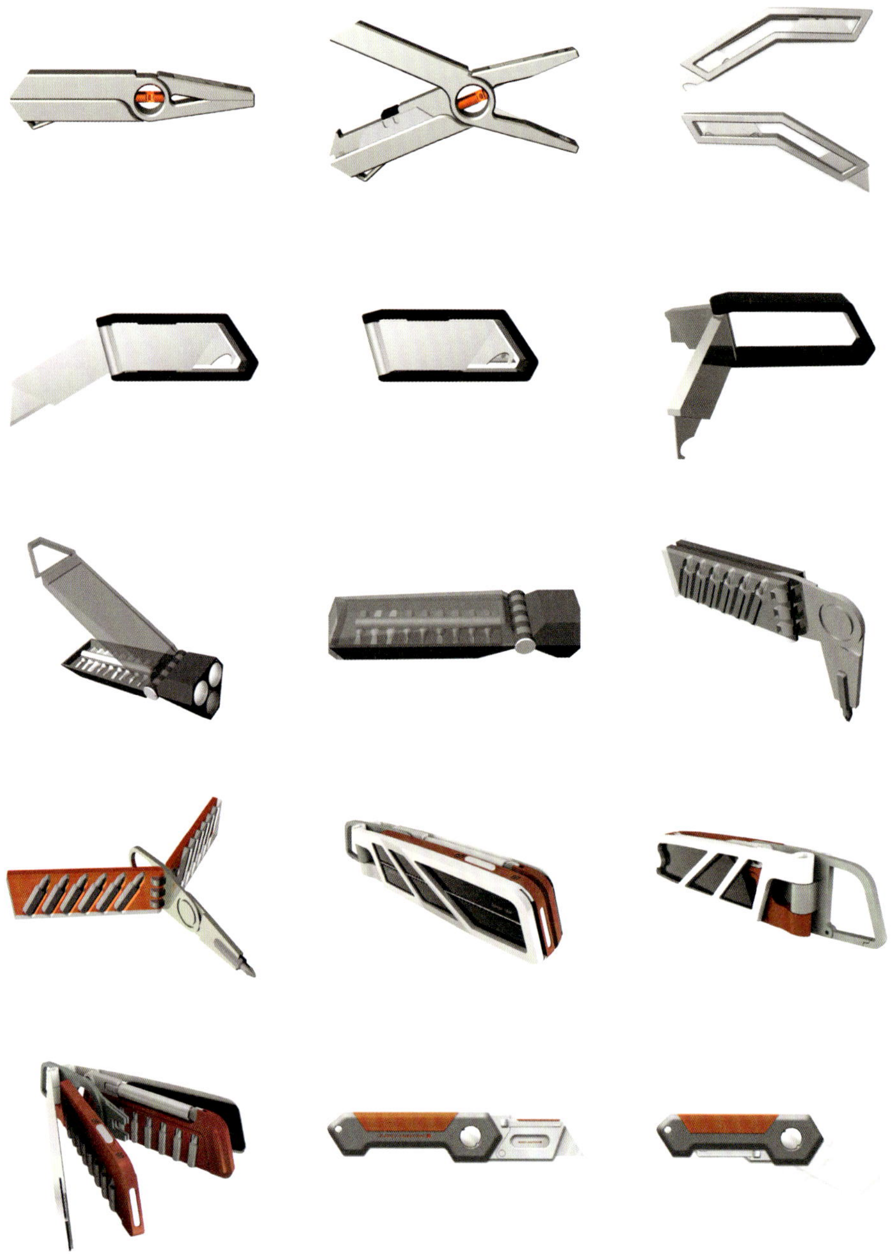

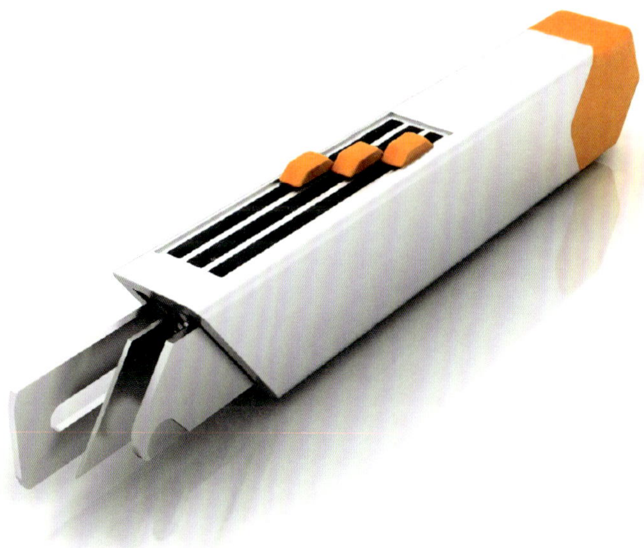

0
1
2
3
4
5
6
7
8
9
10
11
12
13
14
15
16
17
18
19
20
21

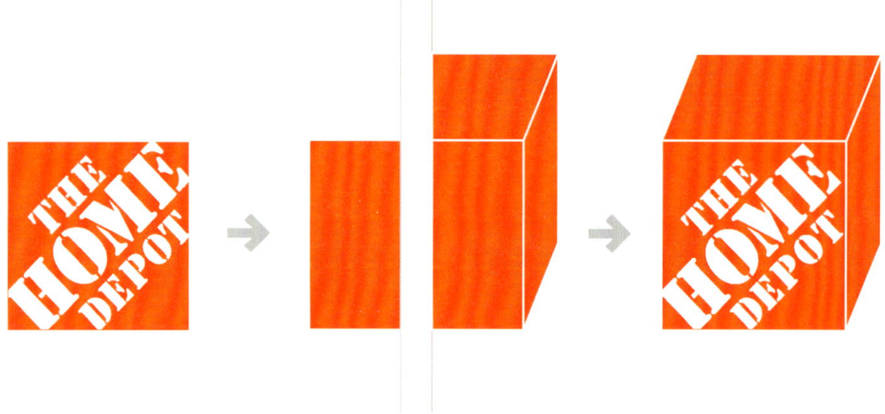
THE HOME DEPOT
THE HOME DEPOT

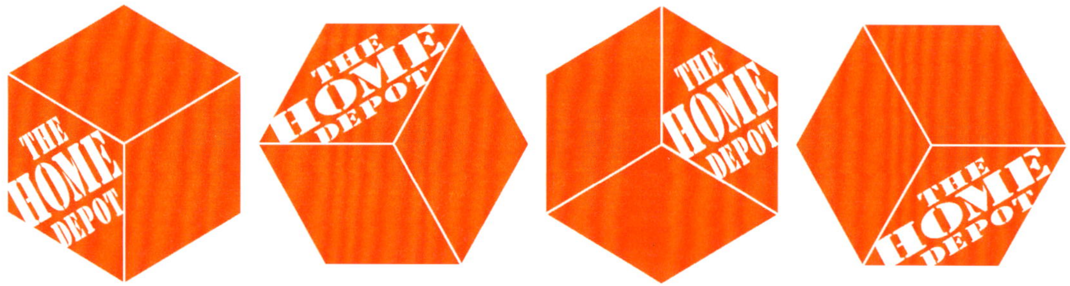
THE HOME DEPOT
THE HOME DEPOT
THE HOME DEPOT
THE HOME DEPOT

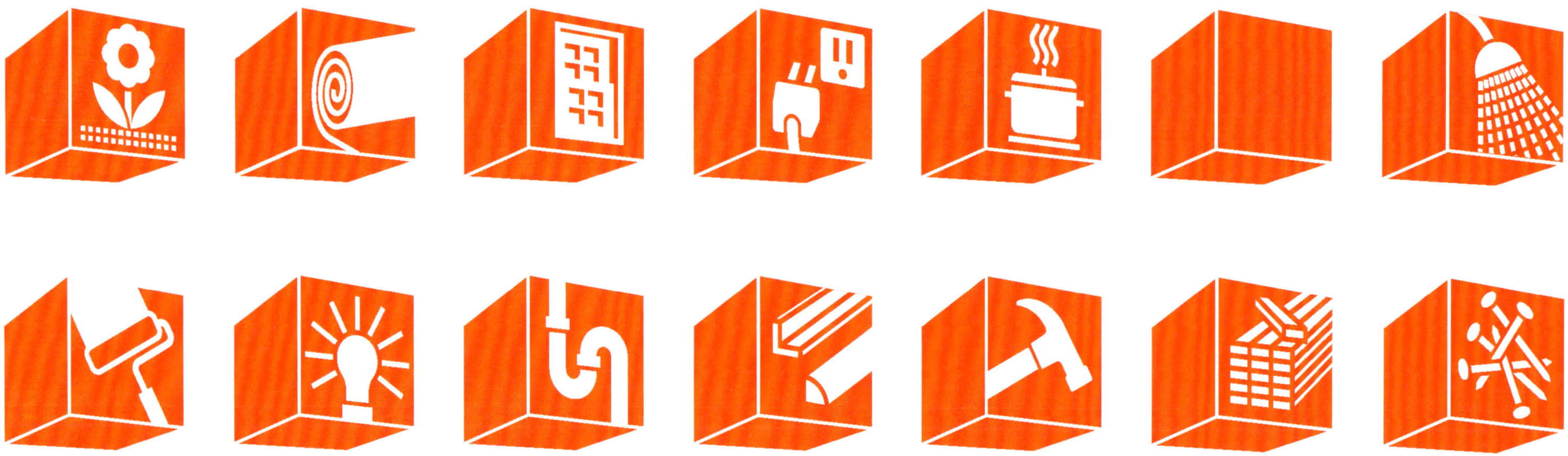

HOME DEPOT WAYFINDING

2007

Working in collaboration with graphic designer Lance Wyman, a concept for an intuitive and informative in-store wayfinding system was developed for The Home Depot to help customers more easily navigate the oftentimes vast store layouts.

The large size of the stores coupled with the volume and variety of products on display meant that Home Depot customers would often end up spending an inordinate amount of time trying to locate the specific items they had come in for. A clear, consistent, easy-to-understand navigation system was identified as the key to making the in-store experience as efficient, informative, and seamless as possible for every customer.

A series of beautiful, simple, universally recognizable icons was developed to represent each of The Home Depot's major departments. Set against a bold orange backdrop, these graphics, strategically placed at key points throughout the store, helped customers to instantly identify the general area that they needed to proceed to, at which point they would encounter secondary signage with a finer-grain level of information that would guide them to the specific items they were looking for.

This program was designed to be rolled out without requiring any significant changes to store layout or merchandising approach: a simple but impactful way of making the shopping experience at Home Depot quicker, easier, and more intuitive.

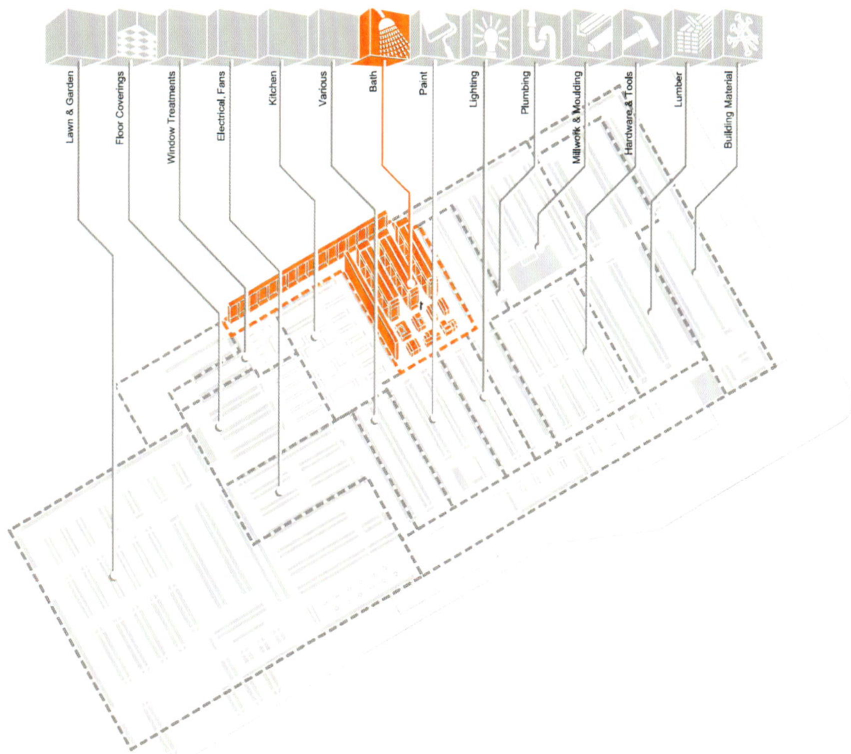

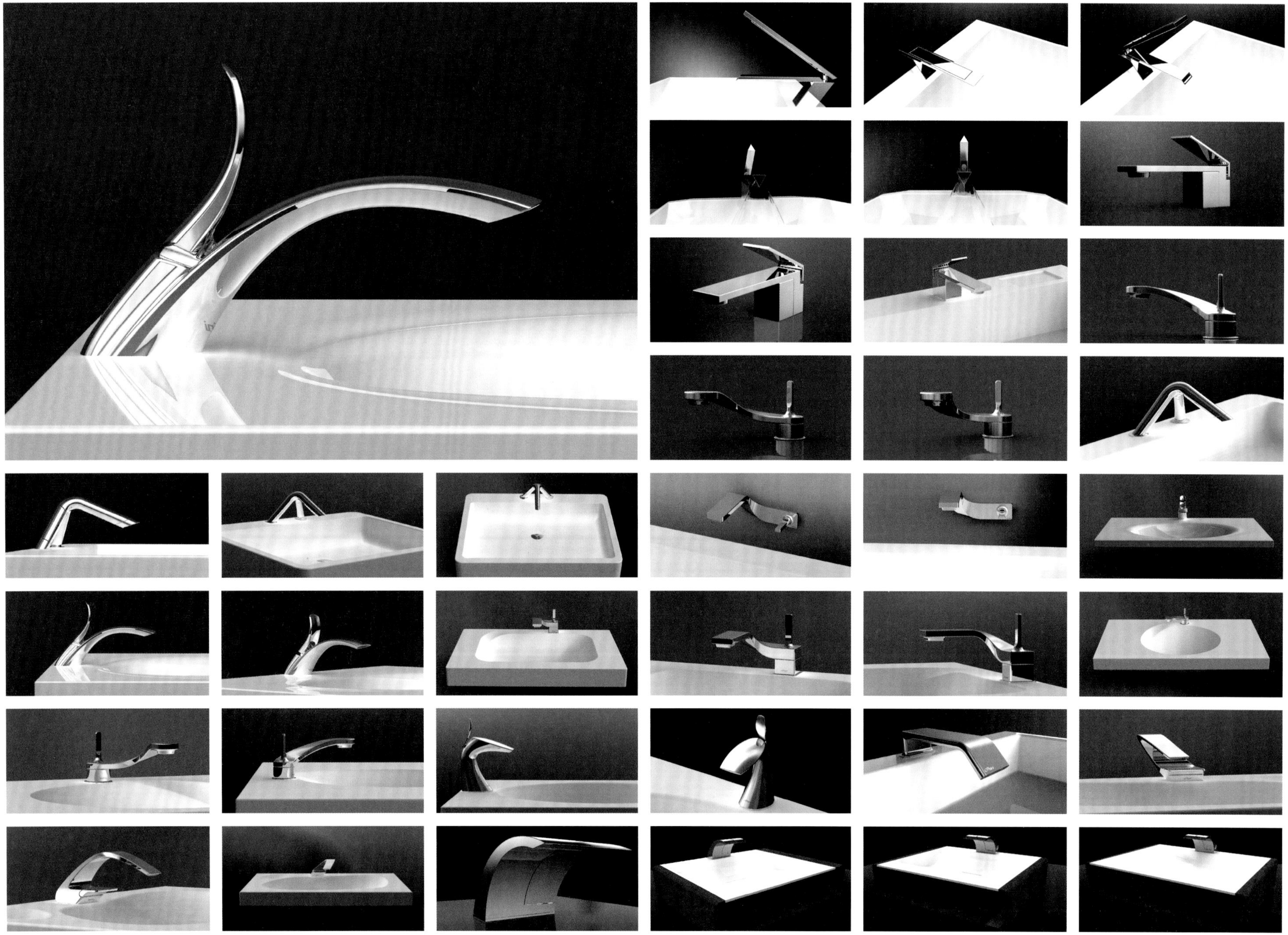

PAINI

2007

A part of a company-wide effort to bring high-design home goods to customers at accessible price points, a partnership was developed between The Home Depot and Paini—the renowned Italian manufacturer of high-end kitchen and bath fixtures—to create a new range of premium faucets that brought Paini's exceptional design quality to a mass audience for the first time.

Working closely with Paini's team of designers and engineers, a range of new bathroom faucet design concepts were developed that combined Paini's engineering expertise with a new level of design excellence and technological capability. The dramatic sculptural forms, inspired by nature, architectural elements, and classic performance automotive design, were unlike anything else on the market, and delivered on The Home Depot's broader promise to provide its customers with world-class products for every home, style, and budget.

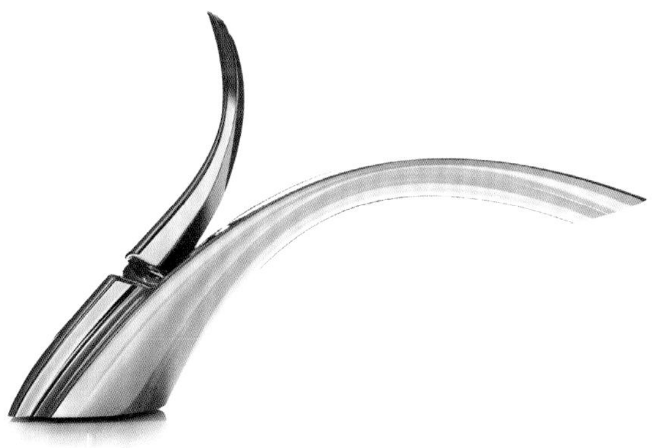

icet

ICET & ECET

2007

The iCet and eCet sinks, a Home Depot-exclusive product range concept developed in collaboration with Paini, were designed to be beautiful, functional hybrids of luxurious style and high-tech design.

The sinks integrated faucet, basin, and shelving components into a single unit, giving the look of a custom wall-mounted sink, with the practical standard plumbing assembly of a traditional countertop unit for ease of installation. The sinks featured a touch sensor for controlling temperature along with easy-to-understand illuminated buttons: on/off for power, hot/cold for temperature, and plus/minus for pressure.

And to ensure both safety and comfort for every user, the sinks also featured concealed LED lights beneath the water stream that changed dynamically to visually represent the water's temperature.

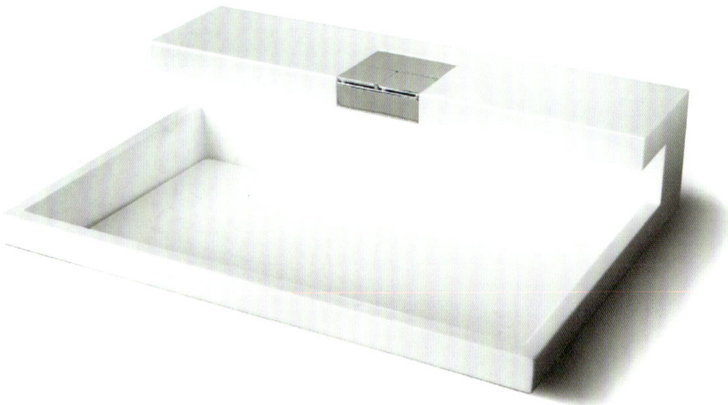

KYOTO KNIVES

2007

Kyoto Knives were a concept product range created for the Home Shopping Network to help target style-conscious home chefs looking for the ultimate in professional-grade cutlery.

This distinctive set of precision-manufactured knives were to be manufactured using surgical grade steel, and were micro-weighted to provide optimum balance for low fatigue. The bold unibody designs were unmistakably contemporary in character, yet also subtly paid homage to the time-honored craft of knife- and sword-making that Japan has long been renowned for.

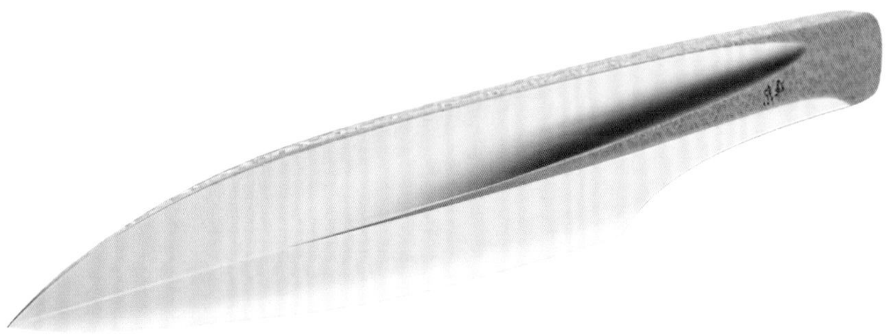

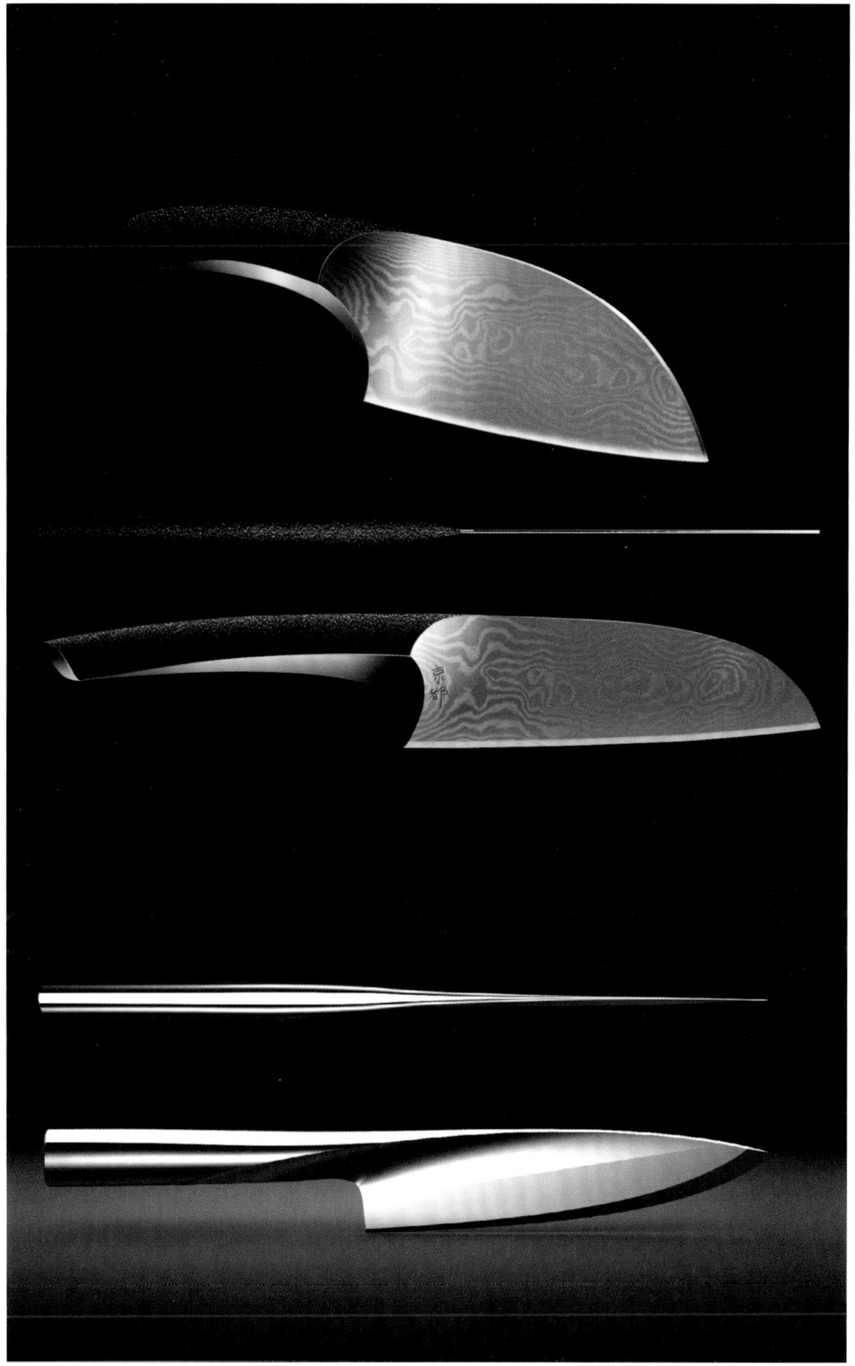

HOMEHERO

2007

The HomeHero range of home safety products were created to address the many untapped opportunities for life-saving innovation in the fire and home safety category.

HomeHero's flagship product, the HomeHero fire extinguisher, was a beautiful, compact, user-friendly alternative to the traditional extinguishers found in most homes. The unit was designed to be compact, lightweight, easy to operate, and attractive enough to be left out on a kitchen counter or other convenient location, ensuring that it would always be on-hand when needed. Graphics on the packaging and the product itself offered clear, concise visual explanations of how to use the extinguisher, helping to shave valuable seconds off of response time in the event of an emergency.

Plans to expand the HomeHero product range beyond the signature fire extinguisher included a full line of innovative, style-conscious, Internet-connected safety devices, including a concept for one of the first fully networked smoke and carbon monoxide alarms—nearly a decade before similar Smart Home devices became ubiquitous. All of these product concepts were designed with a sleek but friendly aesthetic that made them functional and attractive everyday additions to the home, fully living up to the brand mantra of "At Hand for Peace of Mind."

The HomeHero brand gained unprecedented buzz following its launch at The Home Depot. The Home Hero extinguisher won numerous awards and accolades, including the *Time* Magazine Best Inventions of 2007, and the International Design Excellence Gold Award by *BusinessWeek*.

HomeHero
At Hand for Peace of Mind™
Kitchen
Fire Extinguisher
Extinguidor de Incendios para Cocinas
Safety + Confidence + Style
B C
Suitable for flammable liquid and electrical equipment fires
INSTRUCTIONS
Flammable Liquids
Electrical Equipment
LISTED
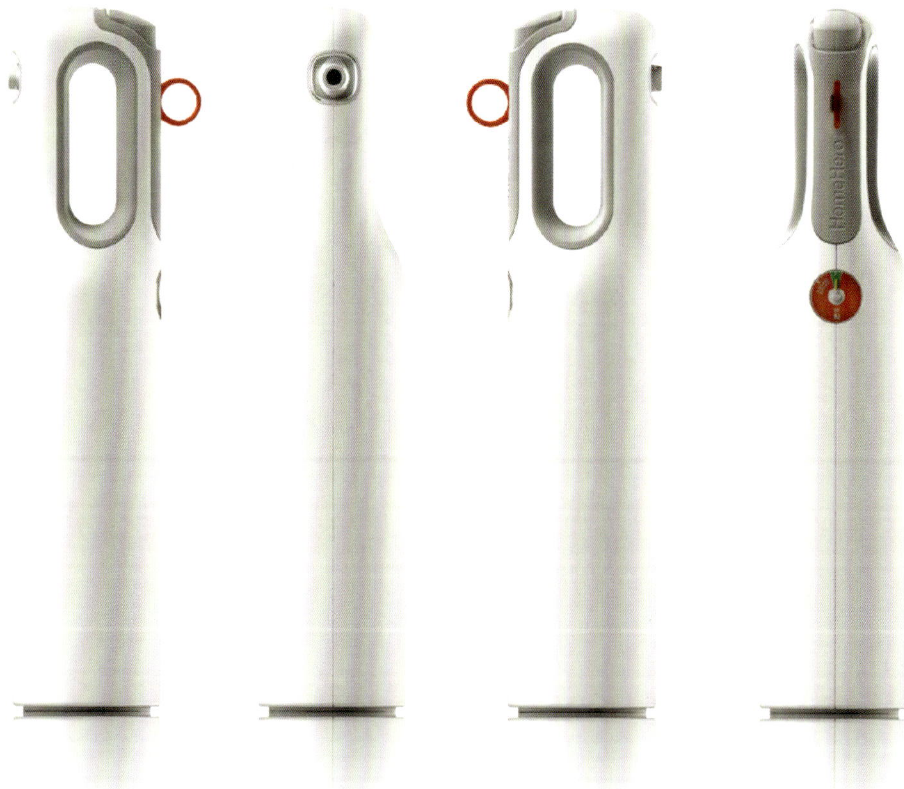
HomeHero

SMOKE AND CARBON MONOXIDE ALARM

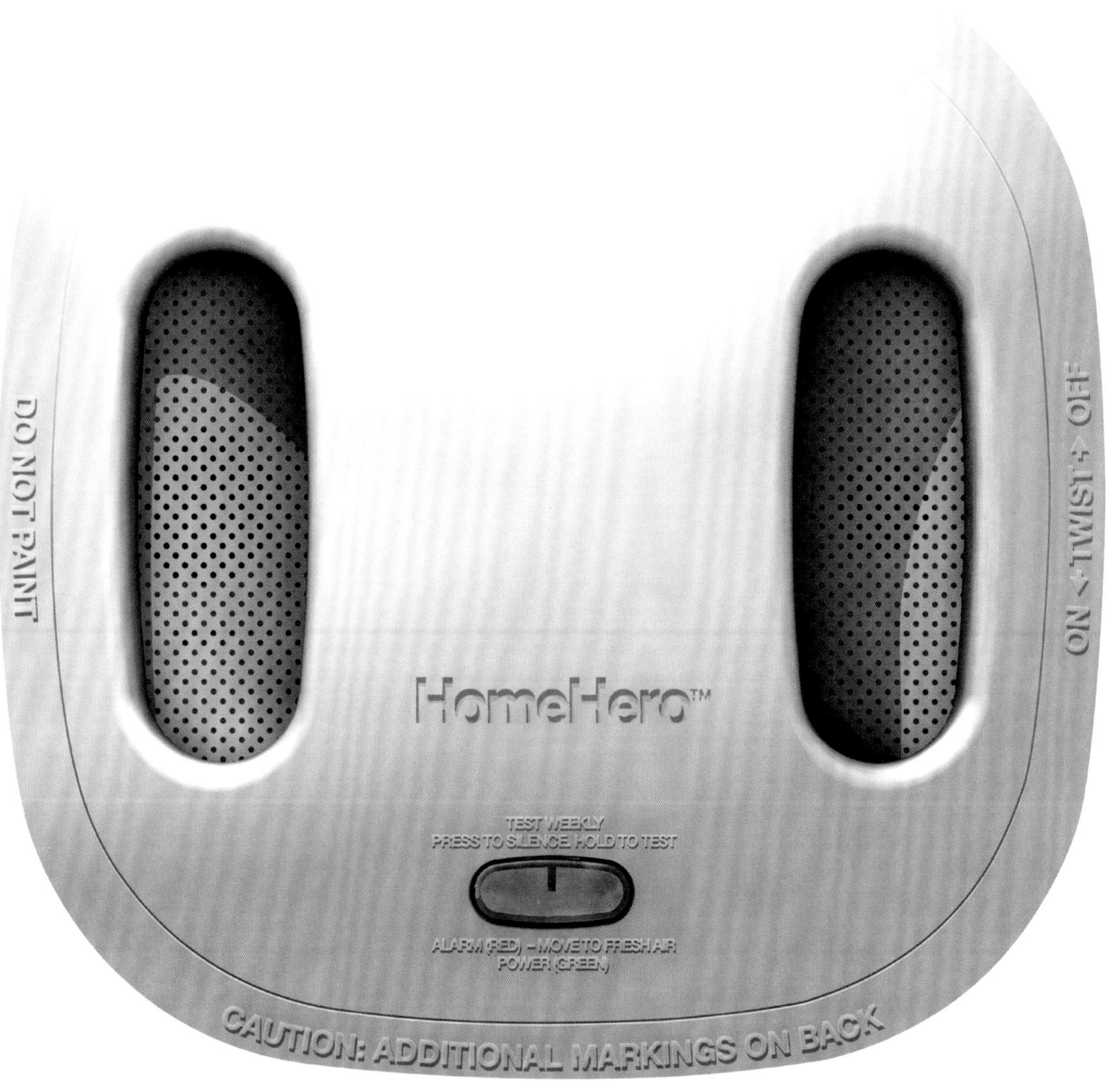

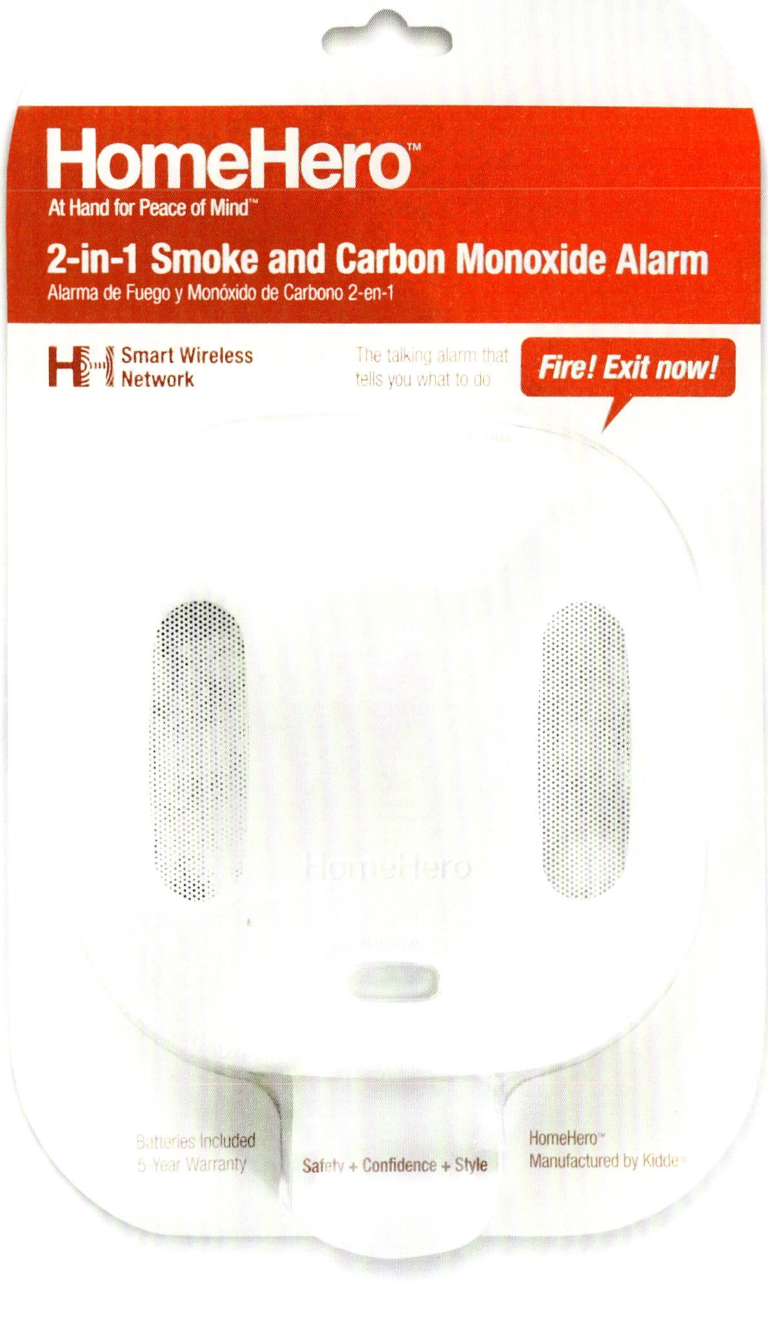
HomeHero™
At Hand for Peace of Mind™
2-in-1 Smoke and Carbon Monoxide Alarm
Alarma de Fuego y Monóxido de Carbono 2-en-1
Smart Wireless Network
The talking alarm that tells you what to do
Fire! Exit now!
HomeHero
Batteries Included
5-Year Warranty
Safety + Confidence + Style
HomeHero™
Manufactured by Kidde

HomeHero
HomeHero

HOMEHERO FIRST AID KIT

2007

After the successful launch of the HomeHero fire extinguisher, plans were developed to extend the brand into the realm of personal health and safety.

The HomeHero First Aid Kit concept, like the rest of the HomeHero product range, was designed to be beautiful, practical, and exceptionally easy to use. Echoing the sleek, ergonomic design of the HomeHero fire extinguisher, the lightweight, molded-plastic carrying case featured a built-in handle for grab-and-go portability in the event of an emergency. Inside, the case was divided into color-coded sections to simplify finding the right tool for any situation. Bandages, burn remedies, antiseptics, CPR tools, and more: each had their own clearly designated compartment, which eliminated the need to search through a cluttered container in the event of any injury.

HomeHero™
HomeHero™
HomeHero™

HOMEHERO CHILD SAFETY BRACELET

2007

The HomeHero Child Safety Bracelet concept was developed to extend a sense of safety and security to the youngest members of the family when they ventured outside of the home. The design proposal envisioned a lightweight, colorful band worn by a child that linked their whereabouts to their parent's phone in real-time, and would automatically send an alert if they wandered too far from a designated area. Years before wearable technologies became mainstream and viable life safety accessories, the Child Safety Bracelet concept provided a vision of a product that would deliver an up-to-the-second peace of mind to parents everywhere.

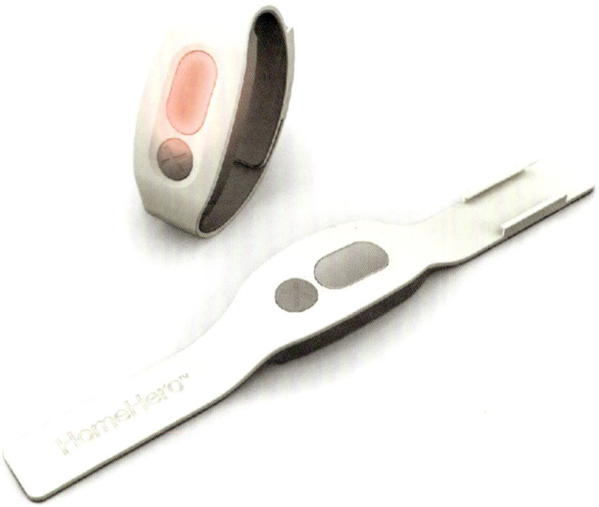

ONE FOUNDATION

2007

When actor Jet Li partnered with the Red Cross to start One Foundation, a charitable organization that raises funds for natural disaster relief worldwide, he needed an iconic logo that would give an instantly recognizable face to his philanthropic group.

The resulting brand mark cleverly incorporated the number "one" into a distinctive smiling icon: a human, friendly, inviting, and, most importantly, universally legible symbol that transcended cultural, geographic, and linguistic boundaries.

李连杰壹基金计划

SPECIAL OLYMPICS MÖBIUS

2007

With over 7,000 athletes from 165 countries coming together to compete over a nine-day period, the 2007 Special Olympics World Summer Games in Shanghai were the largest in the event's history. To mark the occasion, a new brand icon was unveiled for the Special Olympics: one that captured the spirit of resiliency, perseverance, and common humanity that has always been at the heart of their noble mission.

The resulting design took the form of a Möbius loop: a twisting ring whose outside face gradually torques to become the inside surface. The loop, which had no beginning or end and collapsed distinctions between interiority and exteriority, was a celebration of the infinite potential of these gifted athletes, and an implicit reminder that what is on the inside of every person is just as important as what is on the outside.

Realized across multiple scales and applications—in print and digital communications, as commemorative jewelry pieces, and even as a monumental sculptural installation that served as the centerpiece of the opening ceremony in Shanghai—the Special Olympics Möbius was an inspirational, universal symbol that celebrated the skills and accomplishments of these remarkable athletes on a global stage.

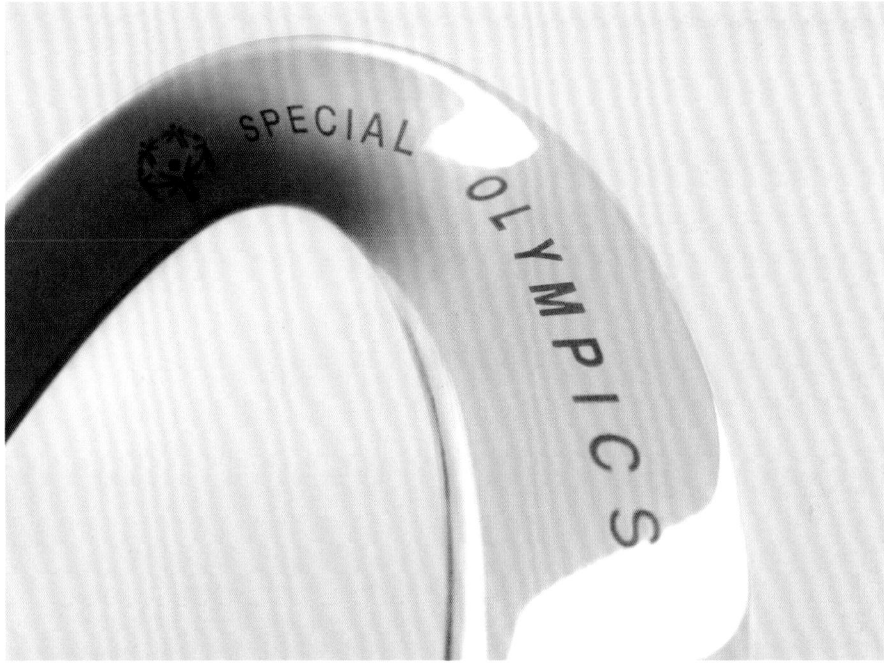

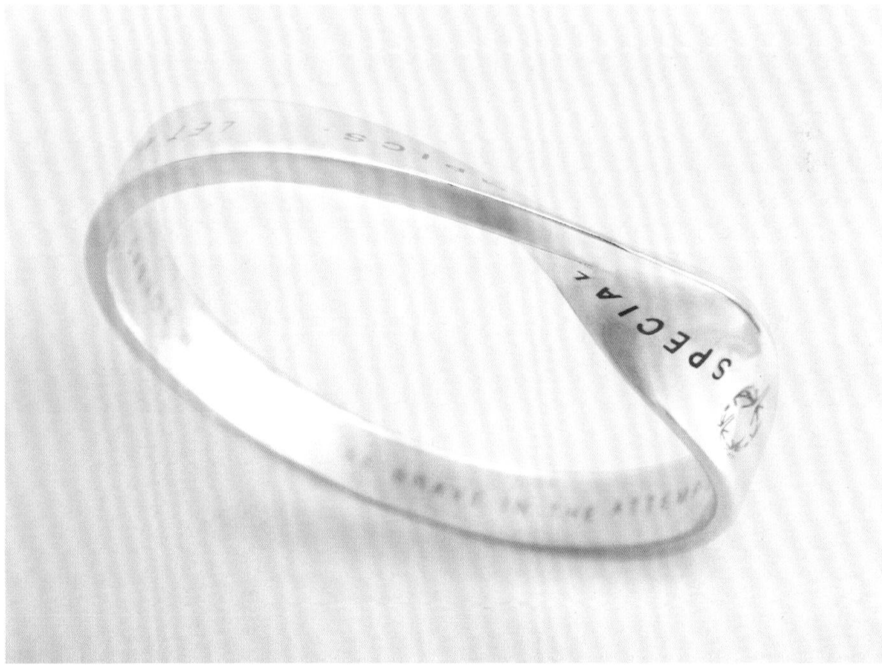

LINESWITHOUTDIVISION
SPECIAL OLYMPICS A WORLD WITHOUT

SPECIAL OLYMPICS

2008

As part of an ongoing effort to help the Special Olympics bring their message of hope and inspiration to the world outside of the biennial games, a new communications strategy was developed that emphasized the higher purpose of the organization, even beyond the world of sport.

"A World Without" encouraged people to imagine a world without the forces that oppress and divide us all: a world without corruption; a world without greed; a world without prejudice. The accompanying campaign featured powerful images of Special Olympics athletes from every background and walk of life, and served as a potent reminder that the core mission of the Special Olympics is about working toward a world in which everyone is able to realize their full potential and be celebrated for who they truly are.

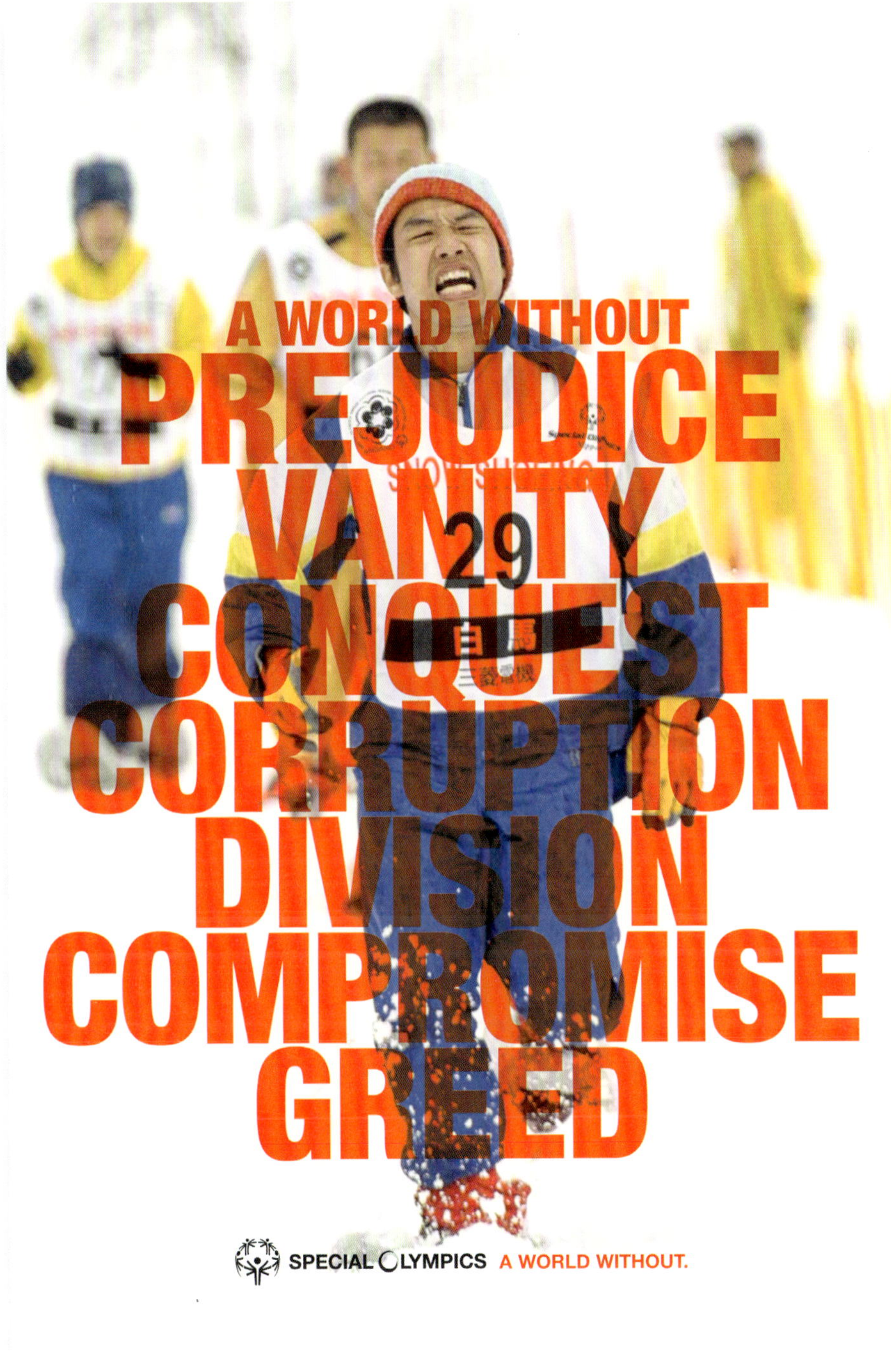

PRIDE
WITHOUT
PREJUDICE
SPECIAL OLYMPICS A WORLD WITHOUT

A WORLD WITHOUT
JUDGEMENT
BIAS
PRECONCEPTIONS
PREJUDICE
SPECIAL OLYMPICS
A WORLD WITHOUT

Victory without Conquest
SPECIAL OLYMPICS
A WORLD WITHOUT
203

PRIDEWITHOUTPREJUDICE
SPECIAL OLYMPICS A WORLD WITHOUT

POWERWITHOUTCORRUPTION
SPECIAL OLYMPICS A WORLD WITHOUT

SPECIAL OLYMPICS & WALTER CRONKITE

2008

In a short film created for the Special Olympics, legendary broadcast journalist Walter Cronkite lent his iconic voice to tell the story of the organization and its ongoing mission of empowerment through sport. As the face of CBS Evening News for nearly two decades, Cronkite covered some of the most consequential news events of the twentieth century, and in the process became, for many Americans, a peerless symbol of truth, decency, and national unity.

Cronkite's heartfelt and impassioned narration accompanied footage of the games as well as interviews with athletes, advocates, and the many individuals who had become involved with, or been touched by, the work of the Special Olympics over the years. Frequently described as the "Most Trusted Man in America," Cronkite's inimitable voice lent authority, gravitas, and, most importantly, humanity to the Special Olympics' noble mission.

DODGE CHALLENGER

2008

Nearly forty years after its introduction, the Challenger, Dodge's iconic 1970s muscle car, retained a devoted fan base and was still widely regarded as one of the automaker's best and most important models. In 2008, following a twenty-five year absence from the market, an all-new version of this great American muscle car was finally set to be released, heralding a new era of power, performance, and design excellence at Dodge.

To help promote the return of this iconic model, a collector's edition print magazine was launched. Modeled after the classic muscle car magazines of the 1960s and 70s, the promotional publication honored Dodge's rich legacy of innovation, and gave car enthusiasts an intimate inside look at this much-anticipated next generation of a true performance icon.

CONTENTS

FONTAINEBLEAU

2008

When the iconic Morris Lapidus-designed Miami Beach Fontainebleau Hotel re-opened in 2008, following an exacting two-year renovation and restoration, it needed to announce that this icon of mid-century style was back, and better than ever. Jeff Soffer, the owner of the Fontainebleau, had a clear vision for the reborn Fontainebleau as a destination that was authentically rooted in its storied past, while still being fresh, vibrant, and unmistakably of-the-moment.

To help bring this vision to life, communications were developed for the resort's reopening that transported Fontainebleau's history of star-studded, mid-century cool into a contemporary context. Beautiful monochrome photography was juxtaposed with vivid, colorful imagery, providing a new expression of timeless cool that respected the past while leading the way to a new future. Sexy, sleek, and dreamlike, Fontainebleau's communications were designed to attract a decidedly sophisticated clientele—the modern-day successors to the effortlessly chic Rat Pack crowd who had made the Fontainebleau the quintessential Miami Beach destination of the 1950s and 60s.

New brand identity systems were designed for each of the nine on-site restaurants, bars, and entertainment venues housed within the resort, as was a proprietary wayfinding system to help visitors navigate the expansive property—a simple way to bring a stylish flair to every visitor touchpoint. In-room communications were also designed to inform and entertain guests, ensuring a seamless experience that lived up to the legendary reputation of this iconic hotel and resort.

FONTAINEBLEAU

FONTAINEBLEAU

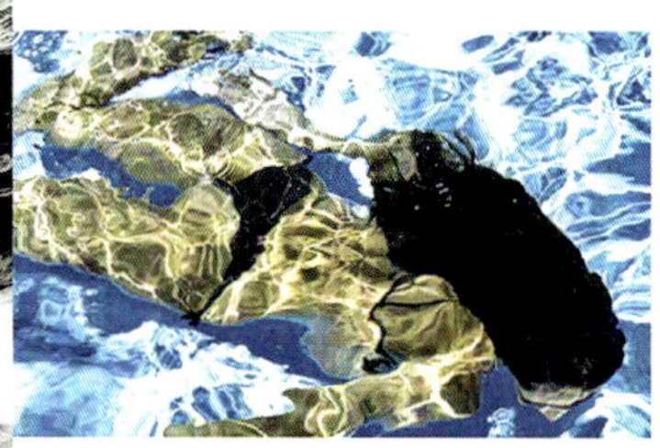

FONTAINEBLEAU

FONTAINEBLEAU

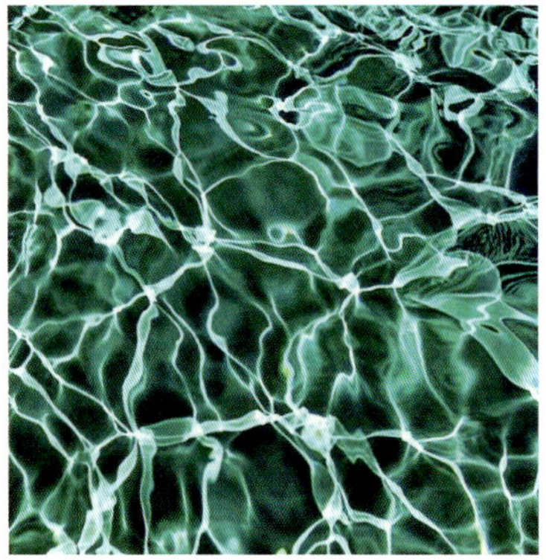

FONTAINEBLEAU
MIAMI BEACH

FONTAINEBLEAU
MIAMI BEACH

FONTAINEBLEAU
MIAMI BEACH

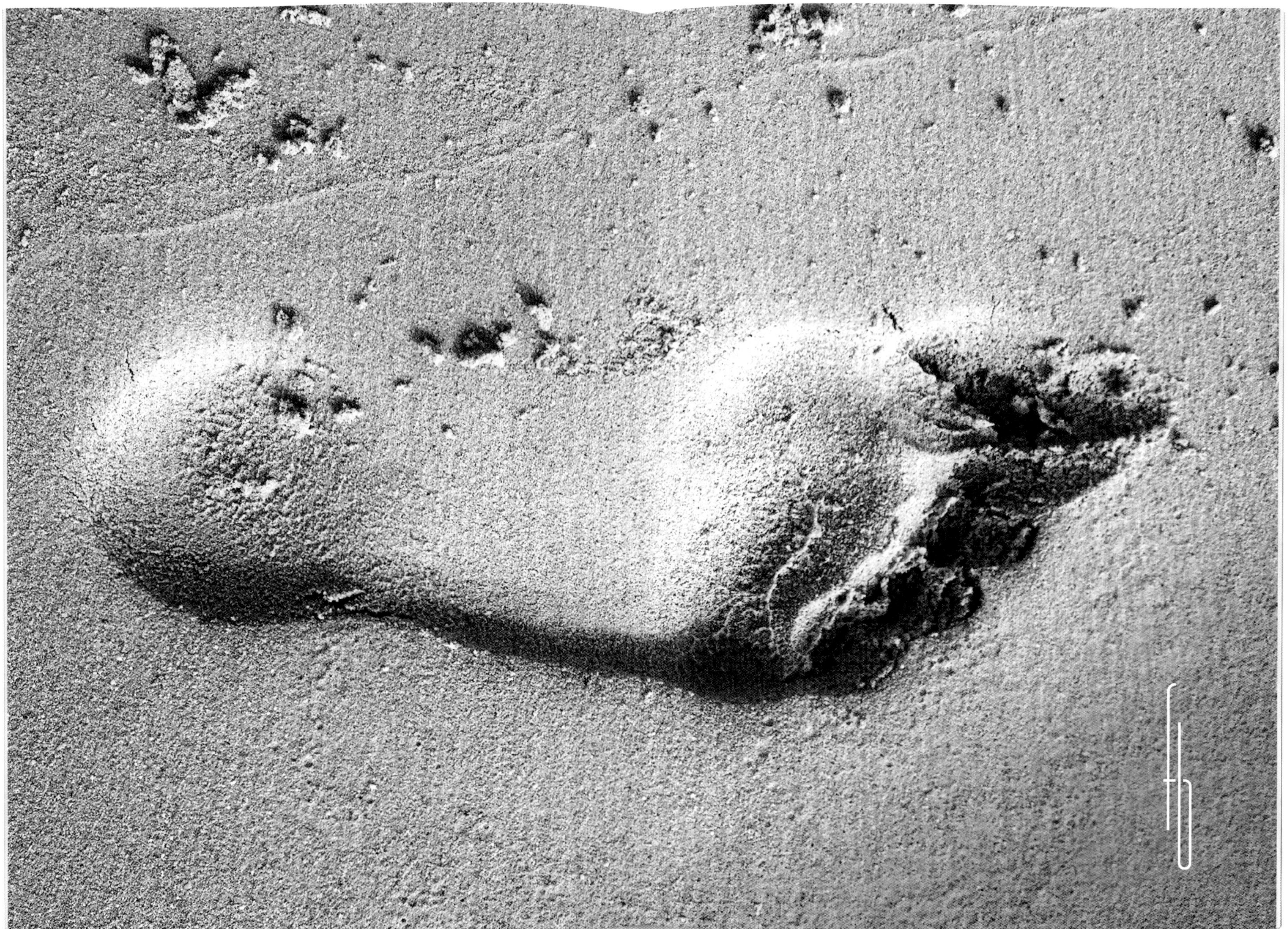

aquamarine

bleau bar

CLOUD9

gelato galleria

monkey

TAI TAM

blade

9

LIV

crush

solo

the buffet

FONTAINEBLEAU
MIAMI BEACH
LAS VEGAS
RESERVATIONS
SET THE STAGE
INVENTION.
EXPRESSION.
THE NEW FB.
EXPLORE THE REINVENTION OF FONTAINEBLEAU RESORTS THROUGH THE FOLLOWING CHAPTERS OF OUR STORY AND DISCOVER THE PART YOU'LL PLAY.
PAUSE
FAMILY + FONTAINEBLEAU = FUN
This summer, reconnect with your family! Two adults + two children from $279/night with FREE or HALF-PRICE dining for kids at Fontainebleau Miami Beach
BOOK NOW
1|2
KEEP UPDATED WITH FONTAINEBLEAU.
Be the first to receive exclusive news and offers.
Email Address
SUBMIT
STAY IN TOUCH
FIND
INSIDE FB
CAREERS
PRIVACY POLICY
© FONTAINEBLEAU RESORTS, LLC 2008. ALL RIGHTS RESERVED.

ATM
V.I.P.
V.I.P.

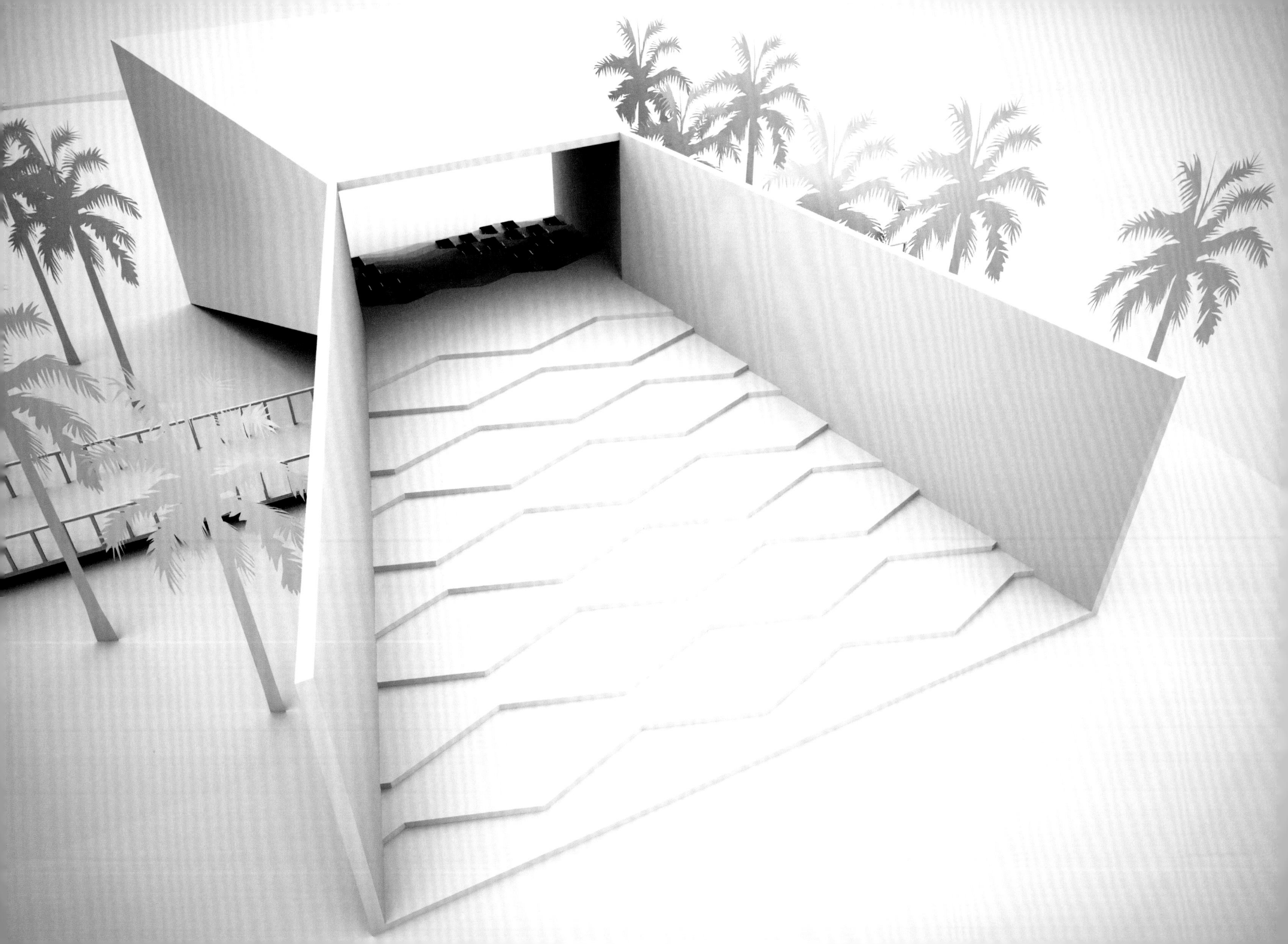

FONTAINEBLEAU BEACH THEATRE

2008

The Fontainebleau Beach Theater was proposed to expand the entertainment opportunities at the Fontainebleau Hotel with a design that fit neatly on the property. In its clever evocation of the classic black bowtie—a recurring motif throughout the Fontainebleau property—the new theatre design organically inserts itself into the grounds, while still being something wholly novel that speaks to the next chapter of this iconic hotel.

The theater's abstracted bowtie form is at once completely legible without sacrificing the technical and acoustic demands of an outdoor theatre venue: a clever nod to the bowties worn by such legendary guests(both real and fictional) as Frank Sinatra and James Bond without being nostalgic, gimmicky, or overly literal. The taut, tailored angles of the theater act as a perfect counterpoint to the sleek, sweeping, curvilinear façade of the hotel itself: unique in form, but completely harmonious within its environment.

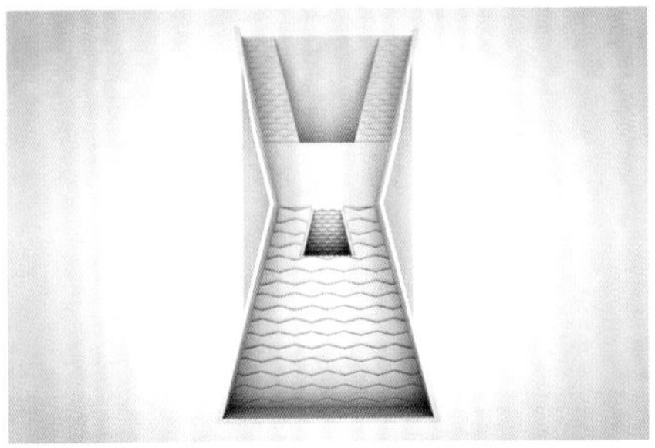

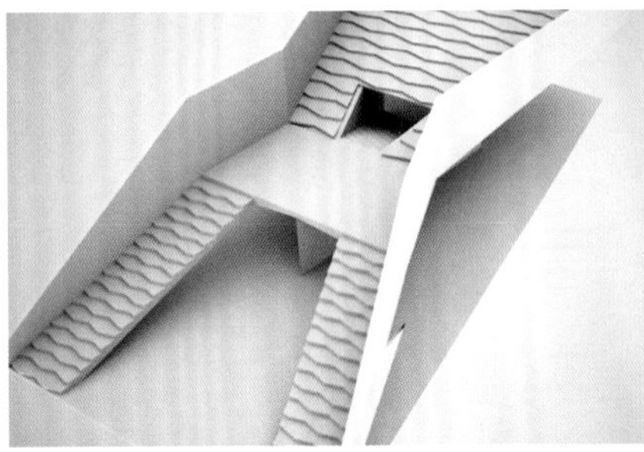

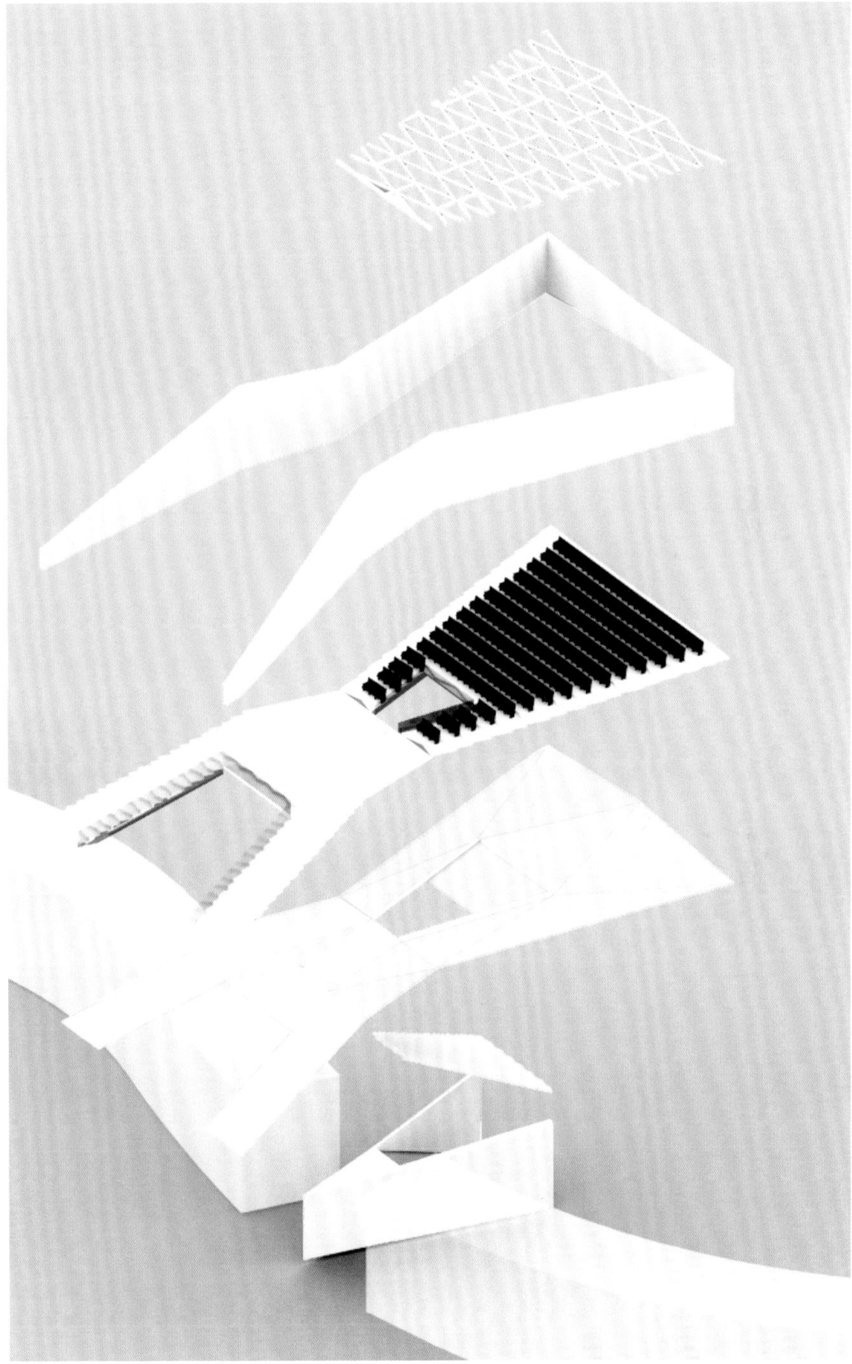

LIV AT FONTAINEBLEAU

2008

When the Fontainebleau Hotel & Resort in Miami Beach reopened in 2008, it introduced what was to become one of Miami's hottest nightclubs in the basement of its lobby: LIV. The club's name was a clear and emphatic instruction to club-goers to let go of inhibitions, have a good time, and be fully in the present.

The sleek identity that was designed for LIV suggested energy, movement, and vibration, and united a fresh, distinctly contemporary spirit with the elegant, mid-century character that the Fontainebleau had always been known for. And in a clever twist that only true insiders would pick up on, the name "LIV" also spelled out the number fifty-four in roman numerals: the year that the hotel first opened.

RESPECT YOURSELF GNC LIVE WELL

GNC

2008

For years, GNC had been a go-to source for vitamins, supplements, and sports nutrition products for athletes and fitness aficionados of all sorts. In order to expand its reach beyond the target market of performance-obsessed, predominately male customers, the brand was relaunched with the mantra "Live Well." This new approach expanded the definition of wellness beyond fitness and performance by providing a wide range of products and resources designed to help everyone achieve their wellness goals. New products were developed to appeal to a more diverse consumer base, while the packaging of existing house brand products was streamlined to become more accessible and appealing to athletes and non-athletes alike.

The retail experience was completely reengineered to be more inviting to a wide range of customers, with lifestyle-oriented products located up front, and performance-based products moved toward the interior where they would still be easily located by long-standing customers. In-store wayfinding systems were also completely redesigned, with imagery-heavy signage enlivening the browsing process and clearly marked pathways simplifying the process of browsing, selection, and purchase.

A subsequent evolution of the "Live Well" campaign encouraged consumers to "Respect Yourself," and reframed health and wellness as an enjoyable act of self-care undertaken to change the way you feel, not just the way you look.

GNC
LIVE WELL
RESPECT
YOURSELF

GNC LiveWell.
NEW STORE
NEW
GNC
EXPERIENCE
50% OFF
50% OFF

Vitapak

GNC VITAPAKS

2008

Vitapaks were a product concept designed for GNC to radically simplify and personalize the process of navigating the often overwhelming range of vitamins and supplements available to customers.

Vitapaks provided streamlined, easy-to-understand, customized health solutions for every person, no matter their needs or goals. Available in eighteen varieties, each Vitapak was designed to address a common set of wellness concerns, allowing customers to mix and match a menu of supplements that was perfectly right for them.

The communications for these and other products remained true to the established GNC brand look of striking black-and-white photography that featured a diverse array of customers, all healthy, happy, and full of life.

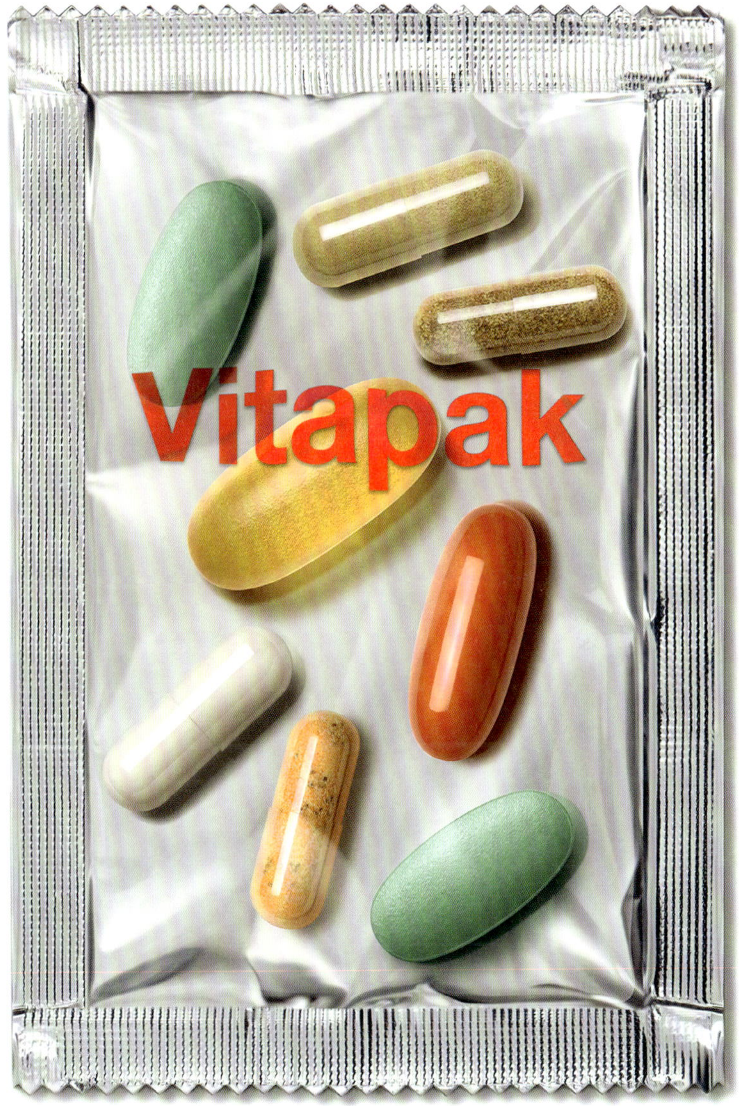

caffeine free
diet
pepsi
0
cal
carb
sug
caffeine free
pepsi
pepsi
diet
pepsi
0
cal
carb
sug
pepsi
max
0
cal
carb
sug

PEPSI

2008

For decades, Pepsi had positioned itself as the drink of choice among the world's youth through superstar endorsements, advertising spectacles that tapped into the pop culture zeitgeist, and a fresh, fun, irreverent attitude that ran throughout all of their brand expressions. To help the brand bring that distinctive energy and outlook into a new era, a comprehensive brand refresh was engineered to represent the contemporary face of Pepsi: joyful, playful, open, and optimistic.

The Pepsi logo—a red and blue globe bisected by a white wave—was gently evolved to represent something even more universal: a smile. The refreshed logo mark had the same shape, the same colors, and the same basic elements, but with an added dimension of genuine, human emotion. The logo was also shifted dynamically depending on which Pepsi product it appeared on: the standard "smile" on regular Pepsi, a thinner "smile" for Diet Pepsi, and a broader one for Pepsi Max. This strategy helped to inject a sense of energy, vibrancy, and life into the "face" of the brand, and bolstered Pepsi's reputation for embracing change and not taking itself too seriously. A new custom typeface developed for the Pepsi word mark also featured a subtle homage to the previous generations of Pepsi logos: a slight wave in the "e" of "Pepsi" that mirrored the geometry and proportions of the classic Pepsi globe.

While still instantly recognizable as Pepsi, the new brand mark was fresher, livelier, and more modern, and today remains one of the world's most iconic and ubiquitous brand marks.

pepsi

pepsi
pepsi max

diet
mtn dew
made from flavors specially blended in the traditional hillbilly style
it'll tickle your innards!
12 FL OZ (355mL)

MTN DEW

2009

Following the comprehensive brand refresh of Pepsi's signature cola in 2008, the rest of PepsiCo's portfolio of carbonated beverages was slated for a similar revitalization.

Mountain Dew, a drink perennially popular with young consumers, received one of the most significant makeovers when its name was updated to Mtn Dew: an abbreviated version that evoked the quick, concise, rapid-fire quality of text messaging, which at the time was quickly becoming the default way in which young people communicated with each other.

The logo and packaging were redesigned to capture this more dynamic, energized style—developed specifically to resonate with the extreme sports aficionados, gamers, and youthful consumers who formed Mountain Dew's core customer base. As part of the revitalization, a series of new characters was developed that cleverly played off of various "mountain man" archetypes and urban legends like Big Foot. Taken together, this work gave a playful edge to the Mountain Dew brand, and solidified Mountain Dew's reputation as the soft drink of choice for young consumers everywhere.

MUG
ROOT BEER
NO CAFFEINE

MUG ROOT BEER

2009

The lovable bulldog mascot who had served as the face of PepsiCo's Mug Root Beer for years got his famous mug made over as part of a sweeping brand refresh that retained all of the attitude and charm of the beloved original, while also elevating Mug's image and broadening its appeal.

Packaging designs were also streamlined and modernized, and featured a clever touch: each can and bottle was topped with foam that seemed to burst out from the top and drip down the sides, immediately cueing the customer to the delicious drink inside and mirroring the overflowing mug held by the bulldog himself.

sierra
mist
lemon-lime soda
with other
natural flavors
caffeine free

SIERRA MIST

2009

For Sierra Mist, PepsiCo's lemon-lime soda, a complete brand image overhaul was developed as part of an effort to give the soda a fresh new look that would appeal to a more mature and discerning customer. Packaging design concepts featured stylized compositions that evoked the clear, crisp refreshment of rolling hills, mountain landscapes, and misty forests, while bright yellow and vivid green lemon-lime hues long associated with the brand subtly signaled the refreshing taste of the soda inside.

Tropicana™ Pulp Free
Pulp Free
Tropicana™
pure premium
100% orange
pure & natural
not from concentrate
100% pure & natural orange juice
pasteurized
59 FL OZ (1.8 QT) 1.75L
Tropicana™ Some Pulp
Some Pulp
Tropicana™
pure premium
100% orange
pure & natural
not from concentrate
100% pure & natural orange juice
pasteurized
59 FL OZ (1.8 QT) 1.75L
Tropicana™ High Pulp
High Pulp
Tropicana™
pure premium
100% orange
pure & natural
not from concentrate
100% pure & natural orange juice
pasteurized
59 FL OZ (1.8 QT) 1.75L
Tropicana™ Pulp Free
Pulp Free
Tropicana™
pure premium
100% orange
pure & natural
not from concentrate
100% pure & natural orange juice
pasteurized
59 FL OZ (1.8 QT) 1.75L

TROPICANA

2009

For decades, Tropicana had long been an integral part of breakfast, but by the early 2000s, the most popular orange juice brand in America was facing increasing competition from products like green juices, smoothies, and enhanced water, and trying to combat a shift in consumer habits away from a traditional sit-down breakfast each morning. Research showed that about half of all consumers assumed there was added sugar in all orange juices, but Tropicana itself had always been made of one hundred percent fresh squeezed orange juice with nothing else added. To reinforce that Tropicana orange juice was pure, natural, and one hundred percent fresh-squeezed, a comprehensive reimagining of the brand identity and product packaging was developed and executed under the leadership of PepsiCo CEO Indra Nooyi.

On Tropicana cartons, the longtime Tropicana symbol—an orange with a straw stuck in it—was replaced by a tall glass filled with Tropicana orange juice that spanned the entire height of the carton. The design took full advantage of all sides of the packaging: the picture of the glass of orange juice spanned the corner of the carton, creating the three-dimensional effect of a fresh glass of orange juice sitting on the table at a 1:1 scale.

The carton's twist cap was reengineered to resemble a halved orange that mimicked the real thing both in terms of color and texture. Made from a special grade of plastic with a soft rebound, it had a gentle, tactile quality and evoked the feeling of squeezing a real orange. Every time this new carton was opened to pour a glass of juice, the symbolic "squeezing of an orange" cued the drinker to remember that the juice they were about to enjoy was one hundred percent natural and fresh-squeezed.

A heartfelt campaign, "Squeeze," was created to both remind consumers of the freshness of the orange juice and to create an emotional link between Tropicana and the simple ritual of coming together with family and loved ones over breakfast every morning.

Tropicana
pure premium
squeezed from fresh oranges
Tropicana
pure premium
100%
orange
pure & natural
Tropicana
pure premium
100% pure & natural orange juice
pasteurized
64 FL OZ (2 QT) 1.89 L
squeeze
it's a natural

squeeze
it's a natural
Tropicana
100% orange
Tropicana 100% orange. pure and natural. squeezed from fresh oranges.

squeeze
it's a natural
Tropicana
100% orange
Tropicana 100% orange. pure and natural. squeezed from fresh oranges.

squeeze
it's a natural
Tropicana
100% orange
Tropicana 100% orange. pure and natural. squeezed from fresh oranges.

squeeze
it's a natural
Tropicana
100% orange
pure & natural
Tropicana 100% orange. pure and natural. squeezed from fresh oranges.

squeeze
it's a natural
Tropicana
100% orange
Tropicana 100% orange. pure and natural. squeezed from fresh oranges.

squeeze
it's a natural
Tropicana
100% orange
Tropicana 100% orange. pure and natural. squeezed from fresh oranges.

Trop50

TROP50

2009

To combat a growing perception that orange juice was an unhealthy option, Tropicana seized on the opportunity to launch a new stevia-sweetened orange juice drink with fifty percent fewer calories, fifty percent of the sugar, and only fifty calories per glass.

Dubbed Trop50, the name eschewed the usual tropes of lower-calorie products: rather than affixing a "Diet" or "Light" to the Tropicana brand name, or focusing on what ingredients were reduced or eliminated, Trop50 simply and transparently asserted what was there: fifty calories in every delicious glass. Packaging was designed to clearly and proudly highlight the health benefits of Trop50 compared to other juices, while the presence of a just-poured glass of orange juice immediately cued customers to the fact that freshly squeezed orange juice was still the main ingredient—just like the original.

Trop50 provided a better choice to the widest possible range of consumers: orange juice lovers, health-conscious non-juice drinkers, and even diet beverage fans could all enjoy this delicious, sensible, genuinely good-for-you option.

SoBe
lifewater
agave lemonade
electrify
passionfruit citrus
b-strong
goji melon
go-ji
orange tangerine
strawberry kiwi
calm-o-mile
pomegranate cherry
zingseng
blackberry grape
enlighten
0 cal
naturally sweetened
fuji apple pear
vitamin enhanced
water beverage
lean machine
yumberry pomegranate
purify
black and blue berry
forti-fight

SOBE LIFEWATER THRILLICIOUS

2008

Fortified with vitamins, herbs, and antioxidants, PepsiCo's SoBe Lifewater was a pioneering product in the now-ubiquitous enhanced water beverage category, but the eventual entry of competitor brands into the marketplace sparked the need for a brand refresh to help SoBe to distinguish itself in an increasingly crowded segment.

Along with a sleeker and more modern brand identity, a colorful, boldly patterned, twisting bottle design was created to evoke a lizard's tail wrapping around the bottle: a distinctive and proprietary form factor that helped to distinguish SoBe in a crowded market of competitors.

Meanwhile, in a landmark campaign developed with the help of James Cameron's animation studio, Digital Domain, the SoBe lizard brand mascot was reimagined as a family of hyper-realistic lizards of various shapes, sizes, and colors, who represented the brand's many flavors and varieties.

The first in a series of commercials that featured the digitally animated SoBe lizards, the "Thrillicious" TV spot featured supermodel Naomi Campbell dancing with the charming, life-like lizards to Michael Jackson's "Thriller" on the twenty-fifth anniversary of the record-breaking album, and became one of the top Super Bowl ads of 2008.

MJJ
Dreamers

thrillicious.com
SoBe
Life
water

Life
water
pomegranate cherry
challengeyourlife
thrillicious.com

thrill
icious.com
SoBe
Life
water

SOBE LIFEWATER LIZARD LAKE

2009

In a collaboration with DreamWorks Animation, a landmark TV spot—one of the world's first 3-D television commercials—was created for SoBe Lifewater at Super Bowl XLIII in 2009. Three-dimensional glasses were distributed in supermarkets across the country in the weeks leading up to the game so that home viewers everywhere could take part in this landmark spectacle.

In the spot, NFL players Ray Lewis, Justin Tuck, and Matt Light, all donning tutus over their football gear, start dancing to a version of *Swan Lake* called *Lizard Lake*—with a special assist from the SoBe Lizards. Midway through their performance, the football players and their reptilian friends are suddenly joined by the animated characters from the then-upcoming DreamWorks film *Monsters vs. Aliens* in a totally unexpected, so-strange-it-works cross-promotional move.

This technologically groundbreaking and unforgettable commercial cemented SoBe Lifewater's reputation as a bold, exuberant, delightfully unusual brand that can always be counted on to lift spirits and bring a smile to people's faces.

lifewater
lifewater

NAKED JUICE

2009

Naked Juice, PepsiCo's range of all-natural juice and smoothie offerings, had built considerable popularity with health-conscious adults owing to its reputation as a sensible, nourishing, on-the-go option made with real ingredients and no added sugar. But in spite of this success, it seemed that there was a missed opportunity to extend the brand's appeal to an entirely new demographic that stood to benefit from its nutritious, delicious, bright, and colorful beverages: kids.

To better engage children, a new crop of colorful characters were created based on Naked's range of fruit flavors. The Drops were undeniably cute and endearing, but, most importantly, each of the characters was carefully conceived to help educate young people about a specific aspect of healthy living. Each of the Drops was an expert in a particular area of wellness, dispensing helpful tips for healthy living in a fun and accessible way. With their playful looks and charming personalities, the Drops were a powerful tool to help children learn about proper nutrition and form good habits early on that can carry them through a lifetime.

naked

PURE
PHENOM
PHENOM HYDRO
100% NATURAL COCONUT WATER
WITH ADDED ELECTROLYTES FOR MAX HYDRATION
QUALITY · POTENCY · PURITY
GNC FORTIFIED
NO SUGAR ADDED
17 FL OZ (500 ML)
COCONUT WATER FROM CONCENTRATE
ORANGE GUAVA PASSION
PHENOM
PHENOM MEGA V
100% NATURAL COCONUT WATER
WITH ADDED VITAMINS AND ELECTROLYTES FOR MAX HYDRATION
QUALITY · POTENCY · PURITY
GNC FORTIFIED
NO SUGAR ADDED
17 FL OZ (500 ML)
COCONUT WATER FROM CONCENTRATE
PINEAPPLE PUNCH
PHENOM
PHENOM MEGA V
100% NATURAL COCONUT WATER
WITH ADDED VITAMINS AND ELECTROLYTES FOR MAX HYDRATION
QUALITY · POTENCY · PURITY
GNC FORTIFIED
NO SUGAR ADDED
17 FL OZ (500 ML)
COCONUT WATER FROM CONCENTRATE
BLUEBERRY POMME
PHENOM
PHENOM ANTIOX
100% NATURAL COCONUT WATER
WITH ADDED VITAMINS AND ELECTROLYTES FOR MAX HYDRATION
+
PLUS A BOOST OF ANTIOXIDANT VITAMINS A, C & E
QUALITY · POTENCY · PURITY
GNC FORTIFIED
NO SUGAR ADDED
17 FL OZ (500 ML)
COCONUT WATER FROM CONCENTRATE
ACAI PEACH
PHENOM
PHENOM ENERGY
100% NATURAL COCONUT WATER
WITH ADDED VITAMINS AND ELECTROLYTES FOR MAX HYDRATION
+
PLUS A BOOST OF CAFFEINE AND B VITAMINS
QUALITY · POTENCY · PURITY
GNC FORTIFIED
NO SUGAR ADDED
17 FL OZ (500 ML)
COCONUT WATER FROM CONCENTRATE
LEMON MOJITO
PHENOM
PHENOM ENERGY
100% NATURAL COCONUT WATER
WITH ADDED VITAMINS AND ELECTROLYTES FOR MAX HYDRATION
+
PLUS A BOOST OF CAFFEINE AND B VITAMINS
QUALITY · POTENCY · PURITY
GNC FORTIFIED
NO SUGAR ADDED
17 FL OZ (500 ML)
COCONUT WATER FROM CONCENTRATE

PHENOM

2009

Developed as a collaboration between GNC and PepsiCo, Phenom was one of the first coconut water beverages to be enhanced with natural flavors and boosted with powerful vitamins and electrolytes.

To help launch this new product, a total branding architecture was developed, including the brand identity, packaging, launch campaign, and, of course, the name of the product itself, which alluded to its phenomenal taste and superior hydration.

A first-of-its-kind product, Phenom opened pathways into new markets for both PepsiCo and GNC, while effectively creating an entirely new product segment, the popularity of which has continued to soar in the years since its launch.

MUJI

2009

Muji, the Japanese home goods retailer, has grown to be a ubiquitous presence all around the world, beloved by customers for its practical, beautiful, well-priced, high-quality designs.

To expand on this tradition of accessible design to new product segments, a series of unique product concepts was developed with the goal of driving last-minute purchases as customers waited in checkout lines. A series of key chains, beauty tools, home office accessories, and toys was designed to be delightful, beautiful, whimsical items that would be truly irresistible must-haves for every Muji customer.

With the smaller apartments of city dwellers in mind, a new range of concepts for space-saving modular furniture was also created to help maximize every square foot. Each lightweight piece folded flat for easy storage, while durable, weather-proof materials were selected to ensure maximum versatility in both indoor and outdoor settings. This new product range embodied Muji's signature design philosophy of "just enough" while still offering beautifully designed, well-made products that were above and beyond anything else at this price point.

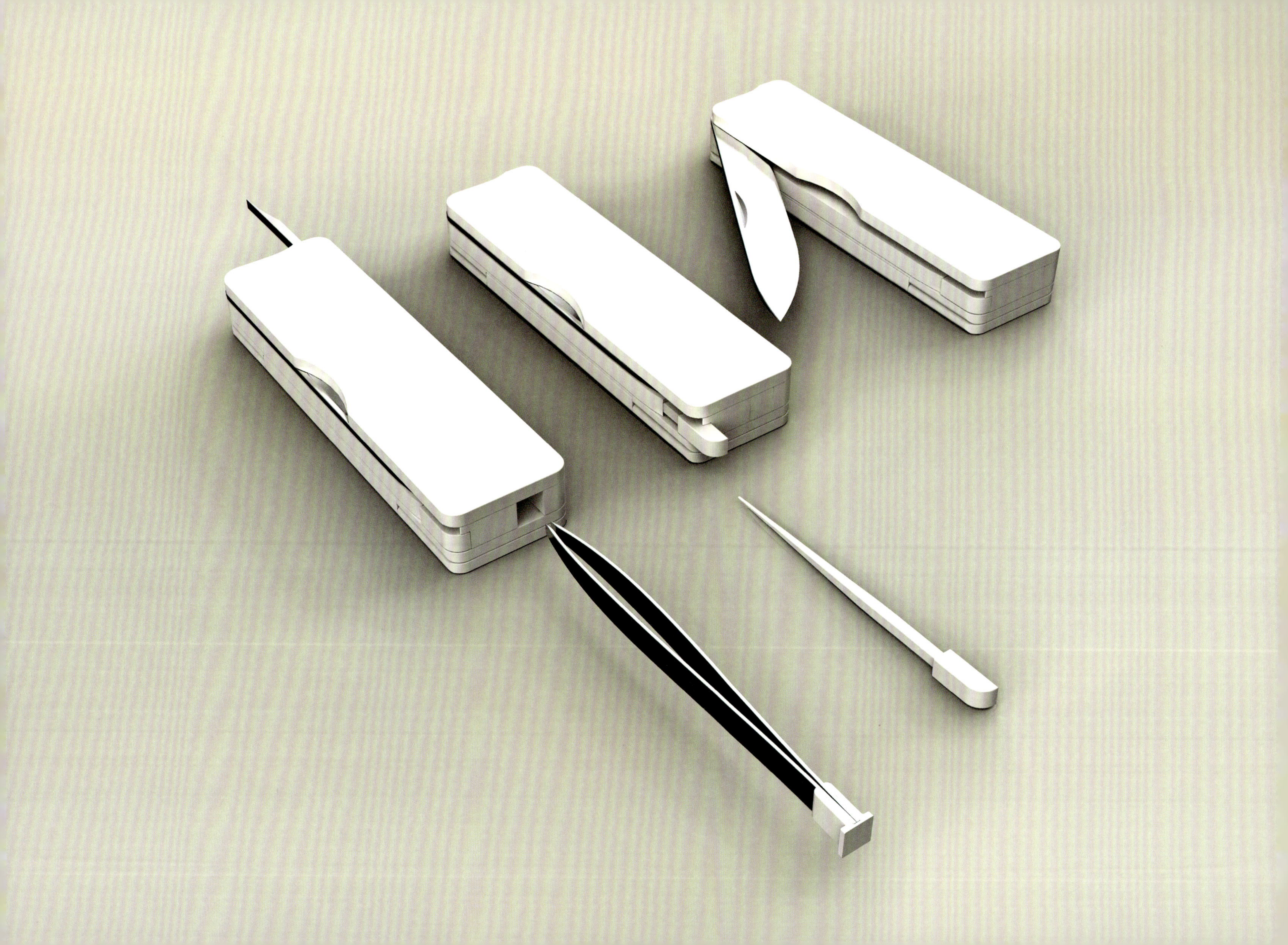

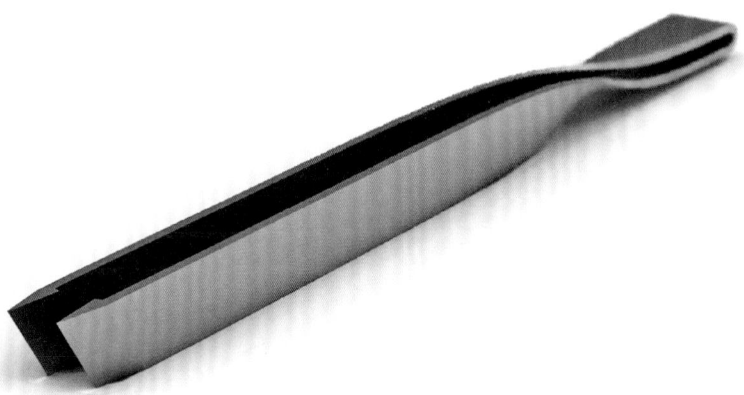

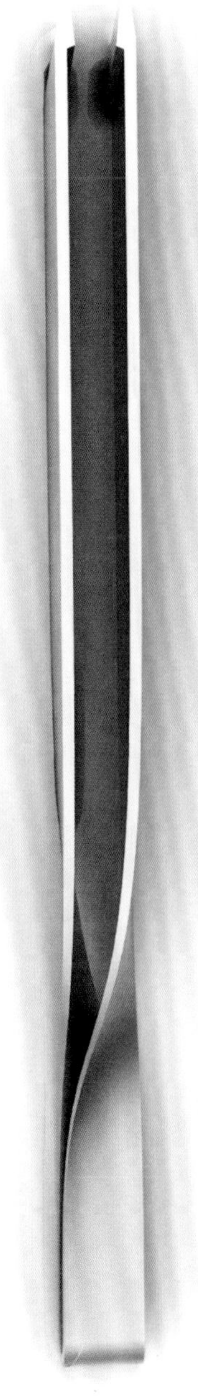

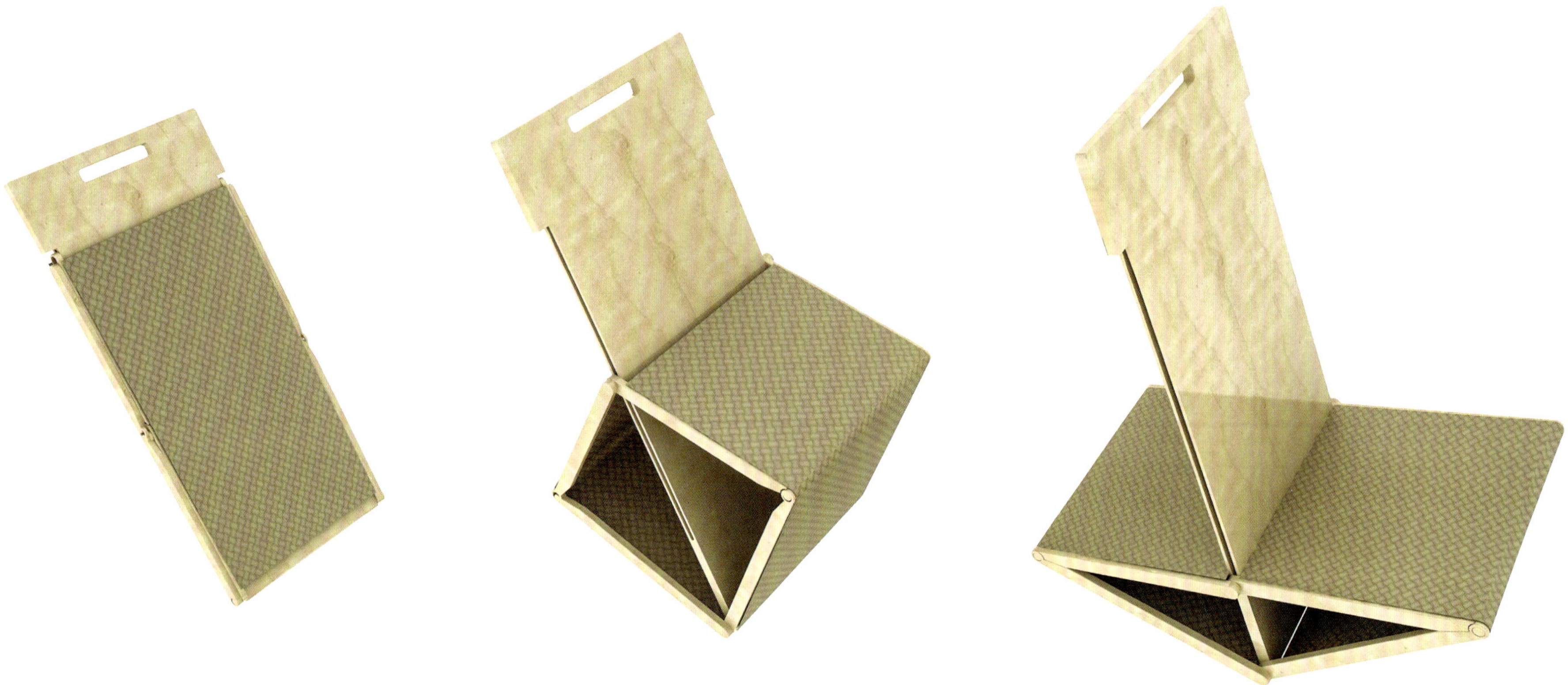

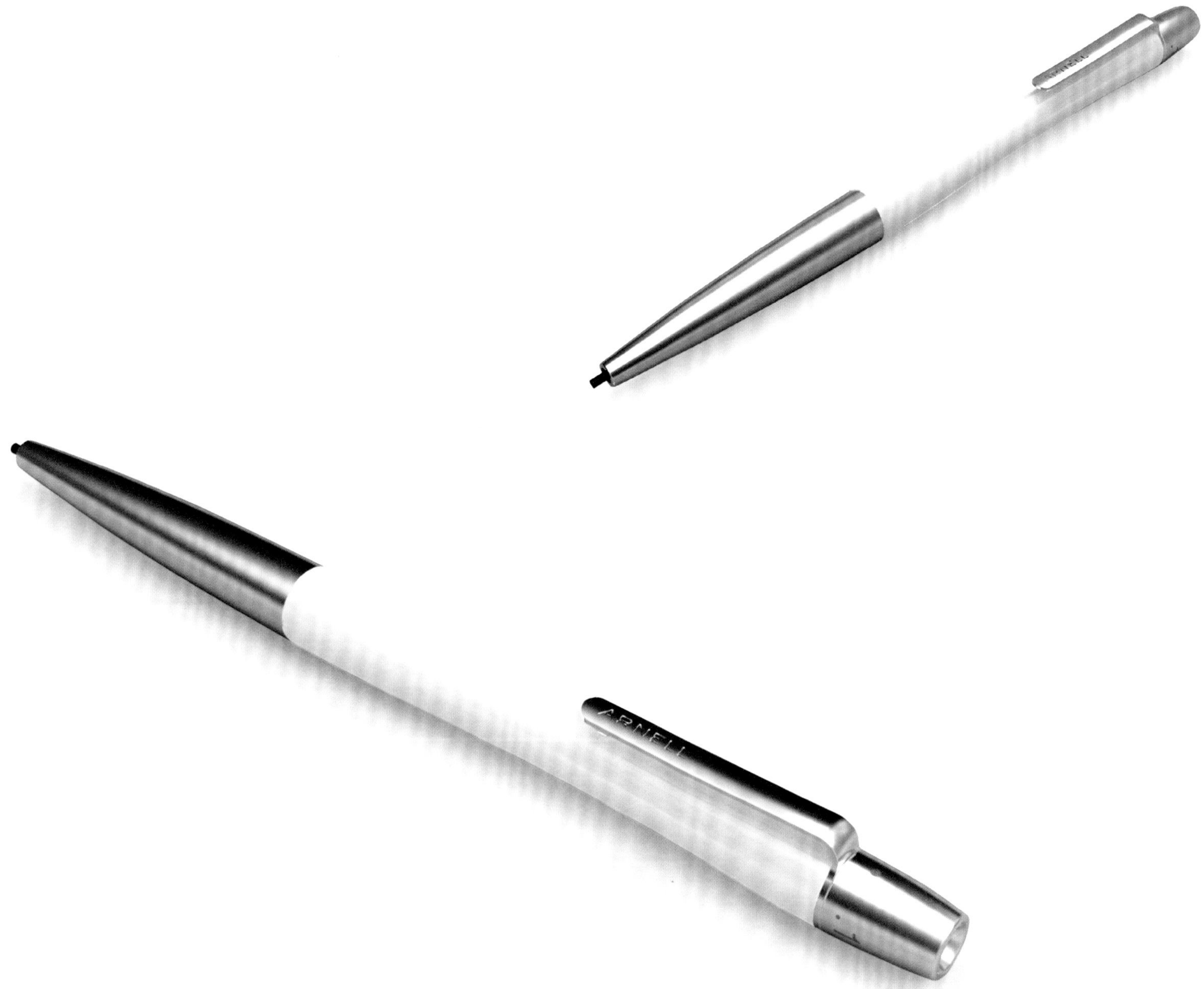
ARNELL

ITO-YA

2009

ITO-YA, a Japanese stationery company started in 1904 by Ito Kazutaro, was founded on the principle of offering innovative, distinctive, and high-quality office, art, and school supplies.

As part of an effort to introduce creative new forms to essential office supplies, a series of design proposals for innovative new writing instruments was developed. For the design of the signature Offset Pen, the subtle shift in the barrel of the pen created a comfortable lip for the index finger to rest upon and a ledge for the thumb to grip while writing. This one small change to the form of a classic ballpoint pen entirely changed the ergonomic experience of using it and was yet another example of the innovation and precision that ITO-YA had become renowned for.

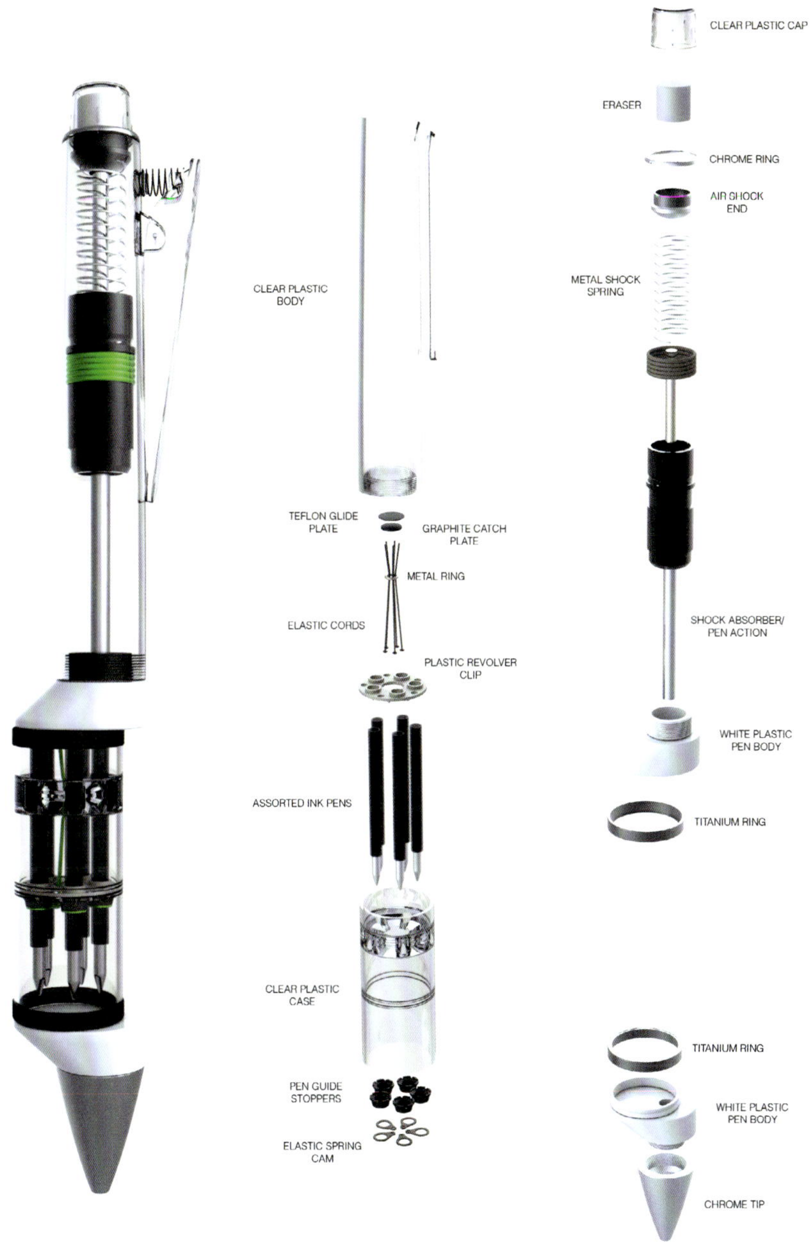

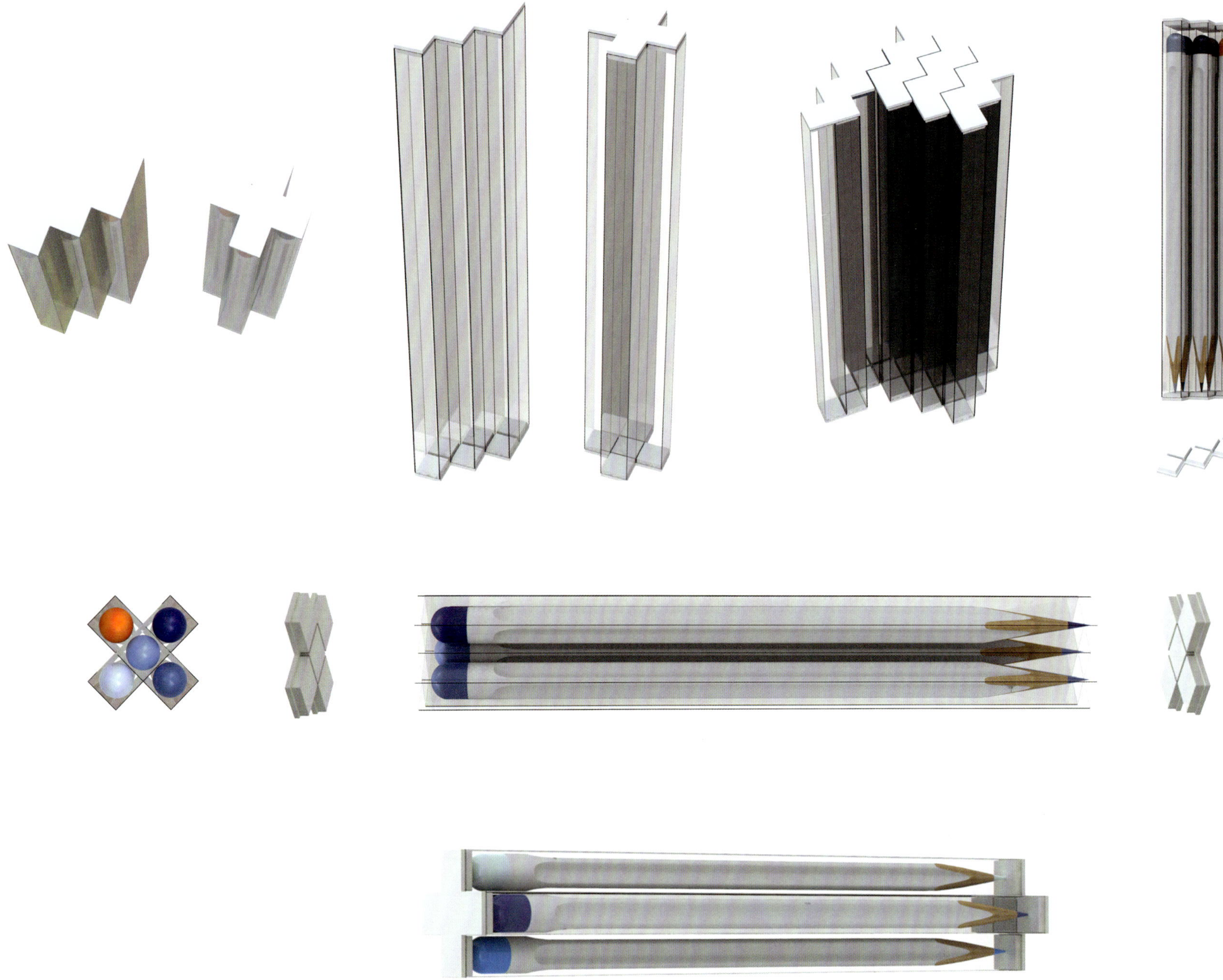

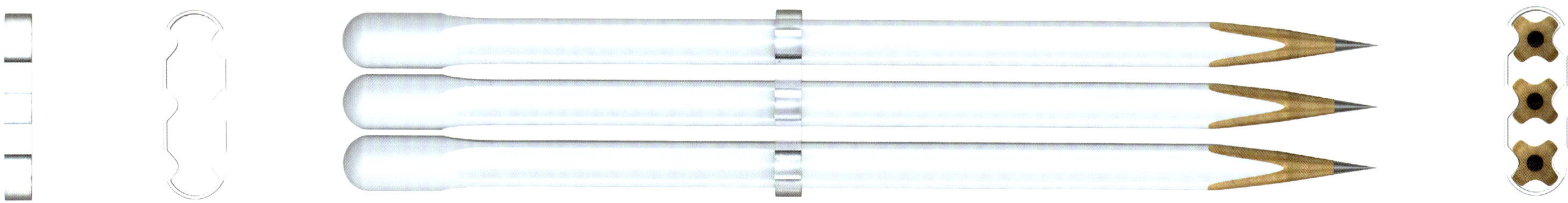

U

UNIQLO

2009

Presented to Kashiwa Sato, then the head of global branding and communication activities at Uniqlo, the "Made for U" proposal envisioned a more personalized consumer invitation that built upon the "Made for All" brand positioning that Uniqlo had used for many years.

"Made for U" served to connect the brand with individual consumers, humanizing the brand message and providing a sense of sincere affirmation to the consumer—all in a way that was an authentic and intuitive extension of Uniqlo's core brand attributes.

Today, many of the core insights of the "Made for U" initiative live on in Uniqlo's U Collection: a range of high-quality, elevated wardrobe essentials designed to suit every style and every budget.

MADE FOR U

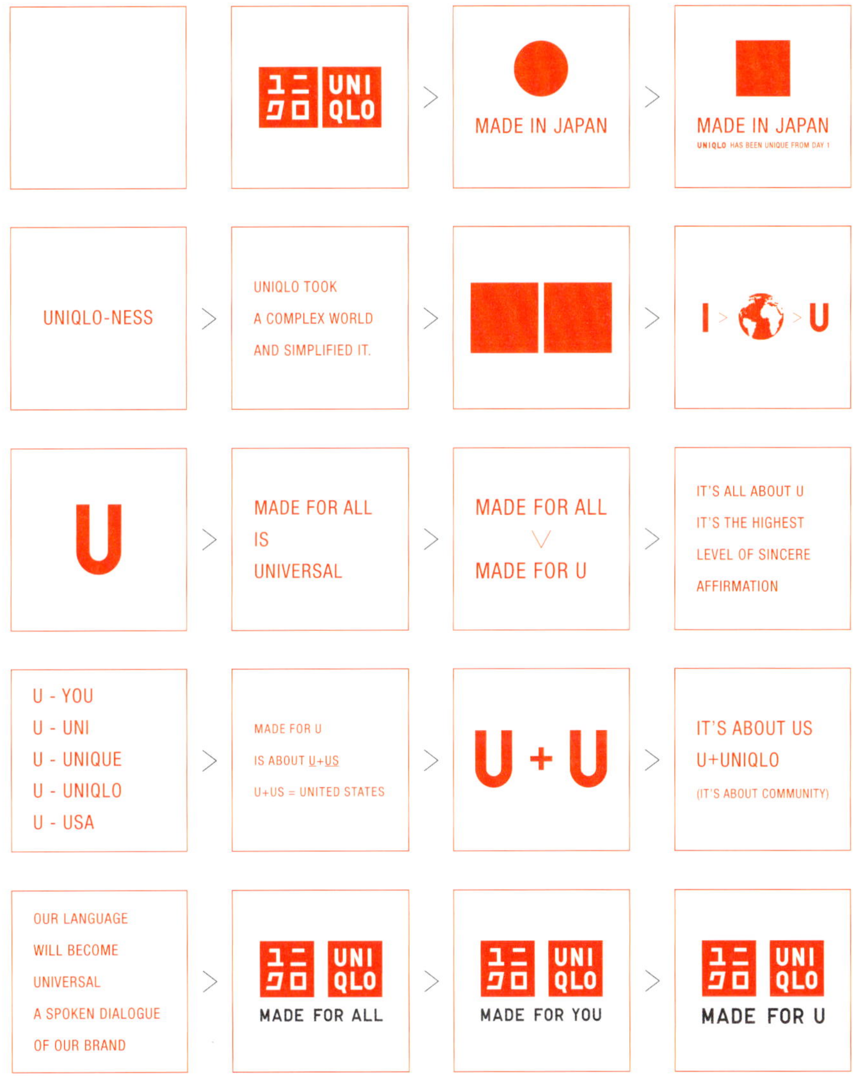

SA SAKE

2009

Despite sake's massive popularity around the world, there has never been a truly global sake brand: a single product that becomes a segment-defining household name. The sake market outside of Japan had historically been dominated by smaller producers, none of which was reliably, widely available from one market to the next. And even those brands that did enjoy more widespread distribution failed to offer clear markers of differentiation to less experienced drinkers who were seeking an accessible entry point to sake.

The opportunity presented itself to develop a premium global sake brand that consumers worldwide could come to know and rely upon. The concept for SA Sake proposed establishing a traditional Japanese sake brewery (known as a *sakagura*) that combined the finest handcrafted brewing techniques with state-of-the-art precision technology to guarantee a perfect product at a mass-produced scale.

An eye-catching brand identity was developed that was inspired by *ensō*—an artistic tradition in Japanese calligraphy that celebrates the irregular beauty of free-form brushstrokes—while a unique aluminum bottle design helped to give the product an iconic and easy-to-spot presence on store shelves. The final result was a product that conveyed a deep respect for the traditions of handcrafted sake while still being accessible and appealing to a vast, untapped global customer base.

SA SAKE

ORIBE

2009

Oribe Canales, the legendary Cuban-born American hair stylist, catapulted to fame in the 1980s when his early work on editorial shoots quickly made him one of the most sought-after hairstylists for the top photographers and designers of the day. Working with brands like Versace, Chanel, and Calvin Klein, he created countless iconic, era-defining hairstyles for supermodels such as Naomi Campbell, Linda Evangelista, Christy Turlington, and Cindy Crawford.

For Canales's namesake line of luxury, salon-quality hair care products, a new brand identity was designed that captured the signature style that he had become renowned for. The sleek, elegant typeface perfectly suited the sexy, refined, elevated look that had made Canales one of the most influential and in-demand stylists of his generation, while grounding his brand in the place where it all started: Miami Beach.

O R I B E
M I A M I B E A C H

GOOP

GOOP

2009

When Gwyneth Paltrow decided to launch a lifestyle brand to showcase her wide-ranging knowledge of cooking, fitness, fashion, design, and wellness, she envisioned a brand that was an authentic extension of her life: a curated selection of objects, experiences, and lifestyle practices that she loved and wanted to share with the world.

For Gwyneth, a distinctive name and identity that would set her brand apart was an absolute prerequisite. After considering a wide range of options, the best solution turned out to be the most obvious one: her childhood nickname. Gwyneth's father had called her "Goopy" as a young girl: an affectionate pet name that riffed on her initials "GP."

As a name, "Goop" was completely unlike anything else on the market and a decidedly unexpected direction for a lifestyle brand. But over the years it has proven to be a unforgettable moniker that, for many, is now a byword for casual elegance and living well.

As part of the brand's launch, a website and e-commerce platform was developed, along with editorial content strategies, brand communications, and a variety of collateral brand material. A range of prototypical product lines were also designed to help illustrate the potential reach of the brand, including home décor, dinnerware, and cookware. Goop has subsequently branched out into the wellness, beauty, and fashion sectors, and its wide range of private label products have become industry-leading lodestars. With a reach that extends across various online and print publications, podcasts, e-commerce platforms, a television show, and numerous social media channels, Goop attracts millions of customers each month and has blossomed into one of the world's most influential lifestyle brands.

Let me win,
but if I cannot win,
let me be brave
in the attempt.

EKS

2009

As founder and honorary chairperson of the Special Olympics, and executive vice president of the Joseph P. Kennedy, Jr. Foundation for more than five decades, Eunice Kennedy Shriver was a leader in the worldwide struggle to improve and empower the lives of individuals with physical or intellectual disabilities.

In 1962, Shriver founded Camp Shriver, a camp for children with special needs that was held on her Maryland farm. Camp Shriver subsequently evolved into Special Olympics Inc., a nonprofit charity which Shriver founded in 1968. That year, the Joseph P. Kennedy, Jr. Foundation helped to plan and fund the First International Special Olympics Summer Games, held in Chicago, Illinois. Since that time, nearly three million athletes have participated in the biennial games.

Following Shriver's passing in 2009, her family founded EKS, the Eunice Kennedy Shriver Foundation, in her honor and to continue her life's work. For this new organization, a bold and iconic brand identity system was designed. The resulting brand mark, a simple, elegant sans-serif type accompanied by a green plus sign, conveyed Eunice Kennedy Shriver's lifelong mission of empowering others to realize their limitless potential for growth and set up EKS to continue the work of her vital advocacy for years to come.

THE NORTH FACE

2009

In a bid to reenergize its retail business, outdoor recreation and performance apparel retailer The North Face launched a competition that invited a select group of architects and designers around the world to develop schemes for a completely new retail design strategy and in-store experience for North Face flagship stores.

Working in partnership with the Milan-based architect Cino Zucchi, a design proposal was developed that brought the North Face's spirit of exploration to life with immersive in-store "landscapes." Dramatically faceted multi-planar surfaces evoked soaring natural terrains, while sleek material finishes and a seamless integration of digital displays and other high-tech elements helped to position The North Face as a true next generation performance outfitter.

Each North Face retail flagship was designed to be an immersive destination where customers could go for the products and specialized knowledge that they needed to tackle new challenges in the great outdoors and beyond.

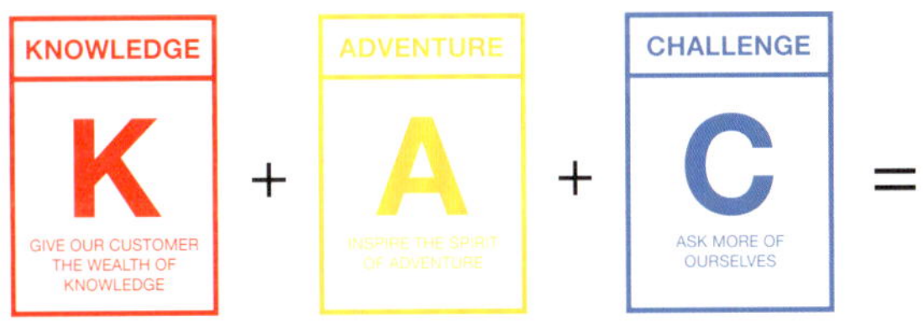

Past is Discovered
Present is Explored
Future is Boundless

THE NORTH FACE

SPORTS
PERFORMANCE
EXPLORATION IS...
ASK MORE OF OURSELVES

ACTION SPORTS

THE NORTH FACE
SUMMIT SERIES

OUTDOOR
GORE-TEX
SUSTAINABILITY

THE NORTH FACE
PERFORMANCE
OUTDOOR
EXPLORATION IS...
ASK MORE OF OURSELVES
AN ICONIC EXPERIENCE
BOUNDLESS

THE NORTH FACE
OUTDOOR
ACTION SPORTS

GEM

GEM

2009

Already a leader in the Neighborhood Electric Vehicle market (NEV), Chrysler-owned GEM (Green Eco Mobility) was well-positioned to expand its dominance in this segment with stylish, safe, more car-like models, while also going after various segments of the electric vehicle market, including passenger cars, golf cars, two-wheelers, light industrial and commercial, military, heavy industry, and other uncharted segments like mobile robots, toys, marine vehicles and more.

Historically, NEVs were flimsy, golf cart-like vehicles with short range, limited functionality, and generic styling. To combat those preconceptions, a range of GEM concept models were developed that featured bold styling, enhanced performance, and a "plug-and-play" strategy that offered customers a variety of add-on accessories to customize their GEMs. Components like storage racks, removable back seats, and flatbed storage compartments were designed to be easily swapped in and out to suit virtually every use case scenario, helping to reposition GEM as functional, viable, and environmentally-friendly secondary mode of transport.

PEAPOD

2009

By the mid-2000s, Chrysler-owned GEM had established itself as the number one electric vehicle brand in the United States—but it was also a brand that almost no consumers had ever heard of. GEM vehicles weren't sold direct to consumers; rather, they were purchased by corporations with large campuses, parks departments, and by other businesses and organizations that needed small, efficient vehicles to transport people within large, confined areas. Using the GEM platform and working with a dedicated team of engineers and designers at Chrysler, the Peapod concept was born: one of the first mass-market neighborhood electric vehicles to be produced by a major automotive company.

Everything about the Peapod was different. It was made entirely from recycled and recyclable materials. Seats, designed like high-tech office chairs, were created to provide lumbar support, superior comfort, and also saved significant space in the interior. And in spite of its small footprint, it still had the interior space to comfortably accommodate a six-foot-eight passenger.

The then-new and groundbreaking iPhone was integrated into the design with an in-app dashboard—years before interfaces like Apple CarPlay or Android Auto were available. Recognizing the growing ubiquity of touchscreen smartphones, the intuitive interface offered at-a-glance information about battery life, trip range, nearby charging stations, navigation tools, music and more, effectively consolidating the usual cluttered dashboard of buttons, dials, and displays into one simple, intuitive, and dynamic user interface.

Working closely with veteran engineers like Arthur Anderson at Chrysler and Katsushi Sasaki at Mitsubishi, the Peapod concept went from napkin sketch to fully functional, production-ready prototype in a little over one year—a fraction of the usual development time in the automotive industry.

Launched on Earth Day 2009, Peapod was an inspired and fully viable design that set the stage for the subsequent generation of mass-market electric vehicles that would enter the marketplace in the years to come.

peapod

no gas.
no emissions.
all electric.
lots of love.
peapod

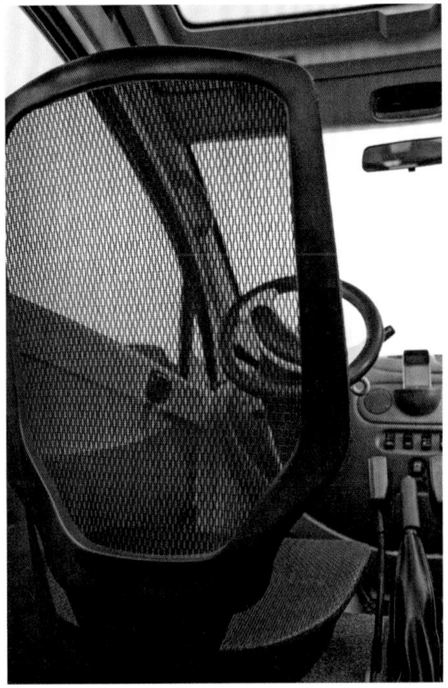

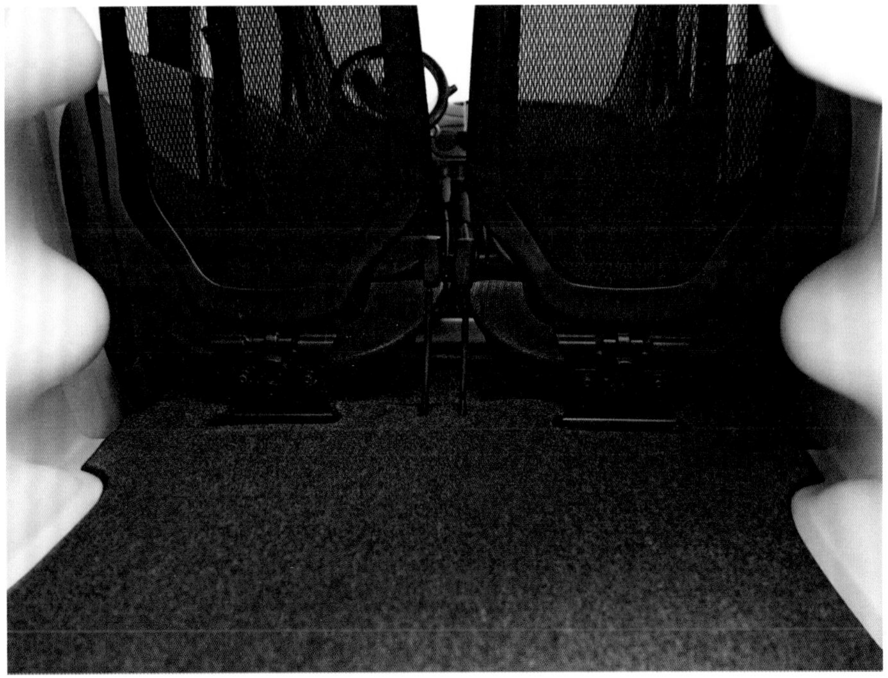

iPod
4:19 PM
peapod
360°
Product Viewer
Media Viewer
More Info
www.peapodmobility.com

iPod
4:19 PM
Back
PRODUCT VIEWER
www.peapodmobility.com

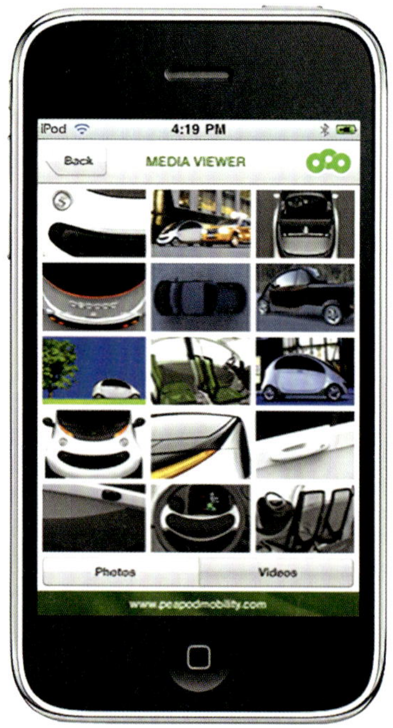
iPod
4:19 PM
Back
MEDIA VIEWER
Photos
Videos
www.peapodmobility.com

iPod
4:19 PM
Done
00:00:00
-01:48:46
www.peapodmobility.com

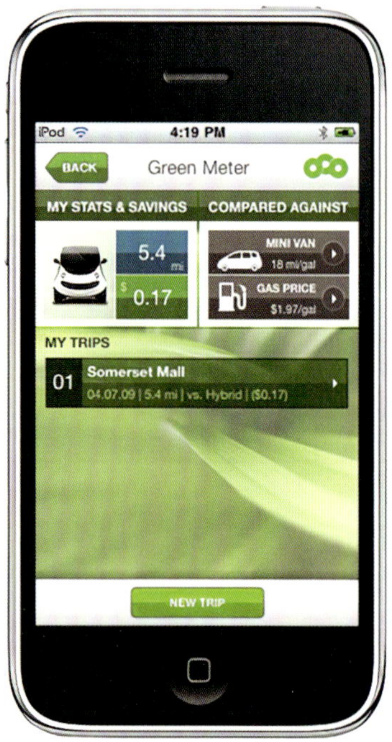
iPod
4:19 PM
BACK
Green Meter
MY STATS & SAVINGS
COMPARED AGAINST
5.4
mi
$
0.17
MINI VAN
18 mi/gal
GAS PRICE
$1.97/gal
MY TRIPS
01
Somerset Mall
04.07.09 | 5.4 mi | vs. Hybrid | ($0.17)
NEW TRIP

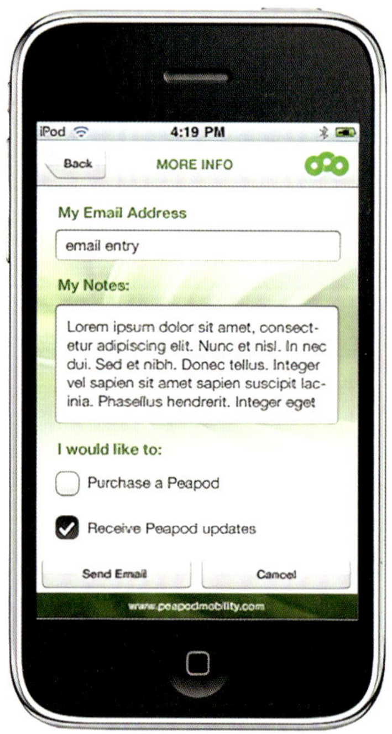
iPod
4:19 PM
Back
MORE INFO
My Email Address
email entry
My Notes:
Lorem ipsum dolor sit amet, consectetur adipiscing elit. Nunc et nisl. In nec dui. Sed et nibh. Donec tellus. Integer vel sapien sit amet sapien suscipit lacinia. Phasellus hendrerit. Integer eget
I would like to:
Purchase a Peapod
Receive Peapod updates
Send Email
Cancel
www.peapodmobility.com

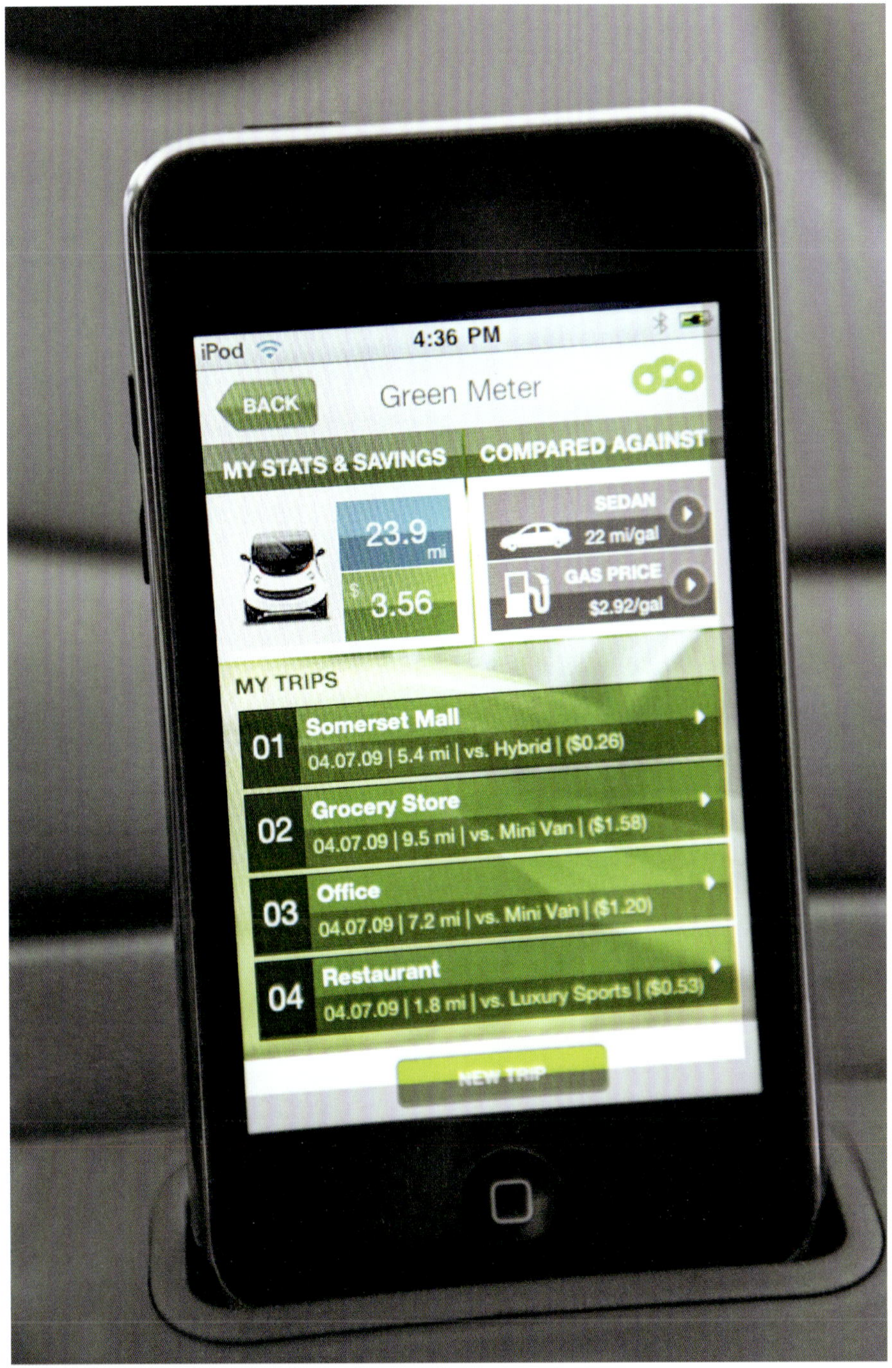
iPod
4:36 PM
BACK
Green Meter
MY STATS & SAVINGS
COMPARED AGAINST
23.9
mi
3.56
SEDAN
22 mi/gal
GAS PRICE
$2.92/gal
MY TRIPS
01
Somerset Mall
04.07.09 | 5.4 mi | vs. Hybrid | ($0.26)
02
Grocery Store
04.07.09 | 9.5 mi | vs. Mini Van | ($1.58)
03
Office
04.07.09 | 7.2 mi | vs. Mini Van | ($1.20)
04
Restaurant
04.07.09 | 1.8 mi | vs. Luxury Sports | ($0.53)
NEW TRIP

4:19 PM
peapod
TRIP

peapod

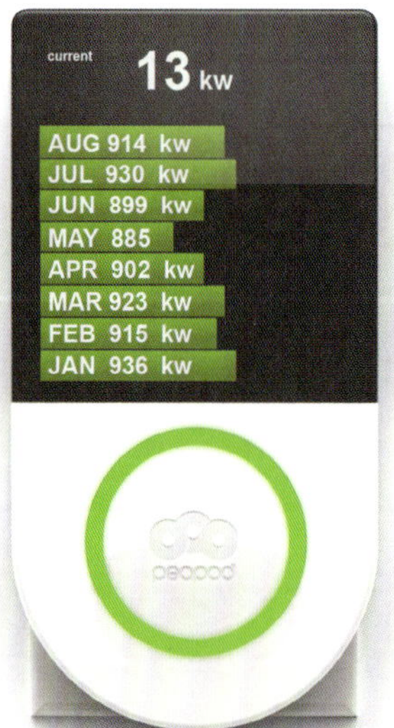
current
13 kw
AUG 914 kw
JUL 930 kw
JUN 899 kw
MAY 885
APR 902 kw
MAR 923 kw
FEB 915 kw
JAN 936 kw
peapod

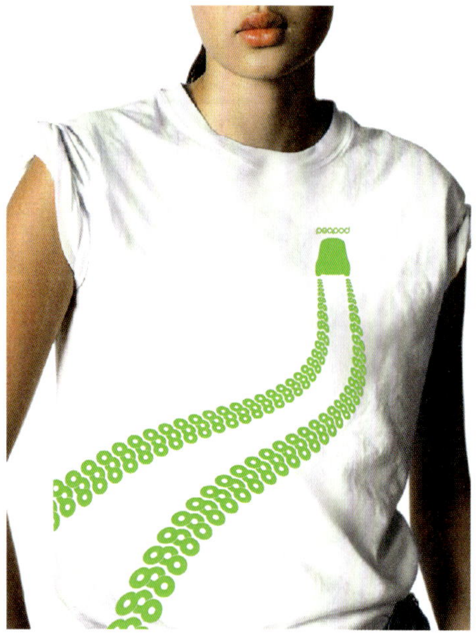
peapod

peapod

no gas.
no emissions.
all electric.
lots of love.
make
love
happen.
LOVE
peapod

peapod
no gas. no emissions. all electric. lots of love.

30 miles for 60 cents
you would be smiling too.....
peapod

DODGE ATV

2009

The Dodge ATV concept was developed as part of a broader brand extension initiative that explored ways of leveraging Dodge's existing automotive engineering expertise and brand equity to capture market share in new segments.

The design proposal for a new Dodge-engineered all-terrain vehicle organically aligned with the preferred recreational activities of many of Dodge's existing customers, while also elevating the rugged cachet that the marque had carefully built up over the years with its iconic trucks and high-performance muscle cars. The muscular styling clearly evoked some of Dodge's classic models, such as the Viper supercar and Ram trucks, while also representing a new era of design and engineering excellence for the brand.

DE TOMASO

2010

By the late 2000s, Italian automaker De Tomaso had been effectively dormant for over a decade, yet the legendary performance marque still had many admirers and devotees. In 2009, businessman Gian Mario Rossignolo purchased De Tomaso, with the intention of reestablishing the brand in the marketplace as a preeminent European luxury sports car maker.

To relaunch De Tomaso, an iconic new brand mark was fashioned by combining the original T brand mark with an abstracted version of the letter D. When mirrored, these two D characters formed a stylized T—at once similar to the iconic original logo, yet entirely fresh and new. This bold new mark helped to launch a new era in the history of this storied automaker: one that was respectful of the past while being firmly focused on the future.

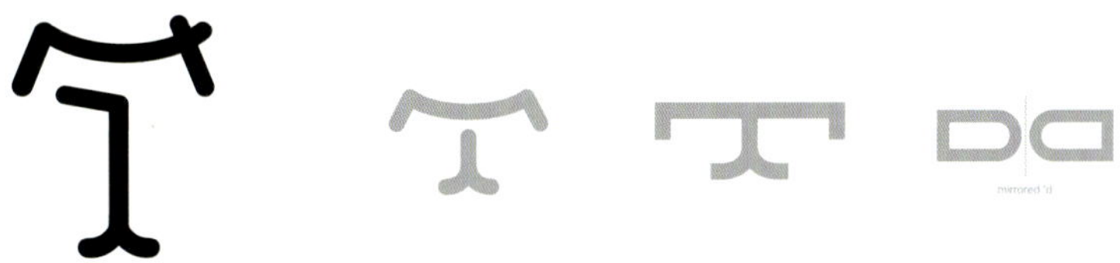

DE TOMASO BLACK PANTERA

2010

Following the 2009 relaunch of De Tomaso, the Italian automaker chose to formally mark its return with a completely reimagined version of its flagship 1970s supercar: the Pantera.

Working in close collaboration with De Tomaso's world-class team of automotive designers and engineers, a fully-realized design concept was developed for this reborn classic: a perfect fusion of muscle, speed, power, and precision engineering. The fluid lines of the new Black Pantera unmistakably referenced the proportions and tailoring of the iconic original model, while also expanding upon its predecessor's distinctive mix of power, agility, beauty, and grace in every way.

On the interior, an innovative full-dash digital display was developed to take the place of conventional instrumentation and controls, while ergonomic seats with razor-thin profiles were designed in an explicit nod to De Tomaso's racing pedigree.

A range of designs for luxury, performance-inspired merchandise—to be handmade in Italy—was also developed. Designed with driver specifications in mind, these items, ranging from apparel, entertainment and motoring accessories, travel luggage, and city guides, enhanced the driving experience and offered an accessible entry point to the De Tomaso lifestyle. Each item was designed to match the sleek design language of the brand, and to reinforce the legendary Italian quality and craftsmanship exhibited in De Tomaso vehicles.

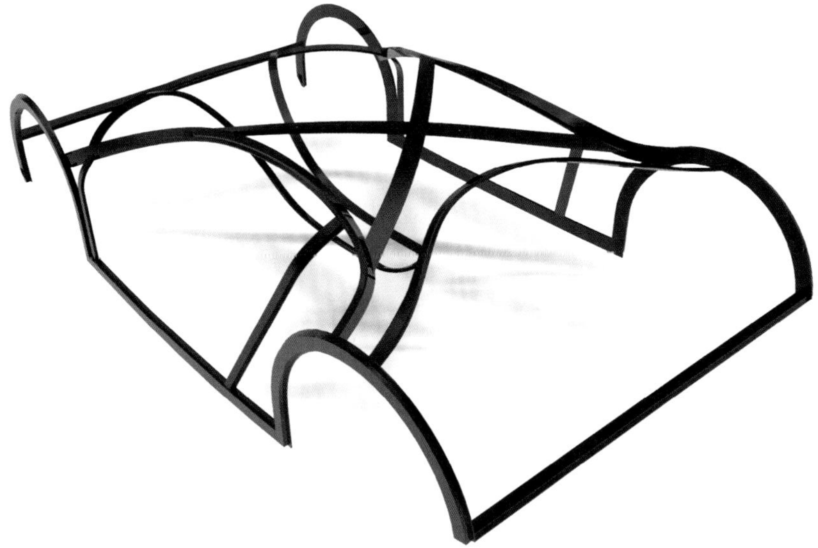

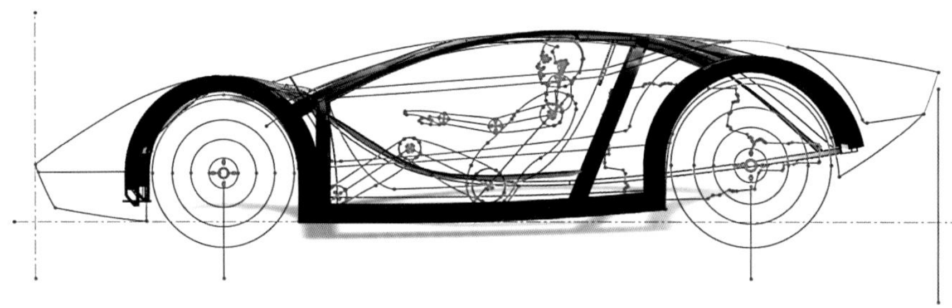
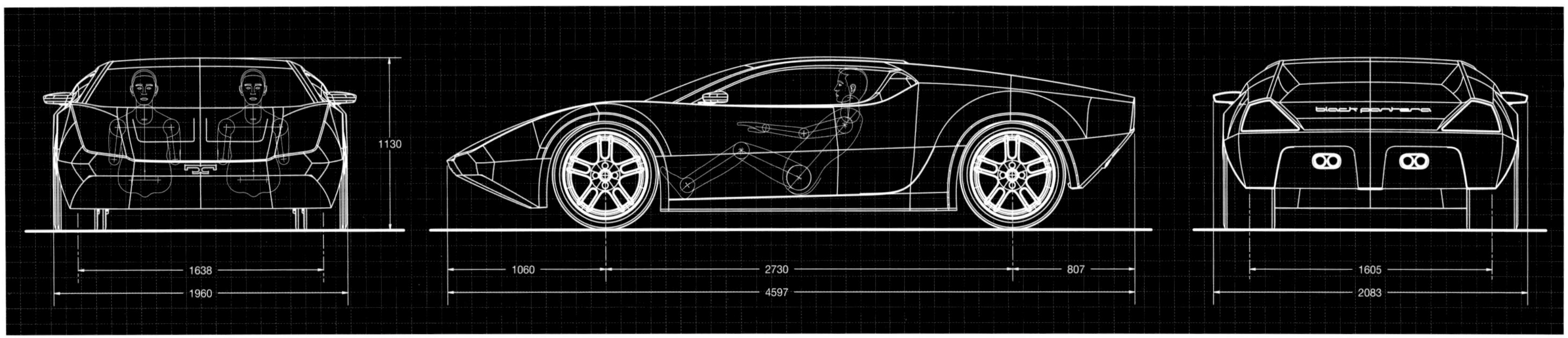
1130
1638
1960
1060
2730
807
4597
1605
2083

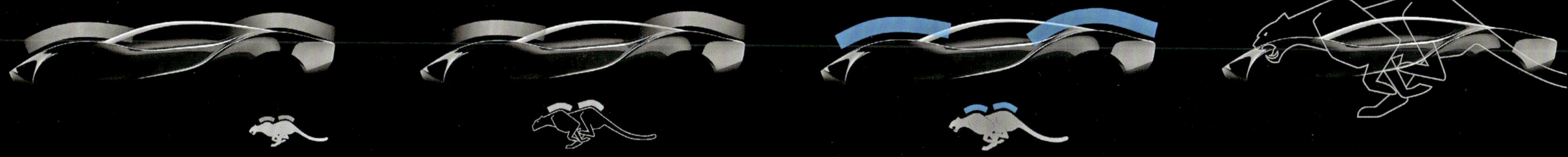

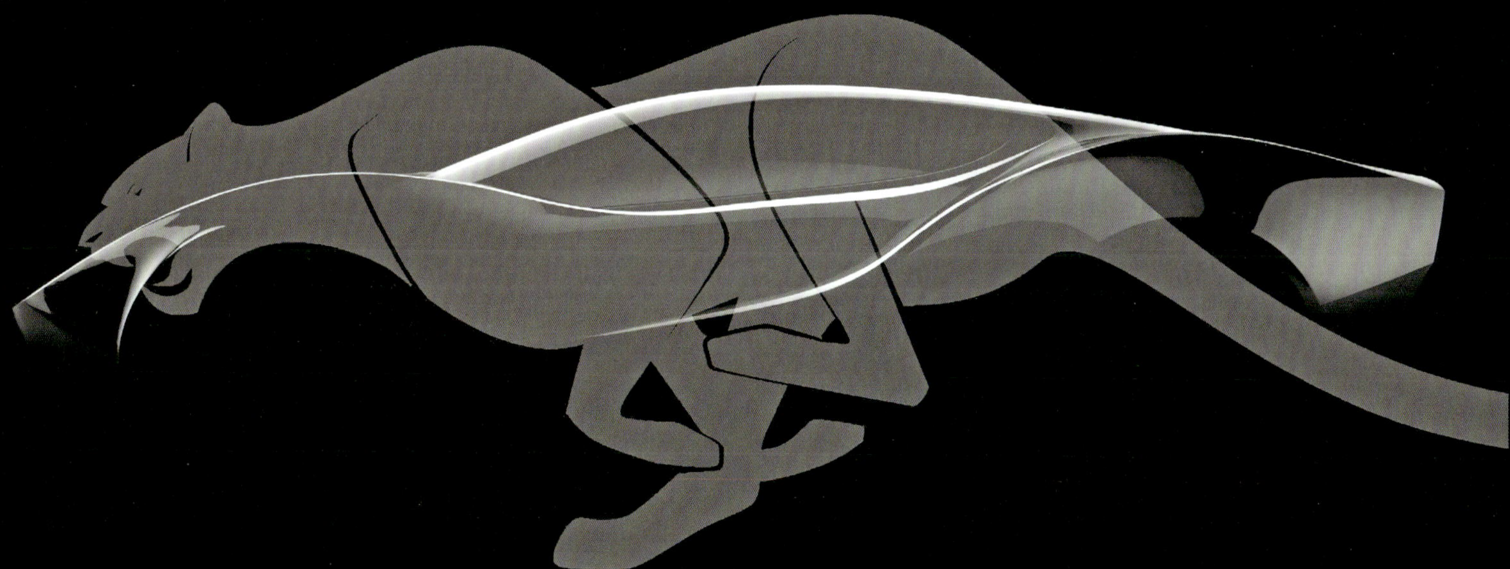

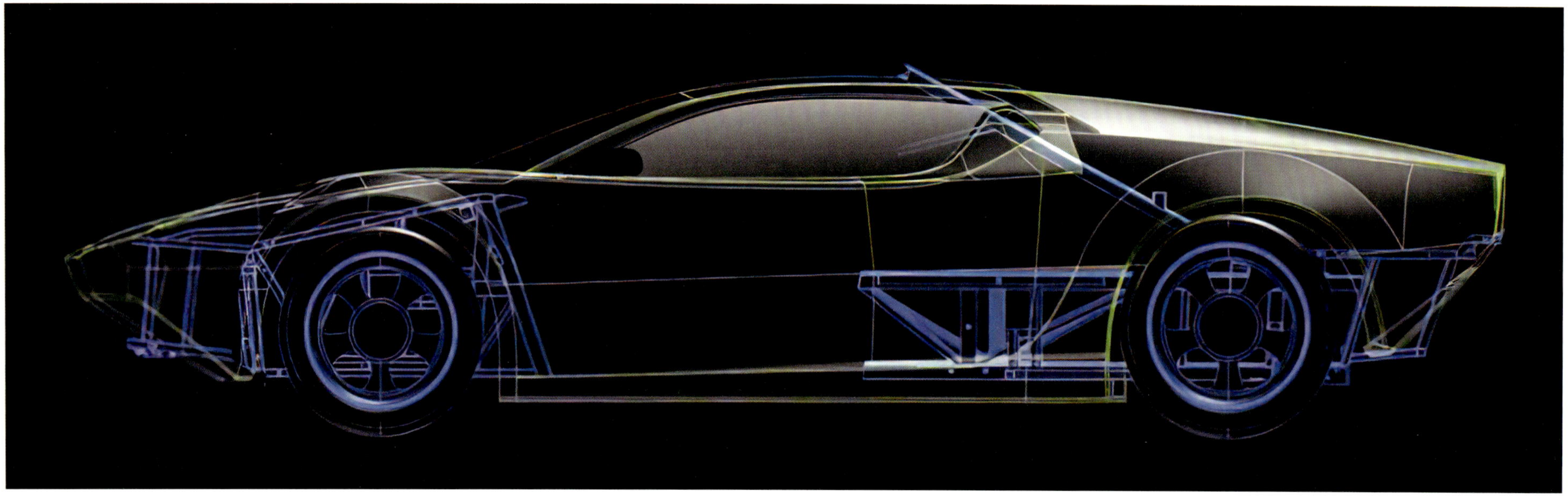

black pantera

0 1 2 3 4 5 6 7 8
N
000 KM/H
0 RPM X 1000
0 10 30 50 70 90 110 150 170 190 210 KM/H
black pantera
002345
21:56
2798 KM
21.5°C
1/1 1/2 0
40 80 100 120
50 100 120 150
0 2 3 4 5

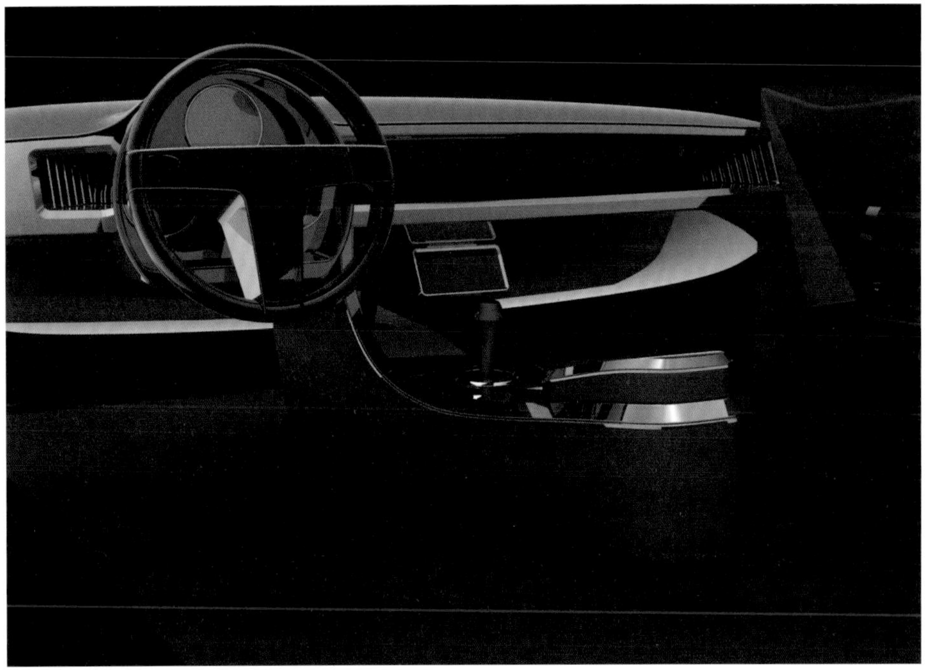
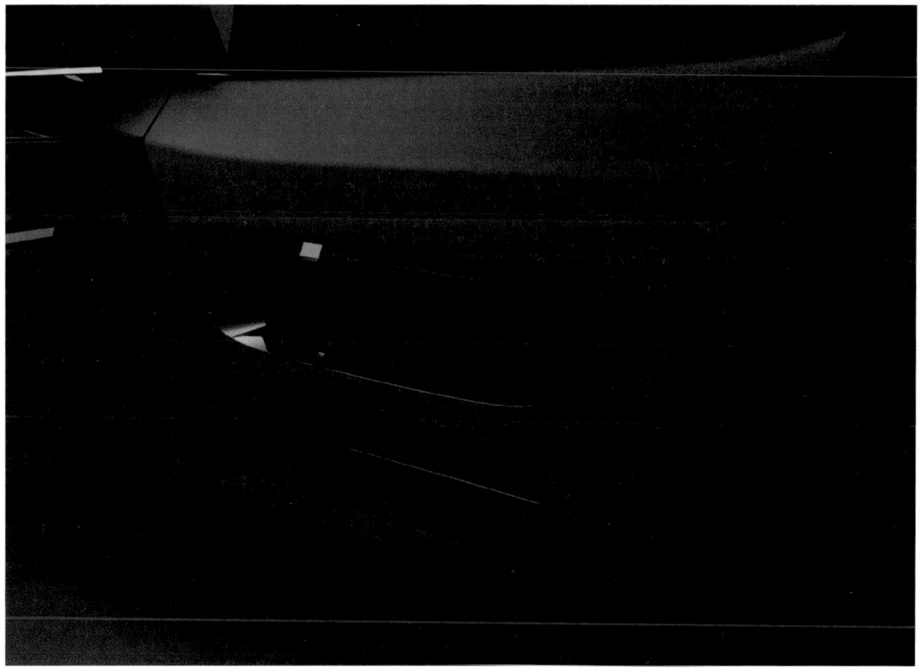

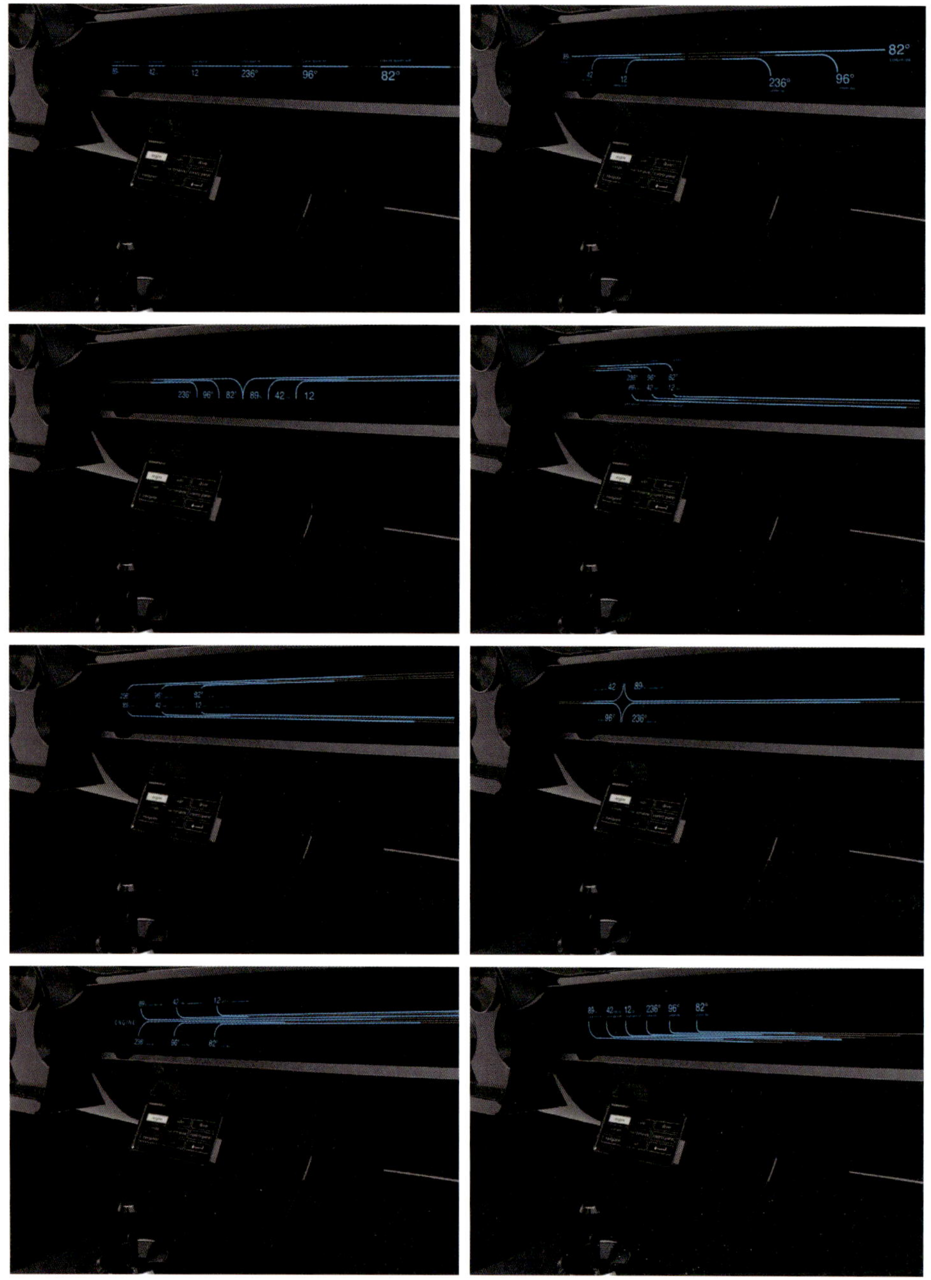

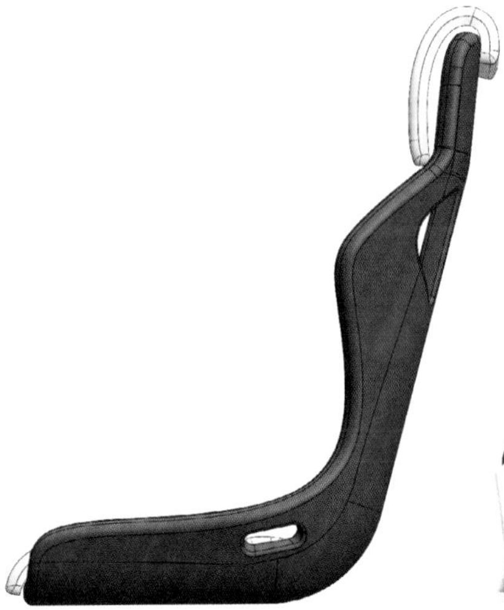

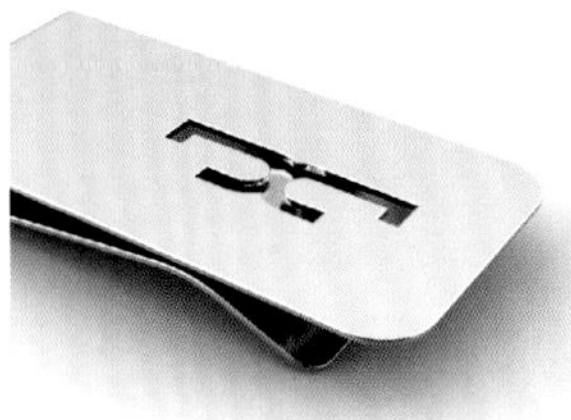
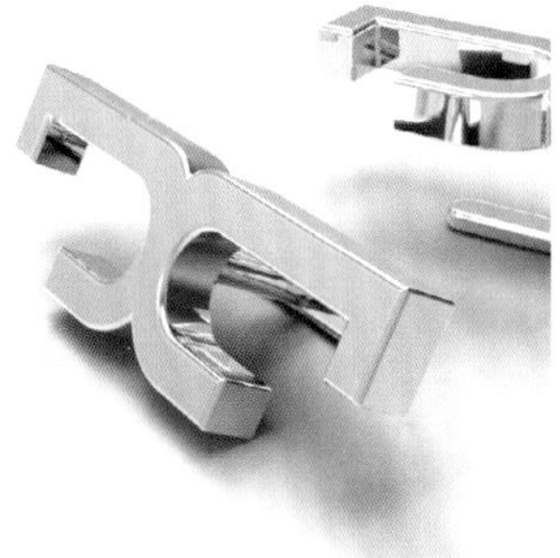
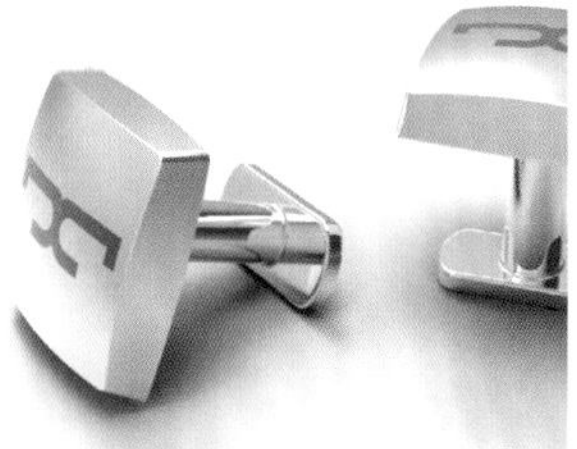

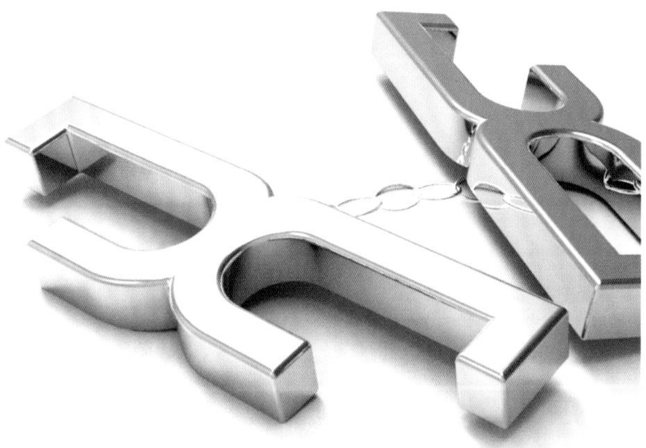

DETOMASO

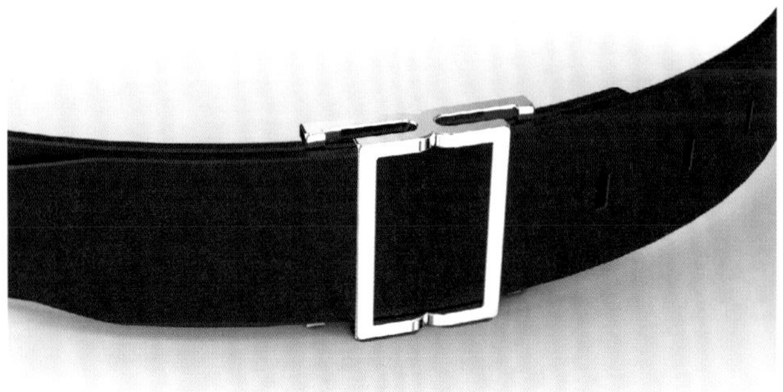

DETOMASO

black pantera

A NEW SPECIES

GIULIO MICRO

2010

Though primarily known for its high-performance racing pedigree, as part of an overarching brand restructuring, Italian automaker De Tomaso developed plans to introduce a compact car that was infused with De Tomaso's exceptional design DNA: the Giulio Micro.

Underpinning the Micro was De Tomaso's UNIVIS system, a proprietary innovation that drastically reduced the time and investment necessary to manufacture a car by minimizing the number of custom dyes needed to a few dozen. With the ability to go to market faster, with lower cost and less environmental impact, UNIVIS allowed for the rapid development of the next generation of Giulio automobiles.

These early design concepts, with their sweeping lines and sensuous curves, defied expectations of what a compact car could look like and helped to establish a new mark of design excellence in this intensely competitive but often overlooked market segment.

micro

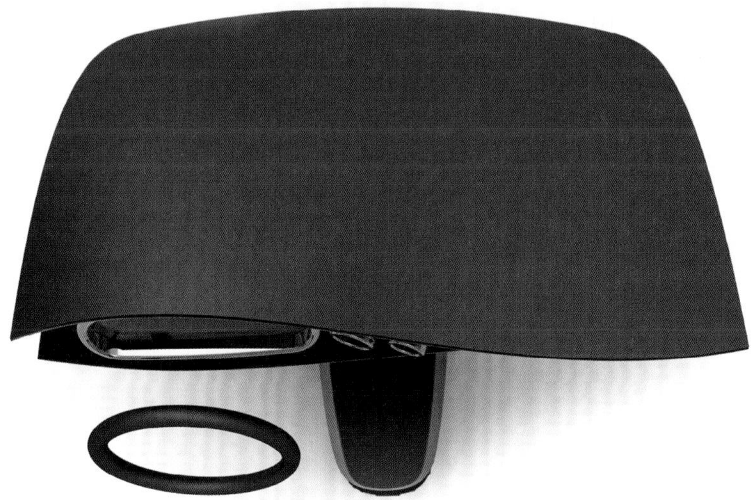

HIGH-SPEED TRAINS

2010

Inspired by the sleek, aerodynamic bullet trains that run on Japan's high-speed train network, a series of design concepts was developed that evolved the form factor of these iconic trains, which had remained largely unchanged since the inauguration of Japan's Shinkansen train network in the 1960s.

The proposed designs featured taut, angular lines and an exaggerated nose that was refined to a long, low, needle-like profile that lent the trains a dynamic, kinetic quality and gave the impression of piercing forward even when at rest. The design concepts also translated aerodynamic and comfort principles from the automotive industry, bringing much-needed innovation to a mode of transport that is set to continue growing in importance, not only in Japan, but around the world.

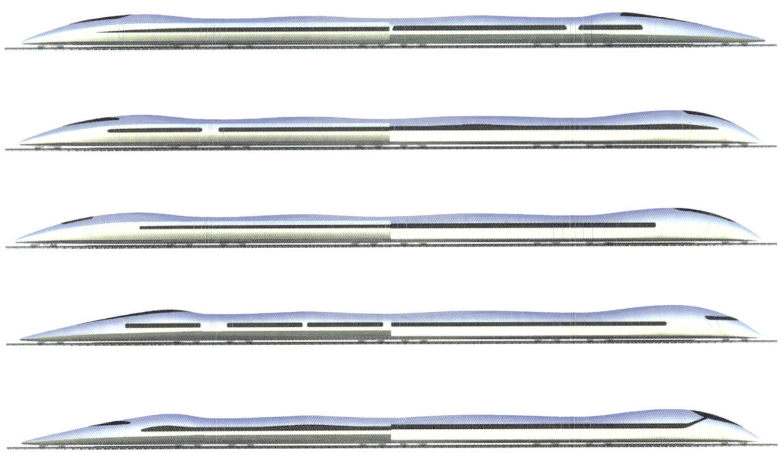

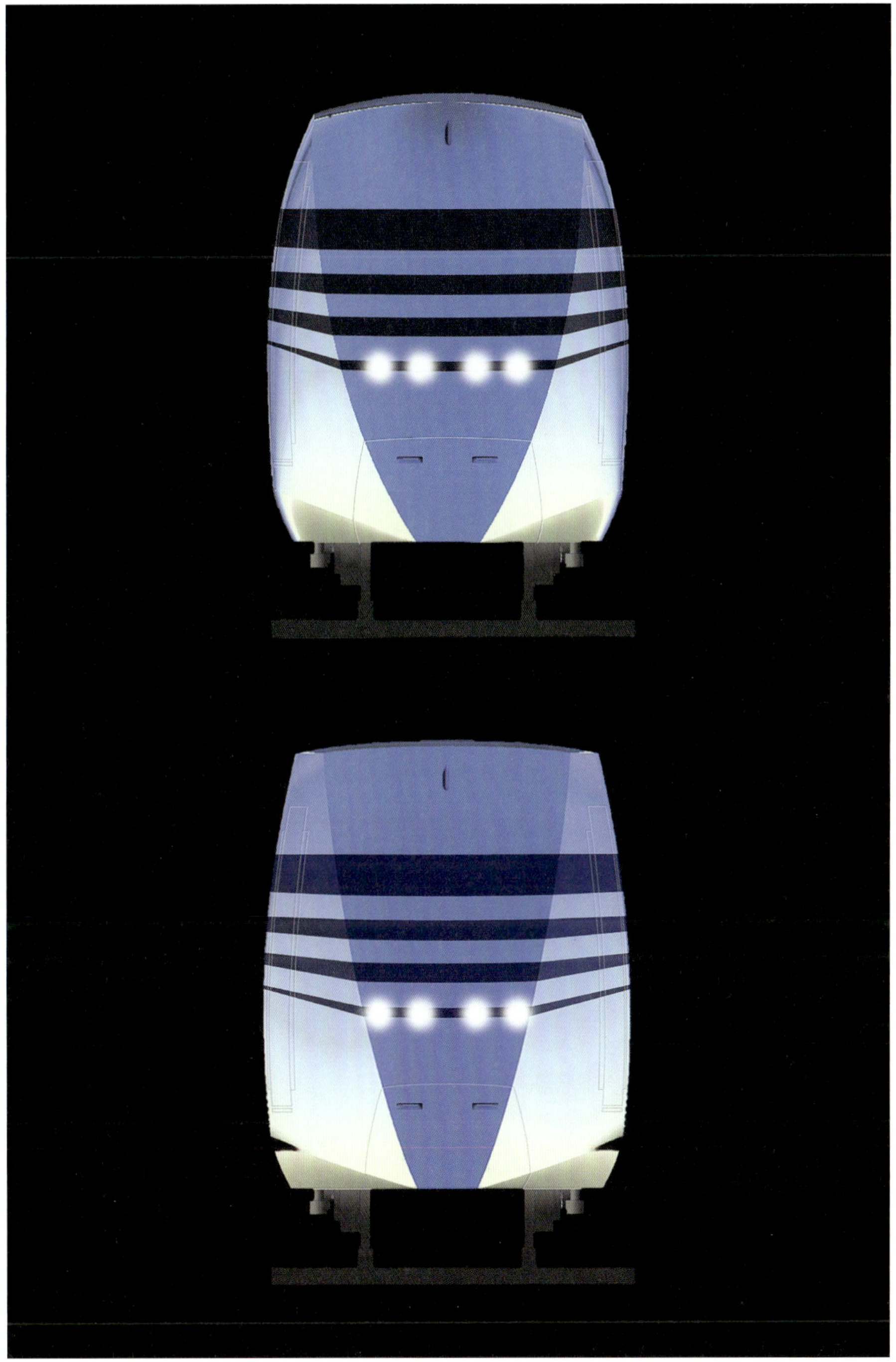

DEVON

2010

Devon was created in 2010 to completely redefine what a luxury and performance brand could be. Inspired by high-performance motor culture, Devon injected some much-needed beauty, fantasy, and sensuality into a sector that typically focused on power and machismo.

At the core of this new ultra-luxury brand was its premier product, the GTX Super Car, which set the one-lap world record at Laguna Seca. The sensuous, sculptural lines of this vehicle informed the design sensibility of the rest of the brand, ultimately leading to the development of a range of luxury timepieces powered by a miniaturized multi-belt motor, a signature fragrance, and a line of designer denim and streetwear.

Along with the branding, communications, and product design for these various brand extensions, plans for a Beverly Hills flagship store were also developed to curate a complete and immersive experience of unparalleled performance-driven luxury.

DEVON

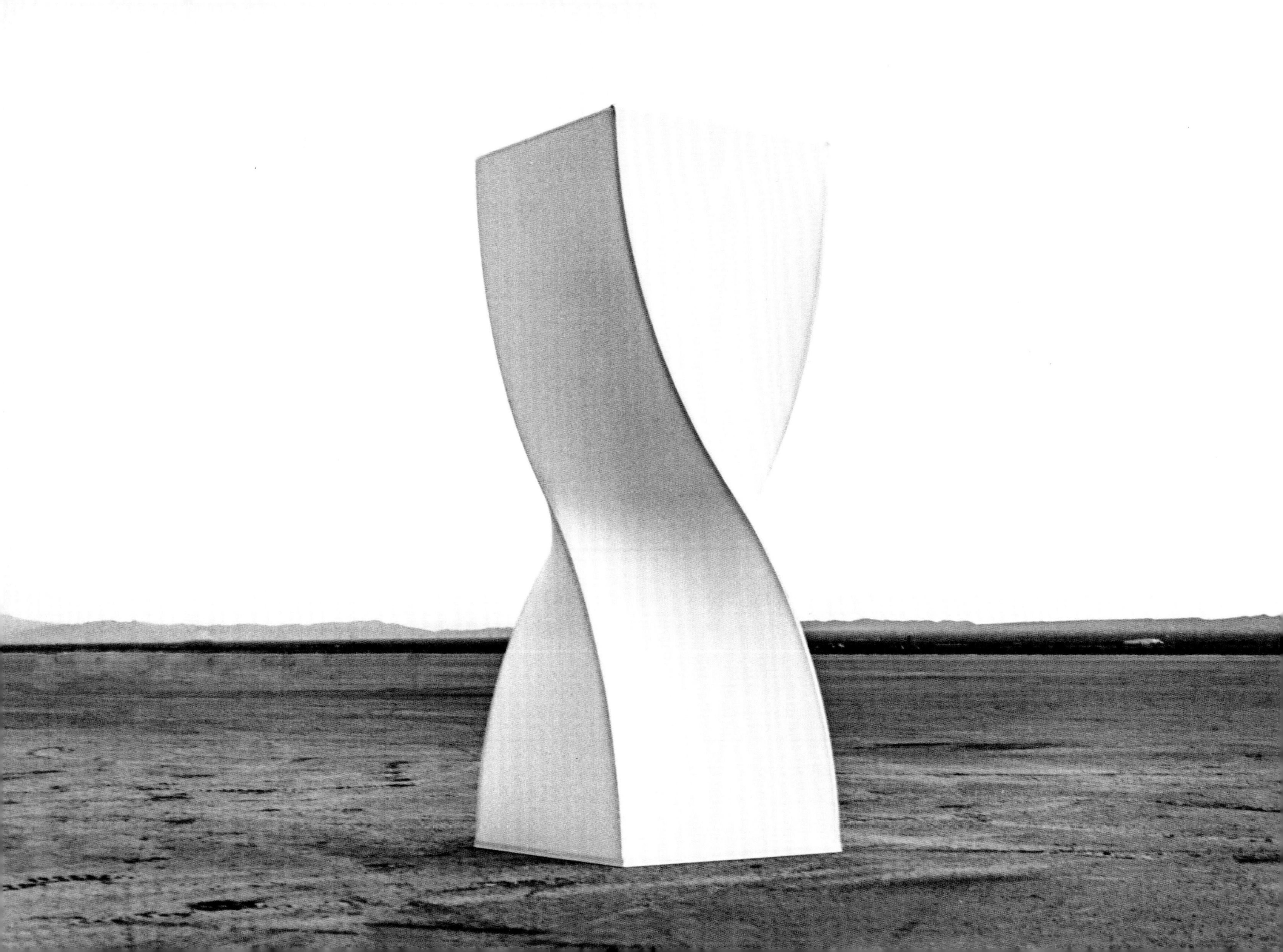

DEVON

DEVON

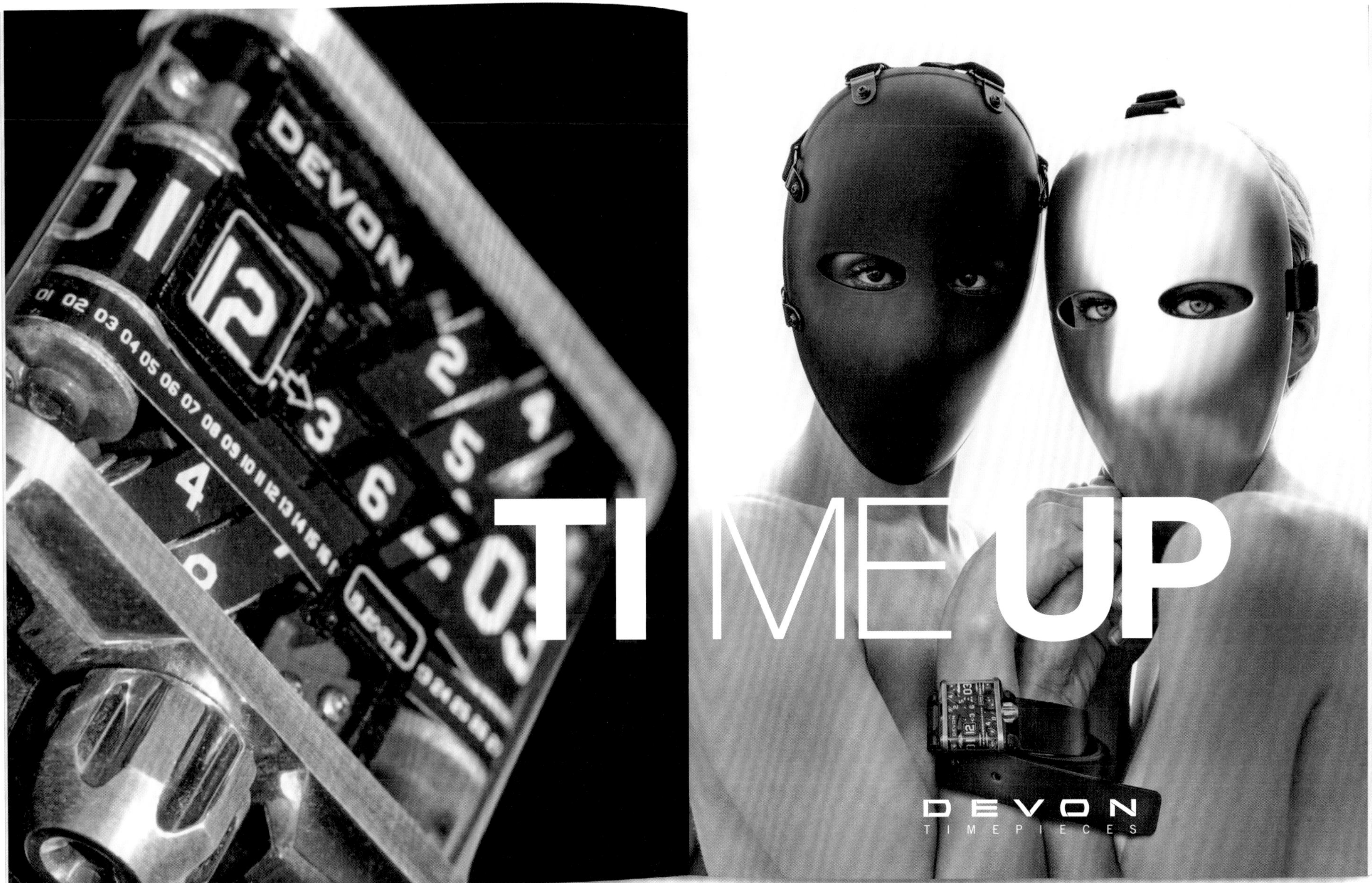
TIME UP
DEVON
TIMEPIECES

SAMSUNG

SAMSUNG LED 9000

2010

For the 2010 launch of the Samsung LED 9000—an ultra-slim, high-definition television—a series of launch communications was developed to introduce this paradigm-shifting product in an unforgettable way. The ads featured the simple but striking image of a woman's hand pinching the wafer-thin display in profile. Even though slim LED television sets weren't entirely new technology, many consumers were still accustomed to TVs whose depth was measured in inches, not in millimeters. The notion that the depth of this particular set could be measured between one's fingers was nothing short of radical. For a product defined by the image it provides, to not show its screen or emphasize its picture quality was a radical gesture that announced just how different this product was from everything else in the marketplace.

The campaign, "Turn on Tomorrow," continued this unexpected shifting of perspective by turning the focus away from the television set entirely and placing it onto the rapt viewers bathed in the glow of the unseen screen, fully immersed in the HD viewing experience.

SAMSUNG

TURN ON TOMORROW

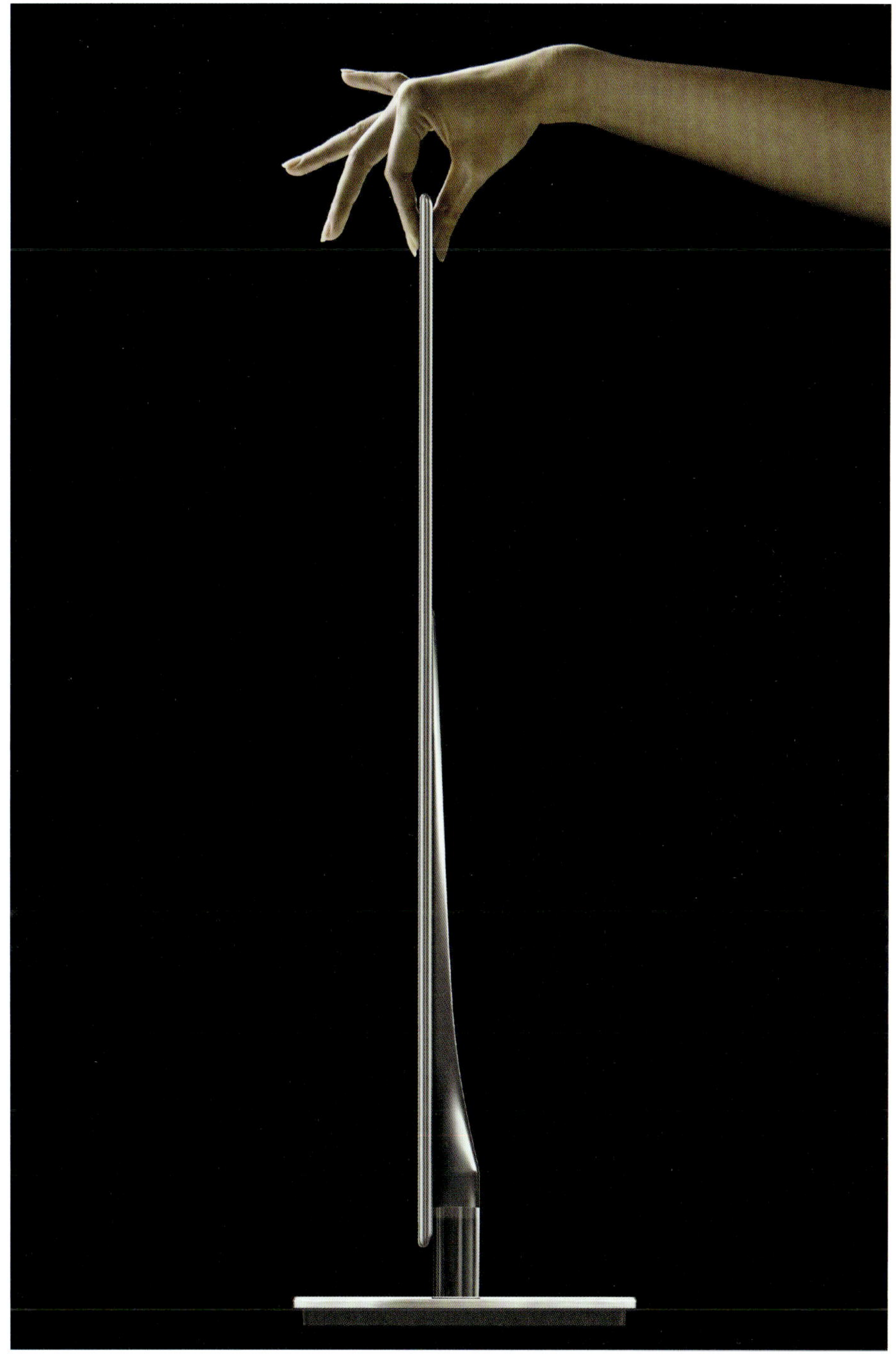

SAMSUNG THE NEW 9000 SERIES LED TV Watched Fashion Week 12:30pm Rue Saint Honoré Paris
THE NEW LED TV 9000
0.3 INCH THIN SIDE PROFILE
ULTIMATE TECHNOLOGY. ULTIMATE BEAUTY. ULTIMATE EXPERIENCE
ONLY FROM SAMSUNG THE WORLD'S MOST WATCHED BRAND
WWW.SAMSUNG.COM
SAMSUNG
TURN ON TOMORROW

SAMSUNG THE NEW 9000 SERIES LED TV Watched Farewell Tour 10:00pm Pacific Heights San Francisco
THE NEW LED TV 9000
0.3 INCH THIN SIDE PROFILE
ULTIMATE TECHNOLOGY. ULTIMATE BEAUTY. ULTIMATE EXPERIENCE
ONLY FROM SAMSUNG THE WORLD'S MOST WATCHED BRAND
WWW.SAMSUNG.COM
SAMSUNG
TURN ON TOMORROW

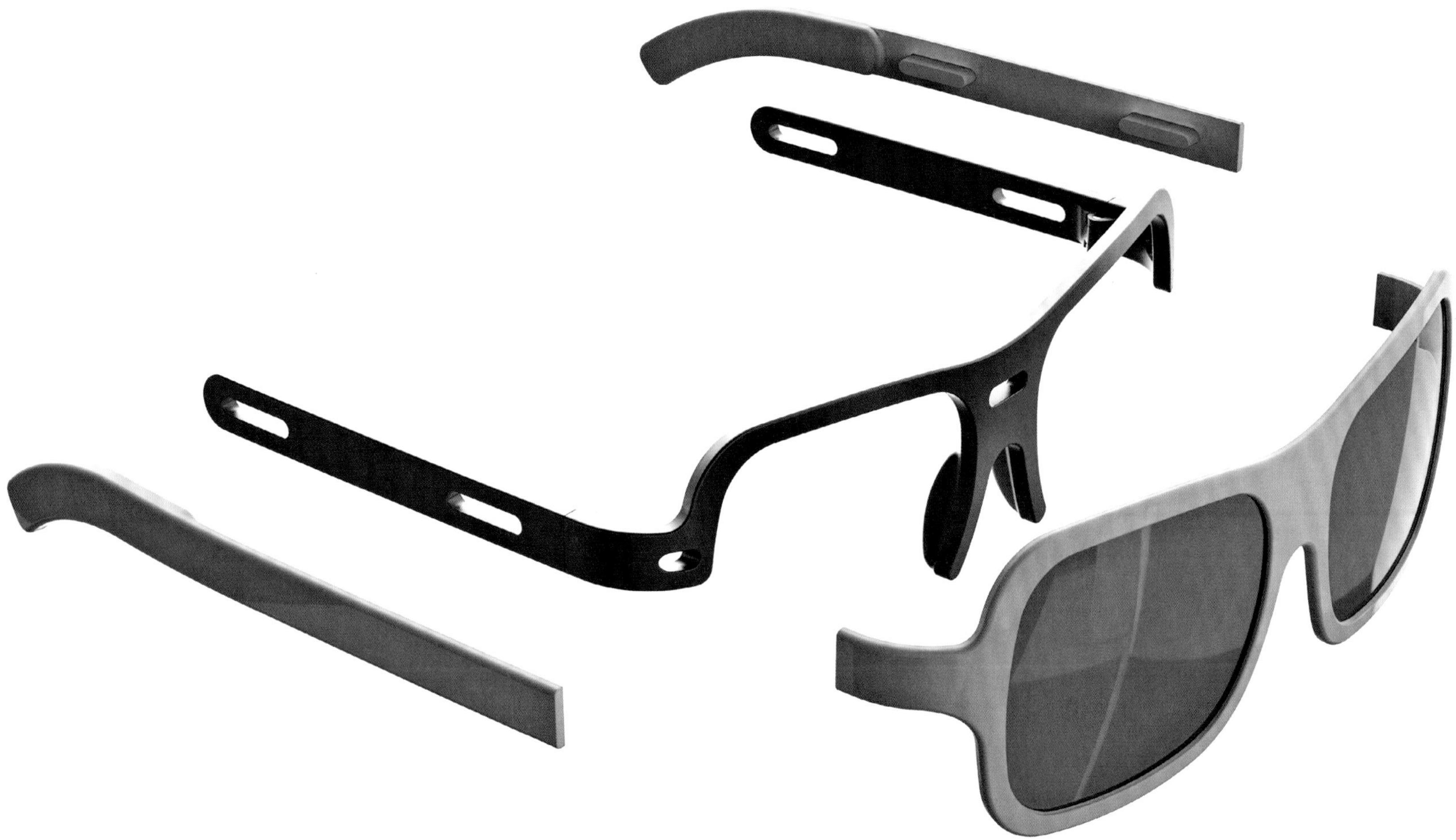

ENTERTAINMENT OPTICS

2010

The 2009 release of the mega-blockbuster movie *Avatar* instantly brought 3-D filmmaking to mainstream attention on an unprecedented scale. The dazzling, immersive world that director James Cameron brought to life using this technology defied filmgoers' expectations about a 3-D experience, and kickstarted the widespread adoption of this technology not only in film, but in gaming, sports, fashion, live television events, and beyond.

The Entertainment Optics (EO) glasses concept was designed to help usher in the next generation of 3-D viewing by offering an enhanced viewer experience that transformed audiences from passive spectators to engaged participants. With EO, the viewer was able to become truly immersed in on-screen worlds by actually inhabiting the identities of the characters that they were watching.

Every pair of EO glasses shared the same basic universal subframe that could be individualized in limitless ways with clip-on frame fronts, customizable temple plates, and interchangeable lenses. The flexibility of the EO subframe allowed for maximum versatility in branded partnership opportunities: limited edition, collectible faceplates could easily be snapped on and off of the universal subframe, while swappable temple fronts allowed for additional brand activation, messaging, and sponsorship opportunities. The lenses, meanwhile, could be easily popped in and out to suit in-theater, at-home, or outdoor entertainment environments. The EO system provided an inexpensive, flexible way to introduce novelty—and increased brand awareness—to every kind of viewing event.

Collaborative partnerships were explored with Disney, Pixar, Marvel, ESPN, DreamWorks, and numerous others in order to leverage their vast troves of intellectual property and enable a unprecedented level of audience engagement across every demographic, and every type of media.

2D
3D
EO

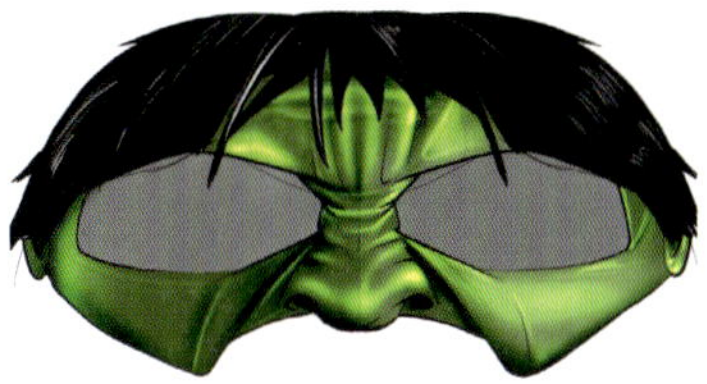

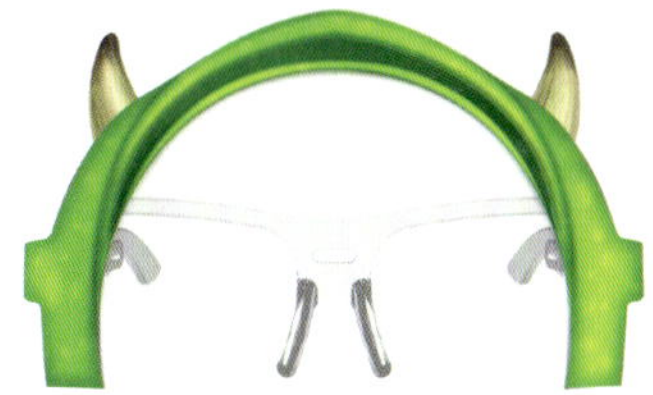

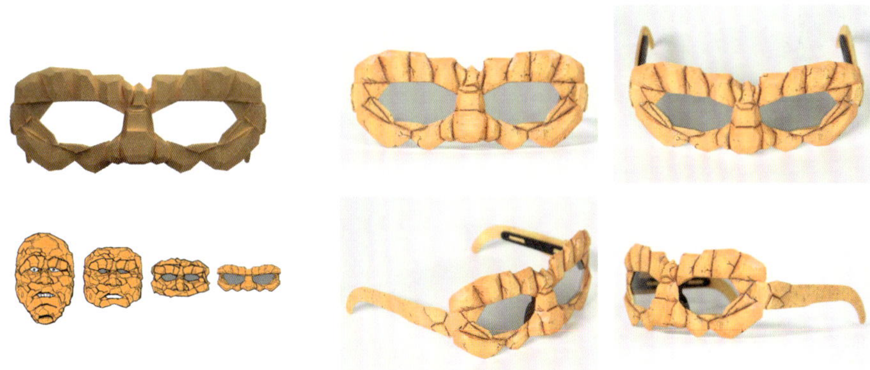

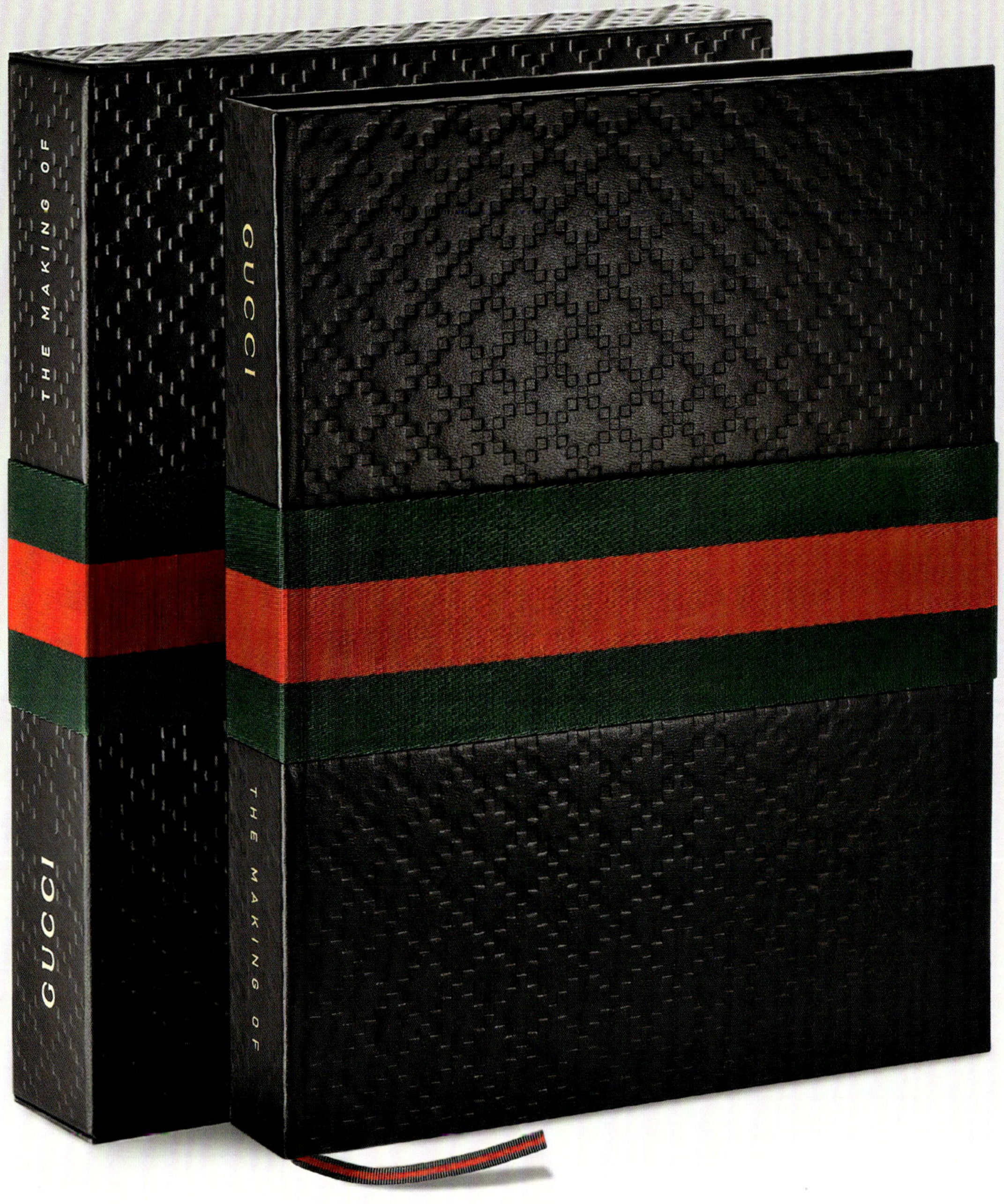
THE MAKING OF
GUCCI
GUCCI
THE MAKING OF

GUCCI

2011

Gucci is world-renowned for its impeccable Italian craftsmanship and perfect melding of classic good taste with adventurous, contemporary style. At every point throughout its long and storied history, Gucci has successfully evolved to define the style of the day, becoming synonymous with a very particular kind of elegance and glamour that is timeless and timely in equal measure.

To celebrate the company's 90th anniversary and commemorate the rich history of this brand, an authoritative publication on the history of Gucci was designed, curated, and coedited. With unrestricted access to the company's vast archives, numerous forgotten or overlooked masterpieces—of fashion, advertising, branding, and design—were finally given the platform and attention they deserved.

Published by Rizzoli International in 2011 to coincide with the opening of the Gucci Museo in Florence, Italy, this exhaustive overview of Gucci's history paid proper tribute to the enormous breadth of work over the past ninety years and set the stage for an equally prolific future as one of the preeminent luxury brands of the twenty-first century.

LIFTBOY

Handbags, suitcases, and luggage are all representations [illegible]

[illegible]

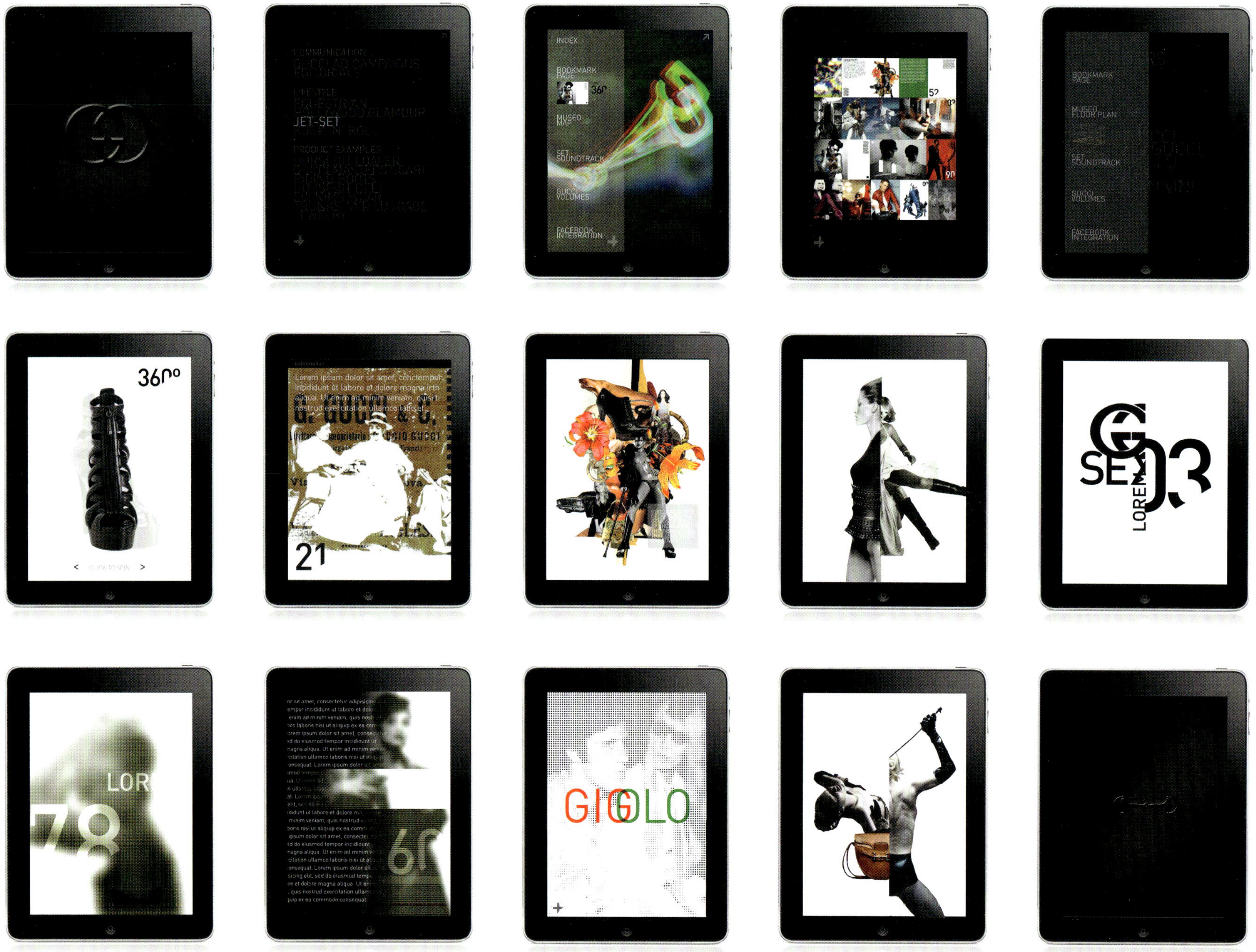
JET-SET
INDEX
BOOKMARK PAGE
360
MUSEO MAP
SET SOUNDTRACK
GUCCI VOLUMES
FACEBOOK INTEGRATION
BOOKMARK PAGE
MUSEO FLOOR PLAN
SET SOUNDTRACK
GUCCI VOLUMES
FACEBOOK INTEGRATION
360°
21
SEX
03
LOREM
LOR
78
60
GIGOLO

GUCCI MUSEO

2011

When Gucci moved its design headquarters to Rome from Florence, its former home in the Piazza della Signoria was reborn as the permanent home of Gucci's history and heritage, the Gucci Museo.

To help celebrate over ninety years of Gucci history, a complete visitor experience program was designed to showcase Gucci's most significant moments in fashion and culture. Traditional artifact-based exhibitions were supplemented by interactive digital displays, intelligent wayfinding, and smartphone and tablet integration that guided exhibition visitors to extended mobile and online content.

A new identity system for the museum was developed, featuring the Gucci Knight: a subtly refreshed version of a classic brand icon. The Knight, holding a suitcase in one hand and a travel bag in the other, represented the origins of the brand, while the shield behind him—a ship's wheel and a rose—represented the careful stewardship of the Gucci family. This traditional symbol, paired with a fresh, modern, serif type, announced the Gucci Museo's identity as an institution that was deeply respectful of its heritage, but thoroughly forward-looking at the same time.

A new range of branded merchandise bearing this identity was also developed to be sold exclusively at the Gucci Museo bookstore that adjoined the museum.

GUCCIMUSEO

GUCCIMUSEO

GUCCIMUSEO
GUCCIMUSEO
GUCCIMUSEO

RUSSIAN STANDARD VODKA

2011

Paying homage to the original Russian Imperial Standard for vodka production set by Czar Alexander III in 1894, Russian Standard's proprietary recipe calls for the highest-quality natural ingredients to be brought together at a state-of-the art distillery in Saint Petersburg, where they then pass through a proprietary distillation process more than two hundred times to arrive at a state of unrivaled purity and smoothness.

To help introduce this premium spirit to the North American market, a campaign concept was developed that helped to distinguish Russian Standard from its many competitors. A series of arresting black-and-white images showed private moments of reflection and quietude—a stark contrast to the provocative style of other premium vodka brands and their elaborate promises of adventure and escape. This quiet, intimate, sensual approach immediately differentiated Russian Standard from its competitors, and cemented its reputation for understated refinement and the utmost quality.

BELVEDERE VODKA

2011

When it first arrived on the global market, Belvedere Vodka was one of the best-kept secrets in the world of spirits. Selective distribution had ensured that the brand—marketed as the world's first super premium vodka—lived up to its exclusive reputation.

The select customer base, who did know about it, began developing its own language when talking about the brand—referring to it as "Belve" when placing their orders. To let the world in on the secret, a campaign was designed that featured provocative, playful, seductive imagery, and the invitation to become an insider and "Make it Belve."

DON JULIO TEQUILA

2012

Since 1942, Don Julio Tequila has been produced at the La Primavera Distillery in Mexico's Jalisco province. Its founder, Don Julio González, began distilling tequila in at age seventeen, and he spent the next forty years improving his craft by tirelessly experimenting with every aspect of the tequila-making process to produce the perfect product.

These stories about González's exacting approach to tequila-making had already taken on an almost mythical status among tequila connoisseurs, which presented an opportunity to elevate the status of this master distiller to that of a true legend. A series of television spots was developed, featuring a young Don Julio roaming the beautiful but desolate countryside of the Jalisco region: a mysterious, stoic, solitary figure who guards the secret to the world's best tequila.

THEORY

2012

Theory was founded on the principle that professional men and women ought to feel comfortable and sophisticated in innovative, highly functional, and stylish clothing that they could wear to work—and everywhere else.

To help bring the label's versatile, sophisticated, easy-to-wear ethos to life, a series of brand communications and new digital experiences was developed, which evolved Theory's message and expanded its global presence. All of these activations were designed to help bring to life the brand's unique air of unstudied elegance that has helped to redefine the uniform for a new generation of professionals.

theory

theory

theory

38

THEORY 38

2012

Theory 38 was developed as a performance-based sister brand to Theory, which complemented the classic American sportswear style of the main line.

Utilizing innovative fabric technology and construction techniques, Theory 38 incorporated revolutionary materials into sleek, sporty clothing that transcended the formerly hard divide between sportswear and lifestyle brands. Elevated enough to wear as part of a daily wardrobe, but high-performing enough to be worn to the gym, Theory 38 was one of the first brands to recognize the potential of high-quality, beautifully designed activewear to become a part of people's everyday wardrobes.

Named for the birthplace of the line and the home of the Theory brand, 38 Gansevoort Street in New York City's Meatpacking District, Theory 38's identity architecture was designed to exude an elegant, elevated, and refined style—as timeless as the main Theory line—but with a bold, progressive twist.

Launch communications featured models striking exaggerated, balletic poses: classically beautiful yet totally contemporary imagery that focused on the freedom, range of movement, and elevated style that Theory 38's clothing offered.

38

theory 38

3

du play
du dine
du swim
du it all
dubai
in dubai, the only question is, what will you do next? will you swim with the dolphins? enjoy a 5 star spa experience while your kids play at the world's biggest waterpark? will you experience the magic of a unique culture and tradition? why choose? du it all.
Ask about our Kids Stay and Play Free Programs. More fun for them. More time for you.
D
Emirates

du swim
du dine
du play
dubai
in Dubai, the only question is, what will you do next? will you eat in one of the many 5 star restaurants? shop in the worlds best boutiques? will you experience a magical culture and tradition? maybe you'll watch the worlds biggest horse race, swim with dolphins, or sunbathe on a white sand beach. or maybe, you'll du it all.
D
Emirates

DUBAI

2013

Offering unique cultural experiences, exciting world-class entertainment, and dazzling global cuisine, Dubai is the ideal gateway to the Middle East for visitors from around the world.

The "DU It All" campaign concept, developed and presented to the Tourism Authority of Dubai, highlighted the infinite range of activities visitors can enjoy while immersed in the magic of this truly world-class city.

Striking images of Dubai's premiere attractions, framed within sleek, modernist, block letter forms, instantly communicated the essence of the Dubai experience: a place where visitors can have it all, see it all, experience it all, and indeed, DU it all.

PETER ARNELL PHOTOGRAPHY 1984-2014

2014

In 2014, Milk Gallery presented a mid-career retrospective of Peter's photographic works at its gallery space in New York's Chelsea district. While his commercial works had been included in various exhibitions over the years, this solo exhibition was the first opportunity to show the world the full breadth of his noncommercial photography practice.

The show, curated and designed by Peter's lifelong friend and collaborator Frank Gehry, featured free-floating, unadorned wood stud wall partitions placed at irregular angles throughout the space. This layout encouraged a meandering approach through the exhibition, creating a dynamic superimposition and juxtaposition of images from every vantage point that defied straightforward chronology and obvious thematic groupings.

Including commercial work, fine art archival prints, and vast troves of never-before-seen personal photographs, the show paid tribute to the prolific scope of Peter's a image-making and honored his relentless pursuit to capture beauty in all its forms.

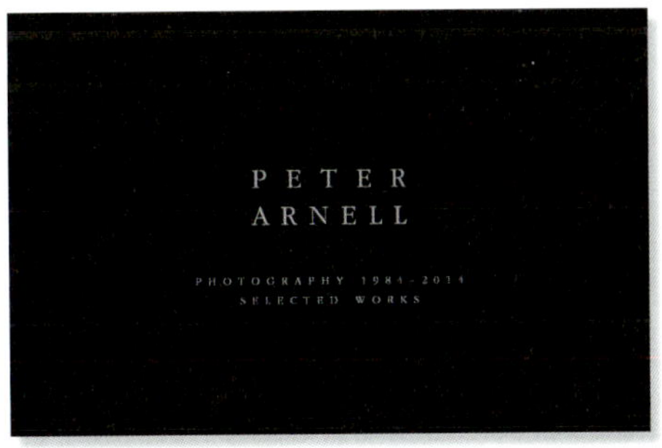

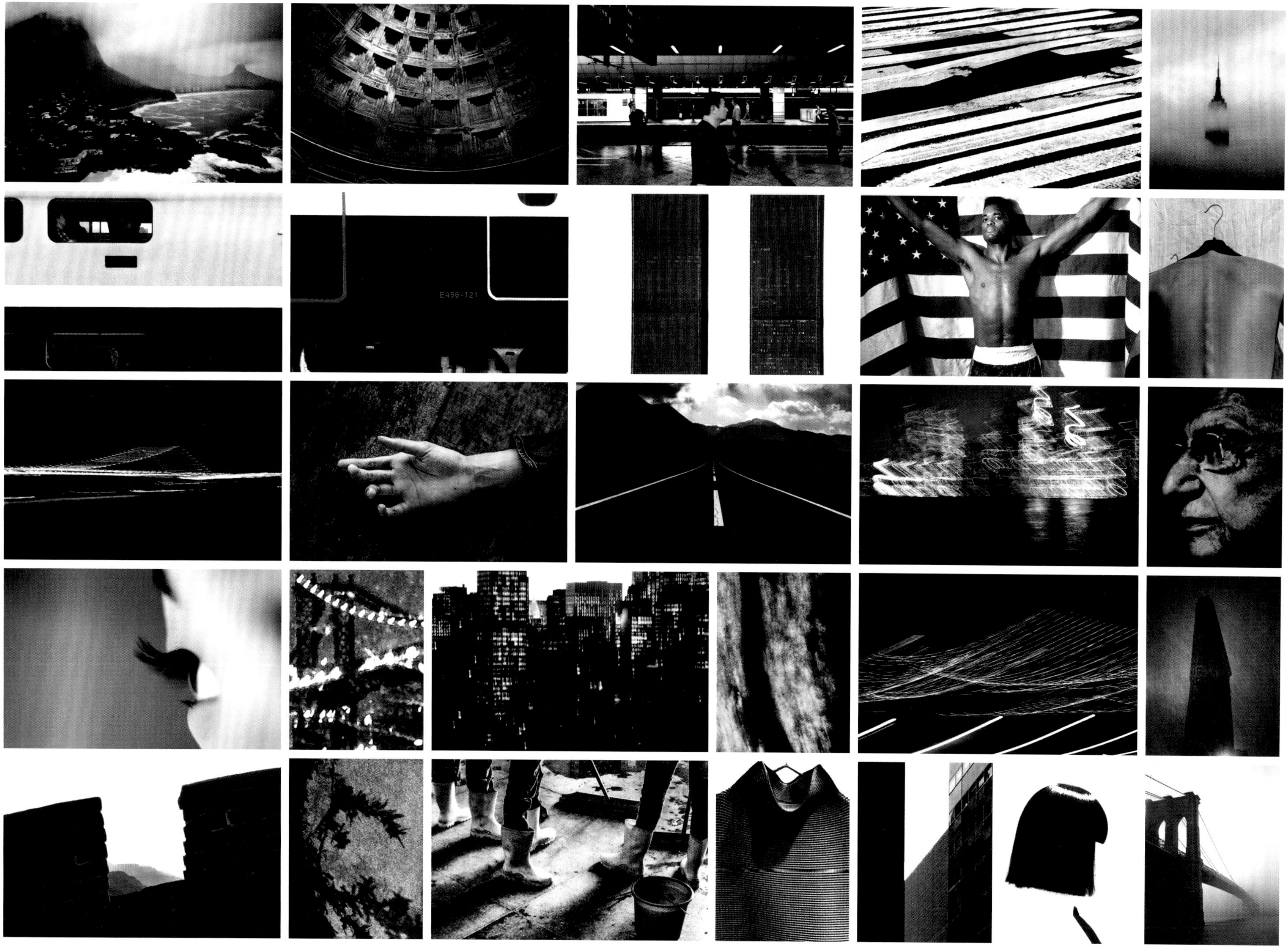
E456-121

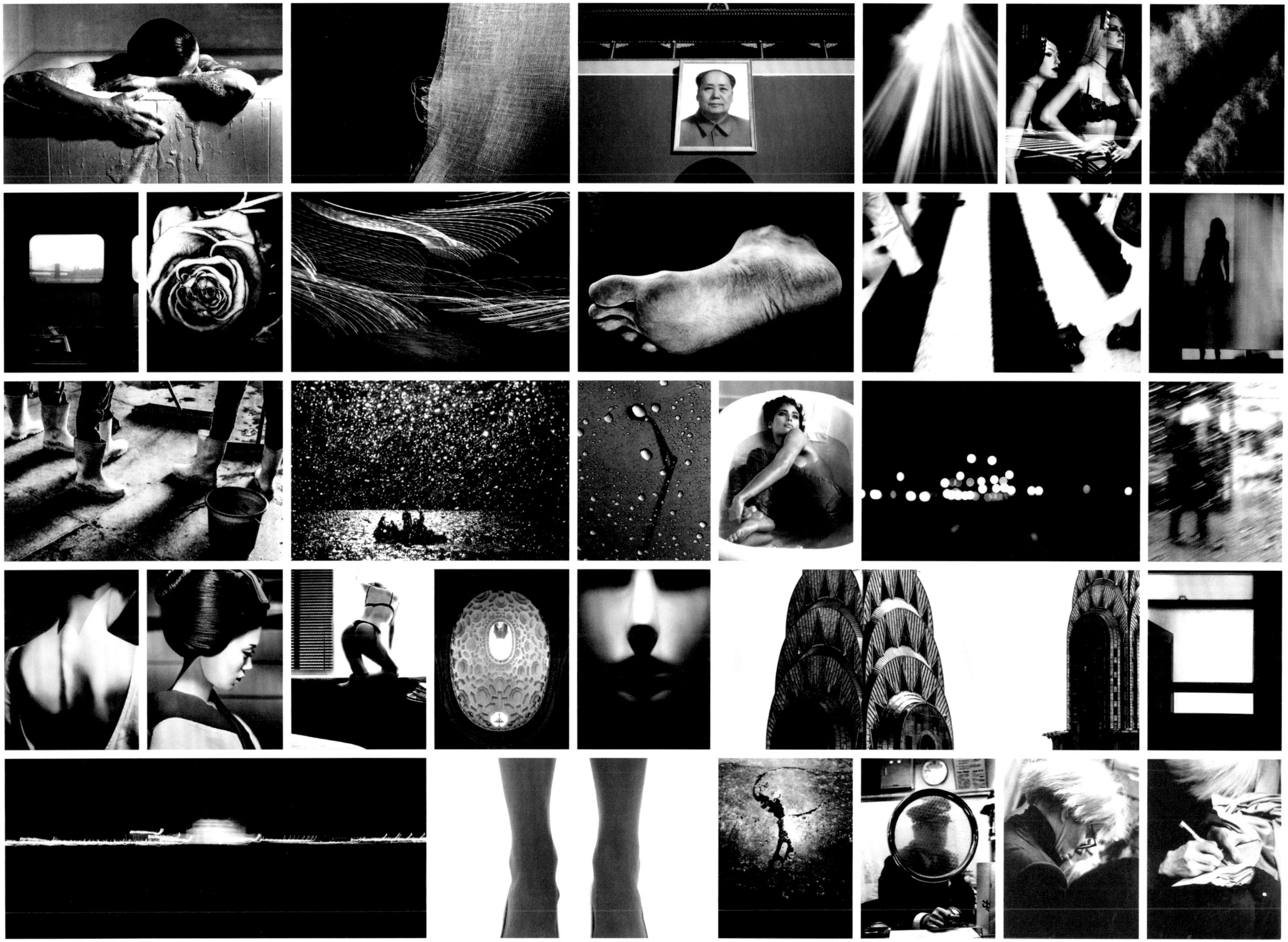

WAIT TO BE SEATED

THAIWORKS

2014

The ThaiWORKS heritage-based product and experience program was developed for the nation of Thailand in collaboration with Thaksin Shinawatra, the country's former prime minister.

Over the years, Thailand had grown into an increasingly popular destination, drawing millions of visitors from around the world. Yet there was still so much about the country that remained misunderstood, overlooked, and underrepresented on the global stage. Thailand's uninterrupted, centuries-long civilization has produced one of the great cultural and artistic legacies of the world, albeit one that has long been under-appreciated relative to its impact.

The ThaiWORKS program was conceived of to respectfully and sensitively leverage that priceless national resource, allowing the true beauty of the Thai people, along with their art and culture, to become better known and appreciated worldwide. By providing a unifying brand vision, single marketplace, and putting the correct operational frameworks in place around production and manufacturing, ThaiWORKS sought to unlock the "soft power" potential of Thailand, and to rightfully make Thai design globally synonymous with beauty, functionality, and sustainability.

A series of sector-spanning integrated initiatives created a vast network of Thai artists, artisans, designers, innovators, and entrepreneurs, and provided a common platform and marketplace to sell their wares, along with the guidance and resources to incubate their burgeoning businesses.

The ThaiWORKS ideology of "Beautifully Functional" was developed to speak to consumers' desire for products of quality, integrity, and authenticity. The program's aim was to spur world-class innovation in all sectors of Thailand's economy through the development of distinctly Thai proprietary products and experiences, creating a new and truly inspired global brand of, for, and by the nation of Thailand.

THAI
WORKS

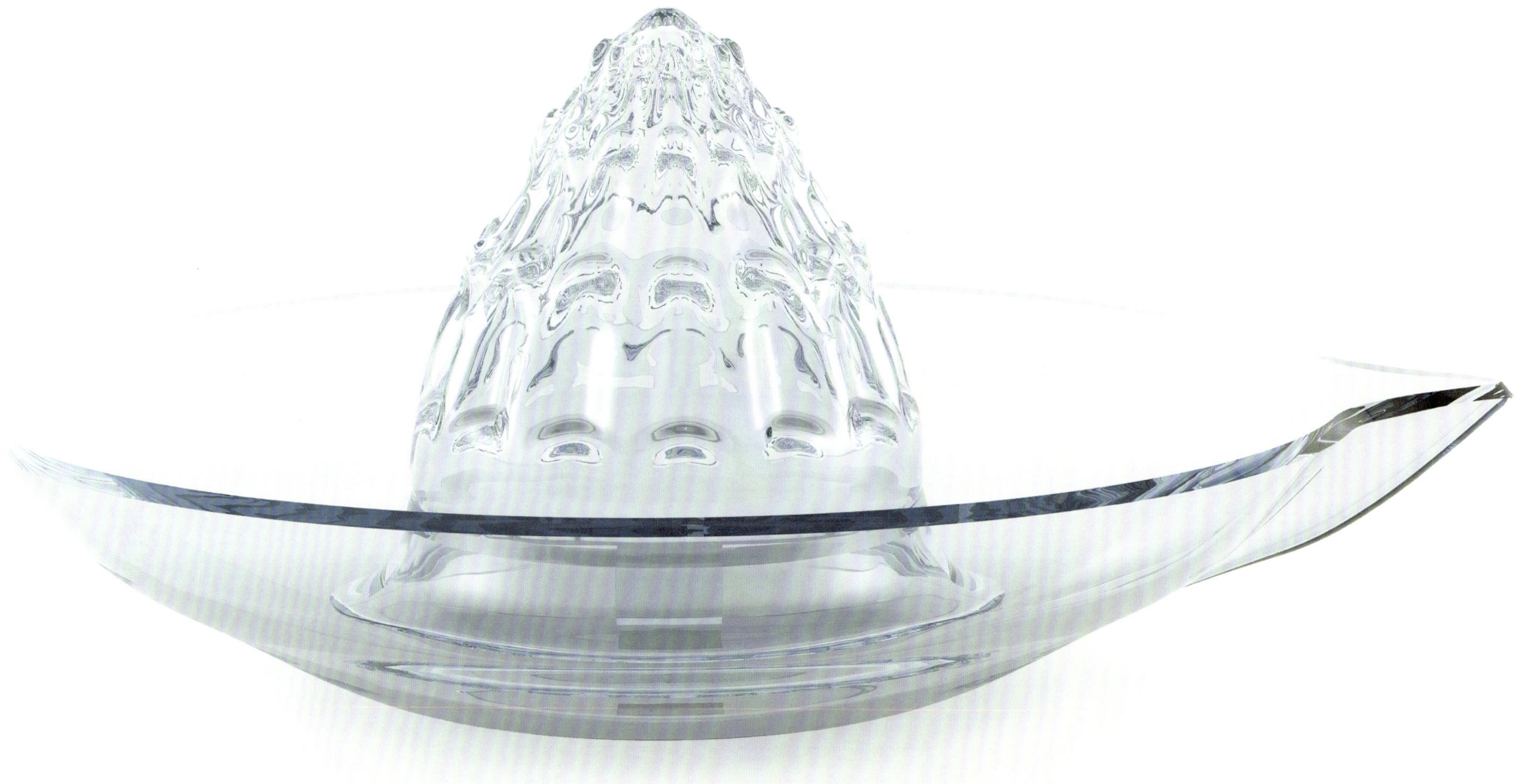

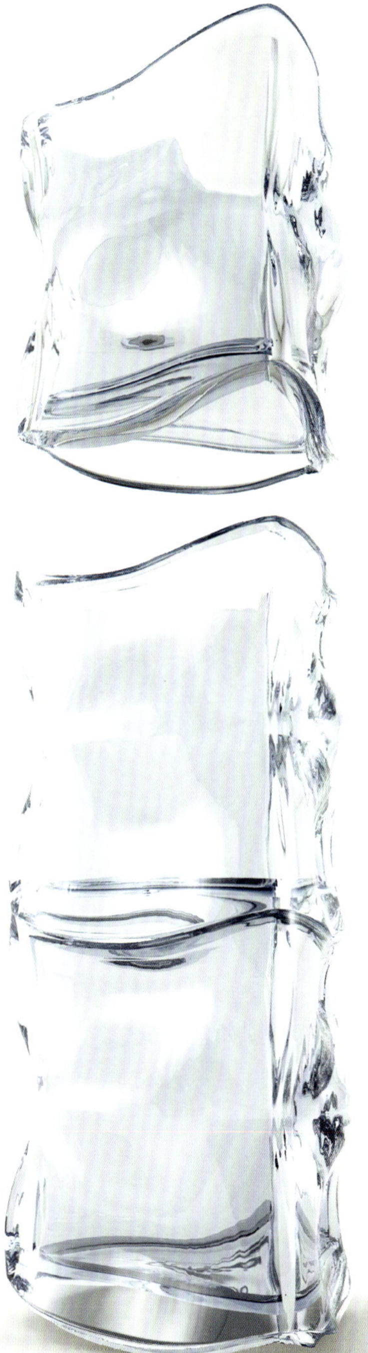

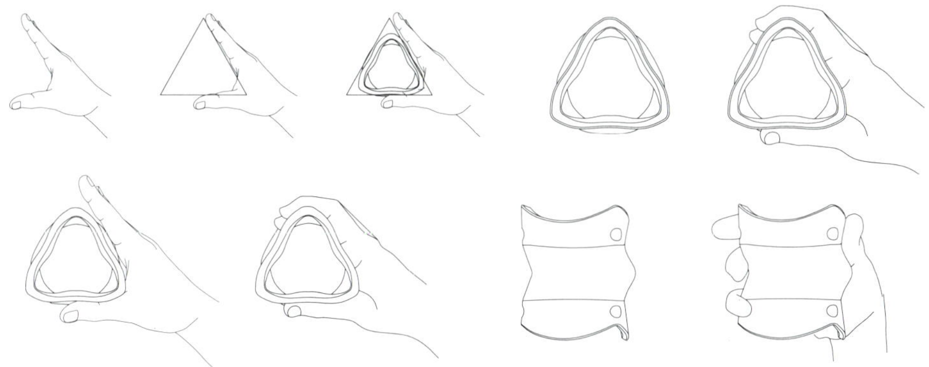

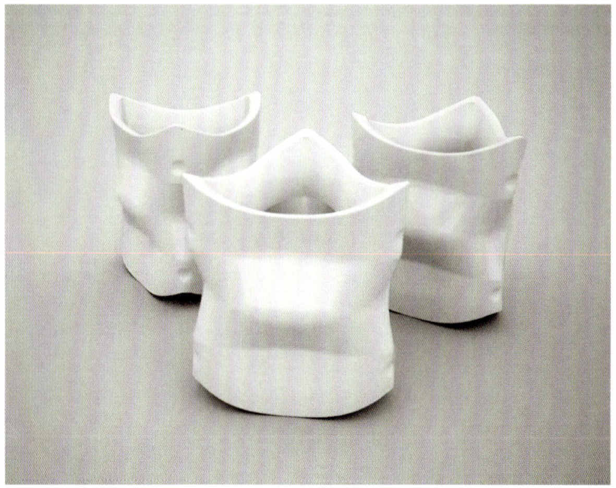

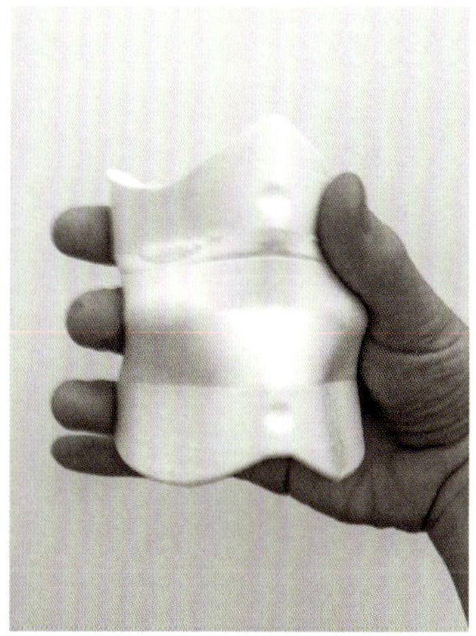

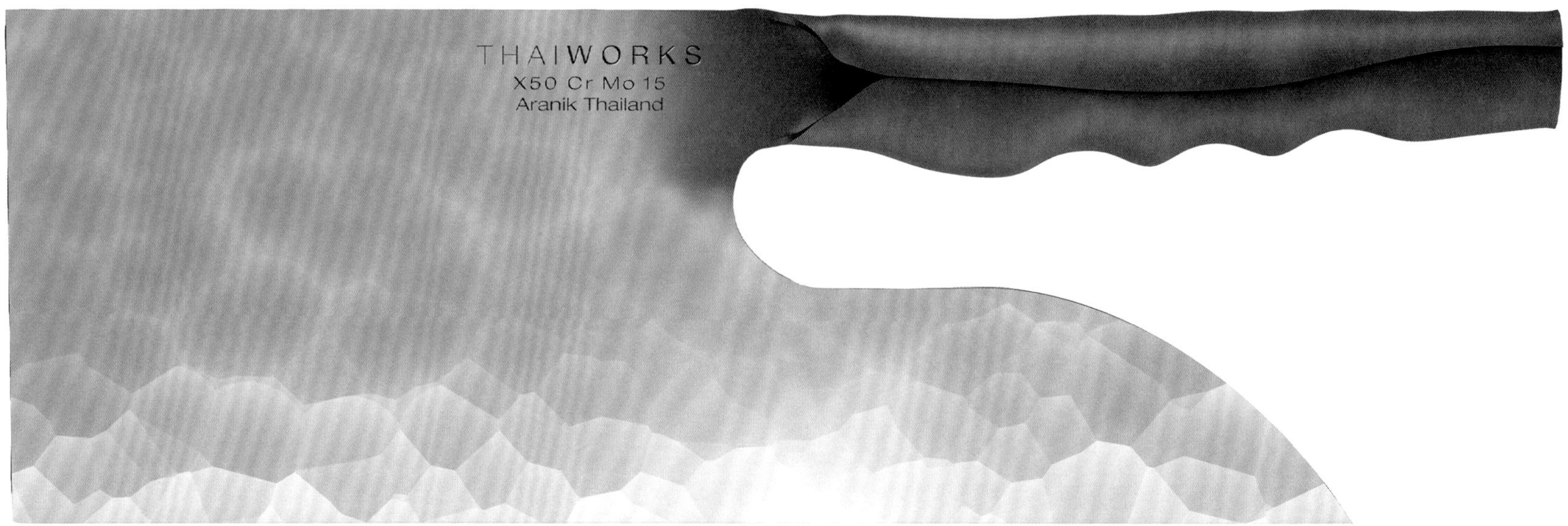
THAIWORKS
X50 Cr Mo 15
Aranik Thailand

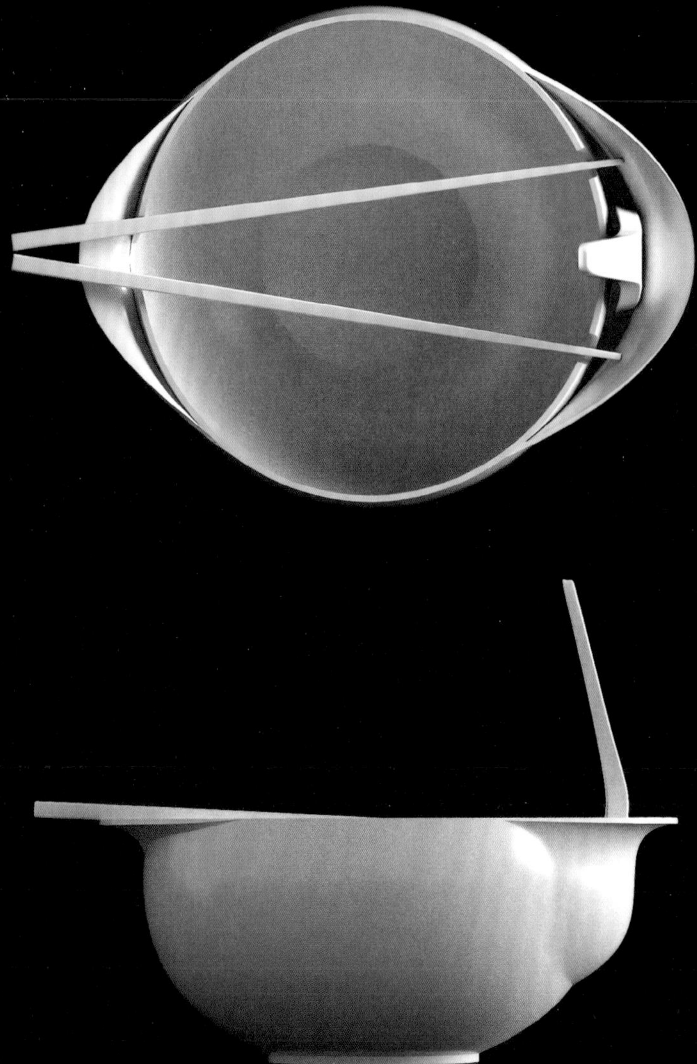

THAIWORKS THAI MASSAGE TOOLS

2014

Thai massage, one of the four branches of traditional Thai medicine, has been practiced for centuries as a holistic healing technique. This ancient method has become popular worldwide for its restorative effects, and for the distinctly Thai spirit of hospitality and generosity that this ritual embodies.

For ThaiWORKS, a set of contemporary massage tools were created to make the distinctively Thai experience of complete relaxation and restoration available as an everyday indulgence in one's own home. Crafted from beautiful and durable natural materials, such as solid hardwoods and Thai textiles, this sleek and contemporary range of tools provided a welcome and accessible introduction to a time-honored wellness practice.

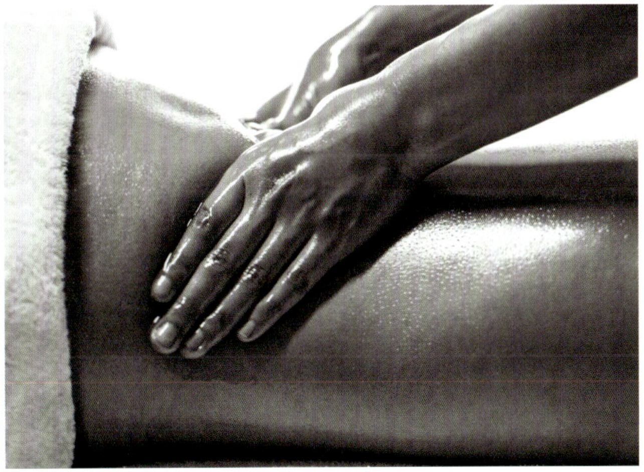

THAIWORKS SANCTUARY AROMATHERAPY

2014

The ThaiWORKS Sanctuary Aromatherapy line, a range of eighteen custom aromatherapy fragrances, was developed as part of the ThaiWORKS initiative. Inspired by the rich fragrances of Thailand, scents like cinnamon, cardamom, clove, neroli, ylang-ylang, banana flower, and jasmine were selected for their beautiful bouquet, and for their soothing and healing properties rooted in traditional Thai medicine.

ThaiWORKS scent pendants were also developed as a beautiful and functional personal delivery mechanism for the Sanctuary line of aromatherapy fragrances. Inspired by the forms of traditional Thai earthenware vessels, the wearable pendants allowed the wearer to enjoy a calming, soothing, personal aromatherapy experience at any time, no matter where they were.

THAIWORKS VILLAGE

2014

ThaiWORKS Village was envisioned as an ecotourism destination that would protect and preserve the outstanding natural beauty of the Chiang Mai province in northern Thailand, while providing the resources for local craftspeople to produce and sell their exquisite, one-of-a-kind wares.

A conceptual visualization of the future Village borrowed traditional forms from vernacular Thai architecture and reimagined them as a spectacular tree-top complex nestled into the dense, lush jungle of northern Thailand, where many of the most skilled craftspeople and artists in the country already operated. The complex was designed to have a minimal impact on the landscape and to be completely sustainable and eco-friendly—both in its construction methods and long-term operational model. Featuring extensive use of local materials and borrowing from traditional construction techniques, the ThaiWORKS Village was designed to be a living testament to the ingenuity and brilliance of Thai craftsmanship.

At the ThaiWORKS Village, local artisans were to be provided with not only the facilities, tools, and financing to produces their wares, but also the scalability models and distribution networks to bring their goods to a global audience. Largely open-air workshops were designed to be accessible to the public, where the artisans would be able to sell their products to customers directly on site. This complete elimination of middleman distribution channels not only provided the largest possible economic return to the artists and artisans, but it also created the opportunity for visitors to truly appreciate the exceptional artistry and passion that went into the making of each unique ThaiWORKS product.

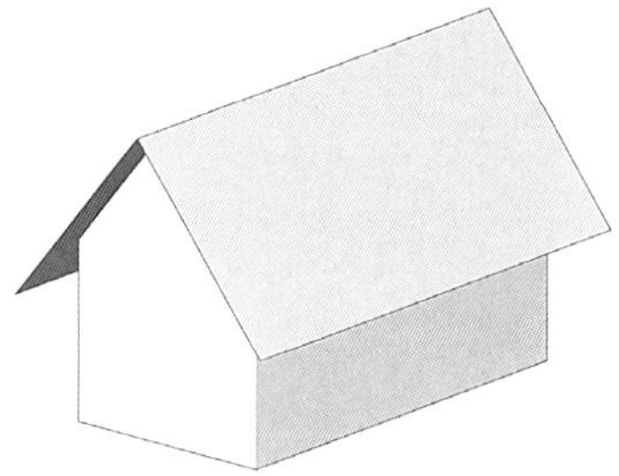

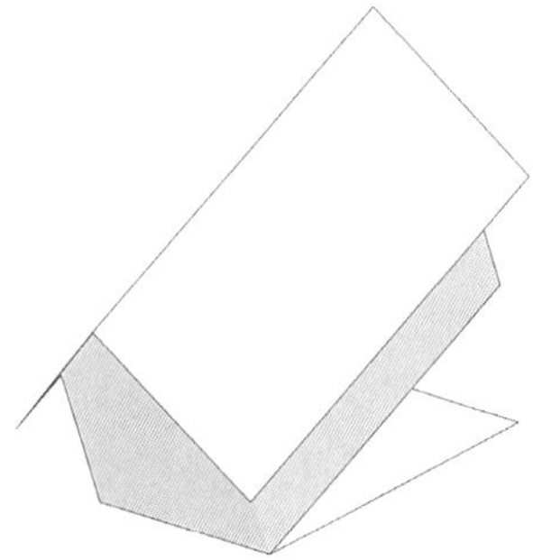

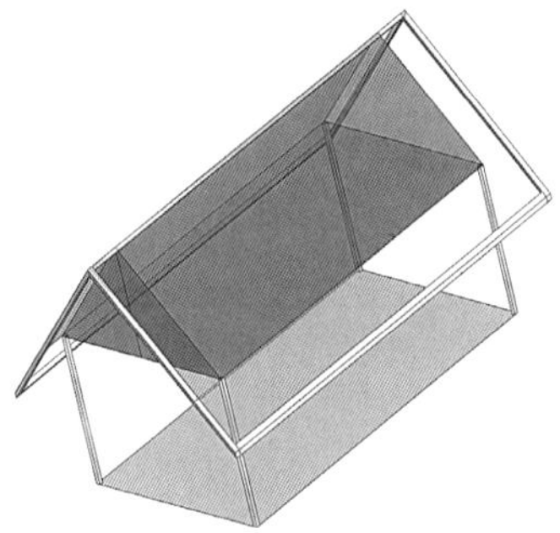

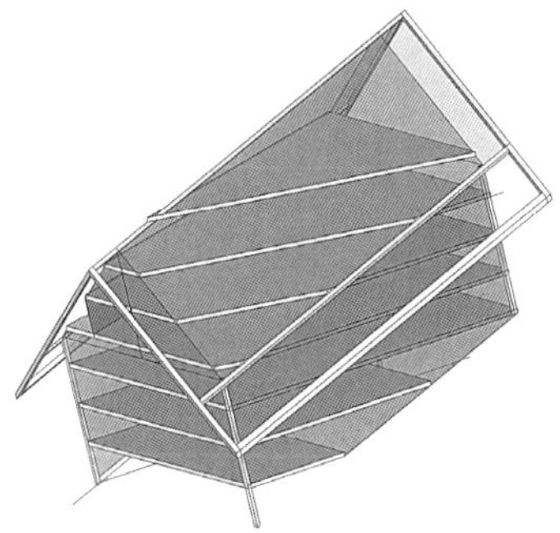

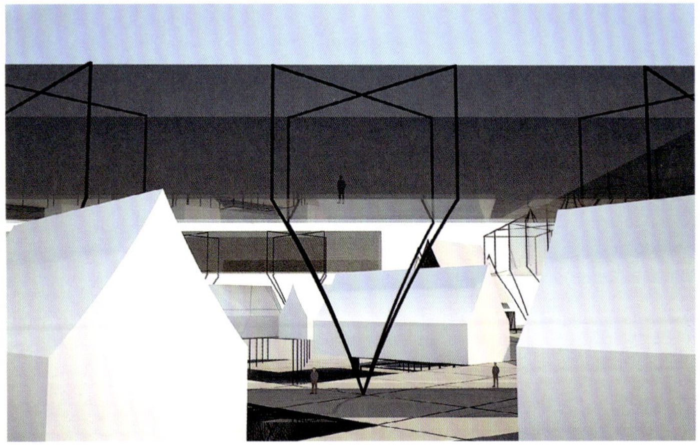

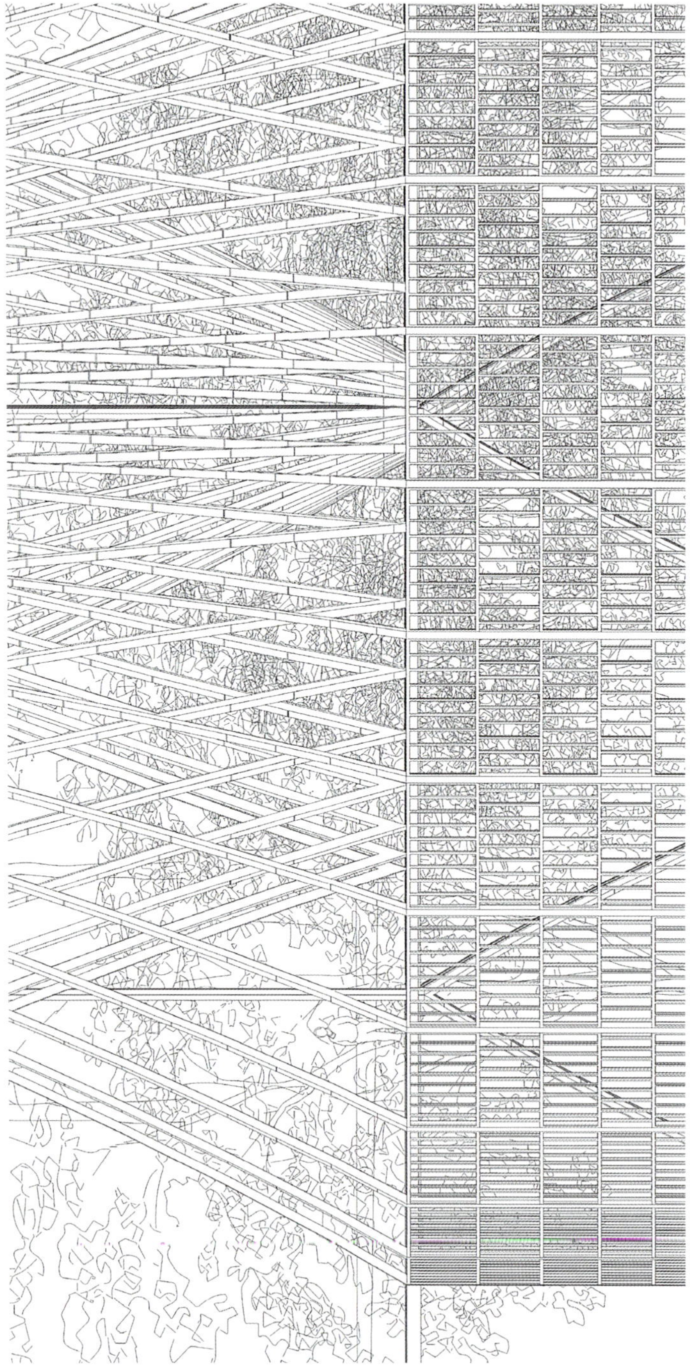
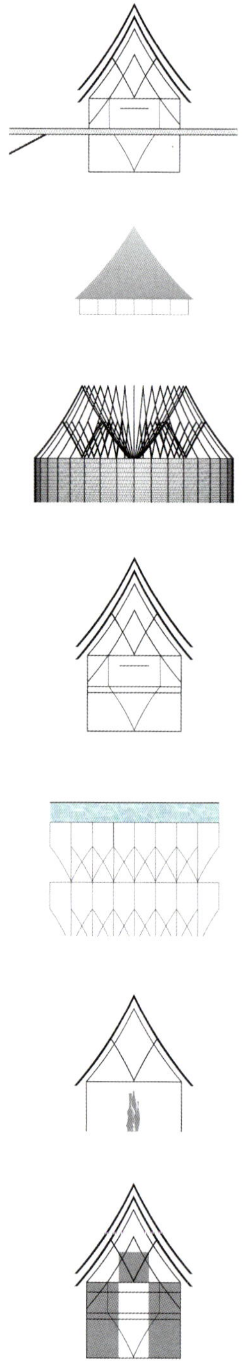

THAIWORKS
VILLAGE
KHAOYAI

THAIMOON SAKE

2014

Owing to its widespread cultivation of rice, Thailand has a thriving domestic sake market—a fact that is largely unknown outside of Thailand. As part of ThaiWORKS' overarching commitment to scale up Thailand's promising but underleveraged soft-culture exports, native Thai sake was earmarked as a product with the potential to become a truly global export.

The concept for ThaiMoon Sake was developed to offer consumers the chance taste an entirely new type of sake—as high quality as any Japanese sake, but with a distinctly Thai flair. Elegant bottle designs inspired by the natural beauty of the Thai countryside and its terraced rice fields provided a subtle link between the final product and its place of origin, and offered an immediate point of distinction from its many international competitors.

TUK TUK

2014

Tuk Tuks—three-wheeled passenger vehicles commonly used in Thailand and throughout much of Southeast Asia—are an essential mode of transportation that millions of people depend on every day. Despite their small size, they're often noisy, fuel-inefficient, and are decidedly not environmentally friendly.

As part of an effort to transform a low-tech polluter into the clean-air, zero-emission, all-electric vehicle of the future, a design concept was developed that completely redesigned and reengineered the Tuk Tuk's unique form factor.

A next-generation vehicle for drivers, passengers, and cargo alike, the all-electric Tuk Tuks offered much lower operating costs, greater cargo and passenger capacity, and dramatically improved durability, all while reducing air pollution in some of the most densely populated urban areas in the world. The new Tuk Tuk was designed to prove that a functional, durable, and cost-competitive vehicle could also be well-designed, technologically advanced, and environmentally sustainable.

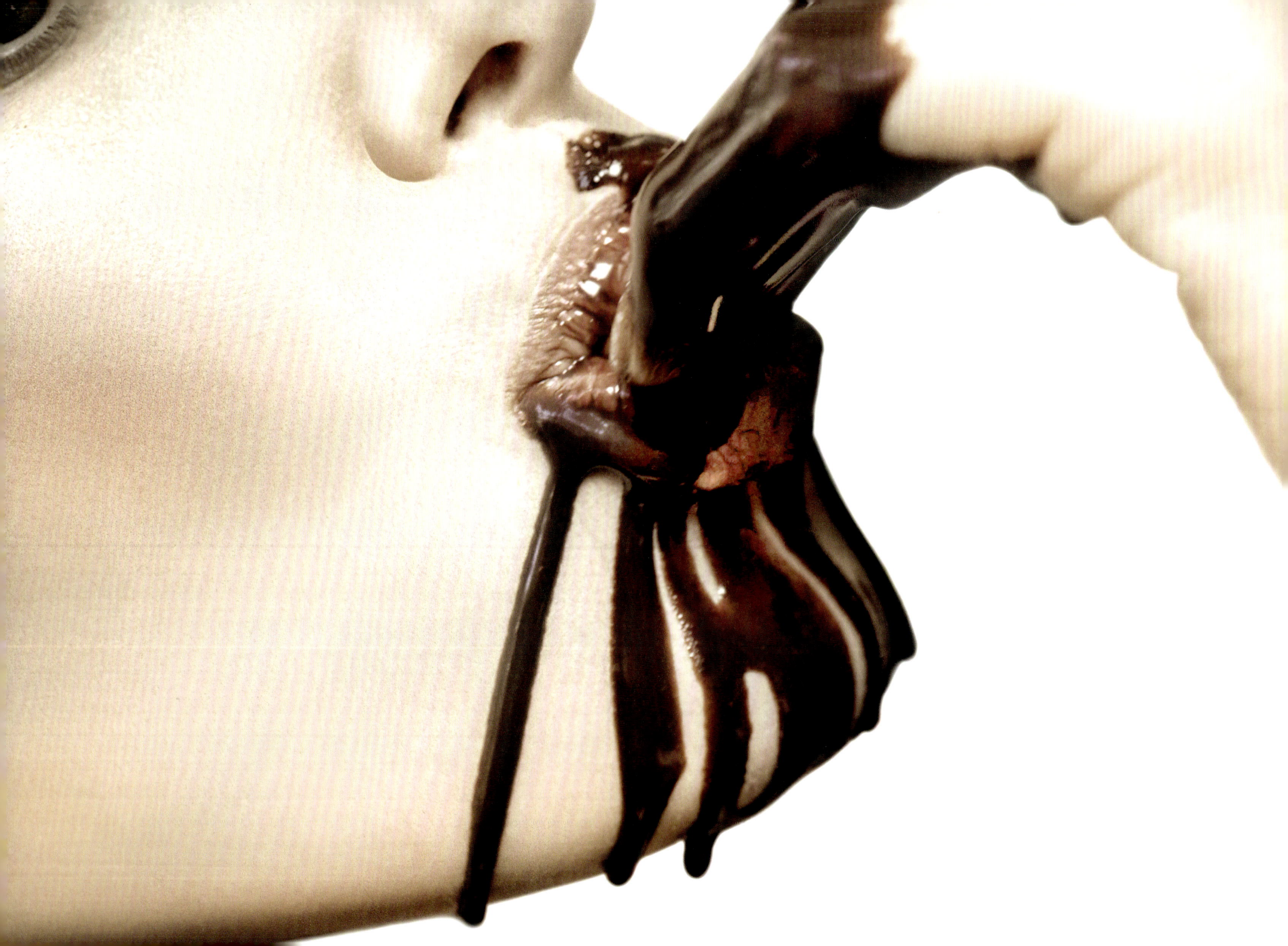

CHEZ BON BON

2014

Originally opened at the Fontainebleau Hotel in 1954, pastry shop Chez Bon Bon was renowned for its elegant pastries and its sleek, stylish retail space. Even years after the original shop had closed, it remained a much-missed Miami Beach institution, and an indispensable part of the Fontainebleau story. When the decision was made to reopen this iconic *patisserie*, an entirely new brand identity was developed to mark this new chapter in the shop's storied history.

The new brand mark featured the iconic bowtie motif used elsewhere at the hotel, along with a custom typeface that captured the exuberant spirit of mid-century design—while still being completely fresh and contemporary. New product presentations, packaging, uniforms, menus, store signage, and merchandise were also designed, helping to give Chez Bon Bon its own distinctive, playful sense of glamour, while still respecting the style, spirit, and history of the Fontainebleau.

DENFORD PARK
PURE & NATURAL HONEY

DENFORD PARK HONEY

2014

Denford Park is a country house and estate in Berkshire, England that boasts beautiful Georgian architecture, exquisitely landscaped grounds, lavish formal gardens, and a working honey farm. An opportunity was identified to commercially market the honey produced by the Denford Park apiary, and to help bring this proposed new product range to life, a brand identity was created that captured the special beauty, quiet elegance, and rich history of this centuries-old estate, along with a series of conceptual packaging designs, which evoked the organic forms of the honeycomb, but in a striking, contemporary way.

DENFORD PARK

PURE & NATURAL HONEY

AL-ULA

2014

Al-Ula is an archaeological site of global significance located near the northwestern region of the Arabian Peninsula. A main stop along the incense trade route in pre-modern times, it boasts vast stretches of incredibly well-preserved ancient ruins and extensive wall drawings left by the Nabataean kingdom some two thousand years ago. But until recently, this spectacular UNESCO World Heritage Site remained virtually unknown outside of the region, and the surrounding area had little to no infrastructure in place to accommodate the modest but steadily growing stream of visitors.

Presented to the Tourism Authority of Saudi Arabia, Epic Arabia was a strategic master plan proposal and a luxury eco-resort concept designed to attract visitors to this historically fascinating, largely undiscovered destination.

The resort's plans reflected a modern interpretation of this region's history as a civilizational crossroads, and drew from both the vernacular architecture of the region and the spectacular, monumental architecture of the nearby ruins. A series of private villas were arranged within a lushly planted, multi-level walled compound that recalled both the labyrinthine character of a traditional *souk*, and the lush calm of a desert oasis.

The resort, directly influenced by the art, architecture, and cultural values of Arabia's history and people, was at once a tribute to the rich traditions that have informed this beautiful, untrammeled corner of the world, and a beacon of a more open and global future.

CAMELUS

CAMELUS

2015

Camelus Automotive was conceived of as the first vehicle for the Middle East. Upon examining the long history of travel and transport in the country, an unlikely inspiration emerged: the camel. The famously resilient pack animal was an ideal namesake for this new brand that promised world-class durability and performance in even the most punishing of climates.

A nod to the rich history of the region, the Camelus name also spoke to a sense of dependability, endurance, and perseverance underreported in even the most extreme conditions. Camels are among the most resilient and 'high-performing' animals in the world and have been an indispensable mode of transport across the Middle East and beyond for millennia.

Capable of carrying massive loads on their backs, and able to go for weeks or, in some cases, months without a direct source of drinking water, camels have been vital means of transport for both passengers and cargo in some of the world's most rugged and extreme locales.

That spirit of extreme durability was paired with a world-class sense of comfort and refinement to create the world's finest, safest, most luxurious multi-purpose utility vehicle: the flagship Camelus Utility Vehicle.

The all-wheel drive Camelus concept beautifully combined an air of utilitarian dependability with elegant, taut, and tailored lines. While clearly referencing the tall, muscular stance of iconic off-road vehicles like Land Rover, Land Cruiser, and Jeep, Camelus also subtly alluded to the anatomy of its namesake with its wide, 'grinning' front fascia and sturdy, top-loaded side profile.

The Camelus flagship vehicle opened the door to an entirely new sector of transportation innovation in Saudi Arabia and offered a new global benchmark in luxury utility vehicles: the ultimate combination of off-road capabilities and on-road comfort and refinement.

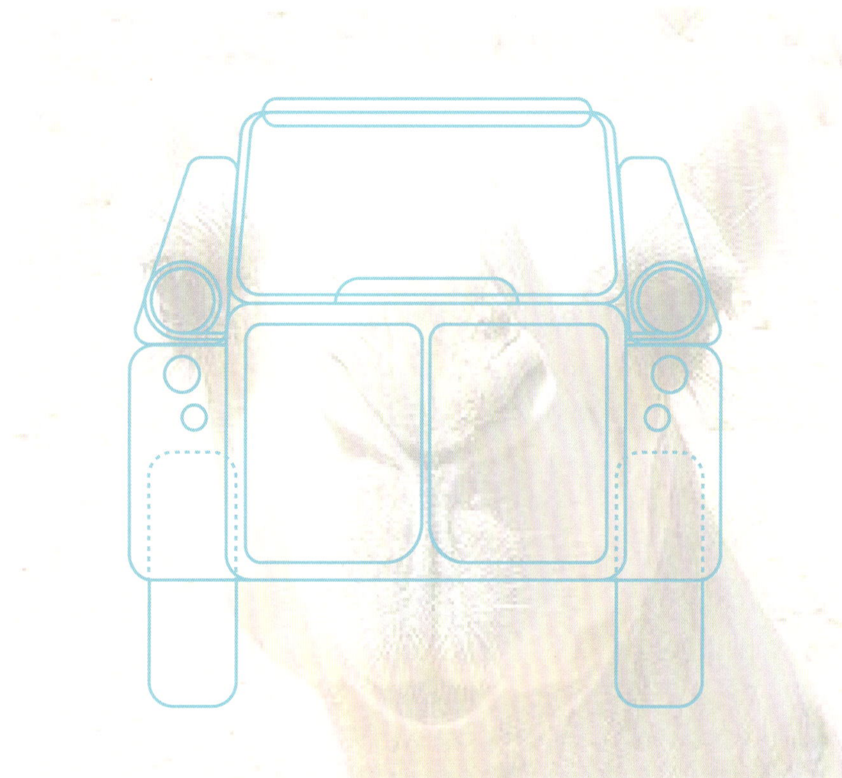

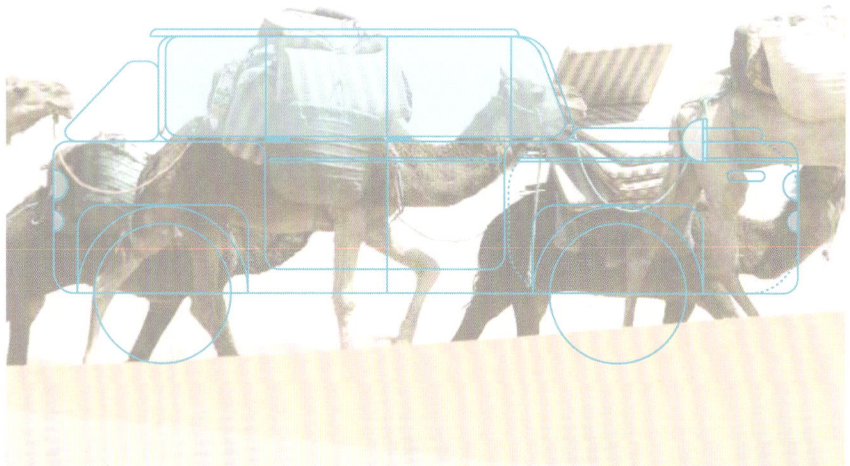

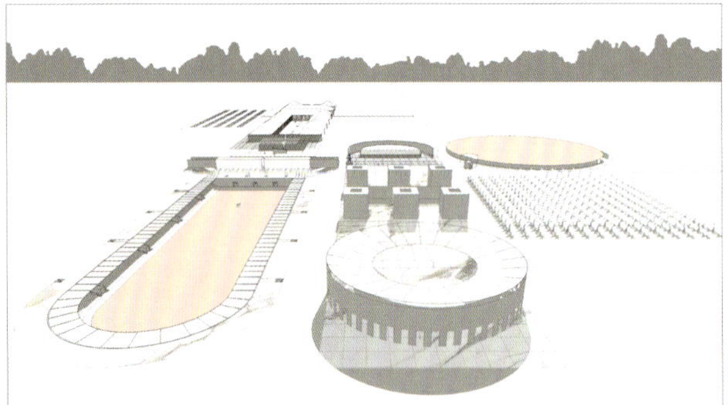
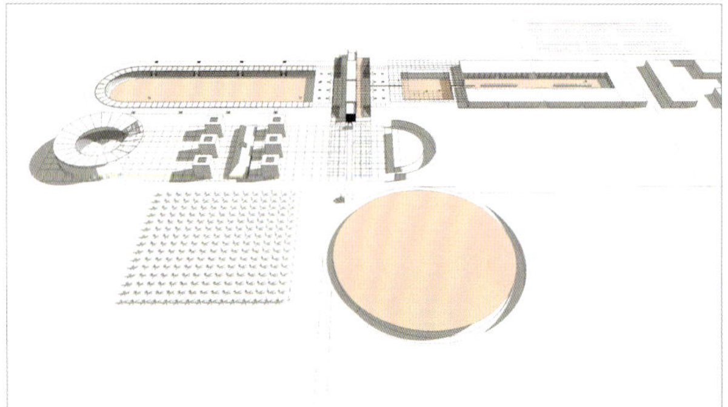

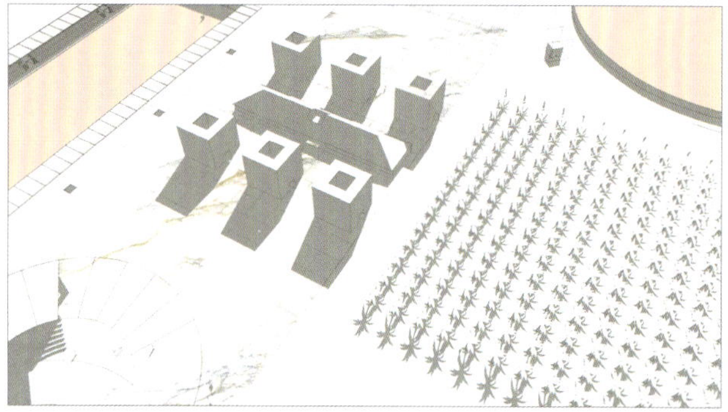

EQUESTRIAN CENTER

2015

Horses have played a significant role in Saudi history for thousands of years. Arabian horses lived among the desert tribes of the Arabian Peninsula, bred by Bedouins who shared their tents, food, and water with them and formed strong connections between man and animal to become a leading player on horse racing's world stage.

Designed as the center of a heritage-influenced resort and destination at Al-'Ula, the center is defined by an architecture that is a testimony to the rich culture from which it arose. The design features classical and timeless geometries, local materials, and ancient construction techniques to create a dialogue between the peninsula's rich past and its modern aspirations. State-of-the-art equine facilities include outdoor, dressage, and indoor arenas supplemented by restaurants, multiple plazas, and a Hippodrome designed to seat up to two thousand people. Together they establish the next generation of equine heritage and culture for the Kingdom of Saudi Arabia, providing a testament to Arabia's rich and profound history for both the country and the world.

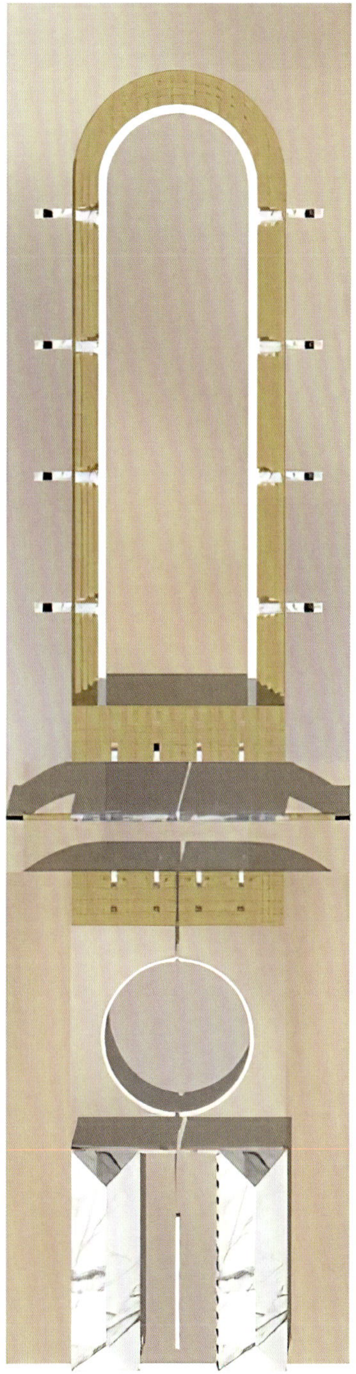

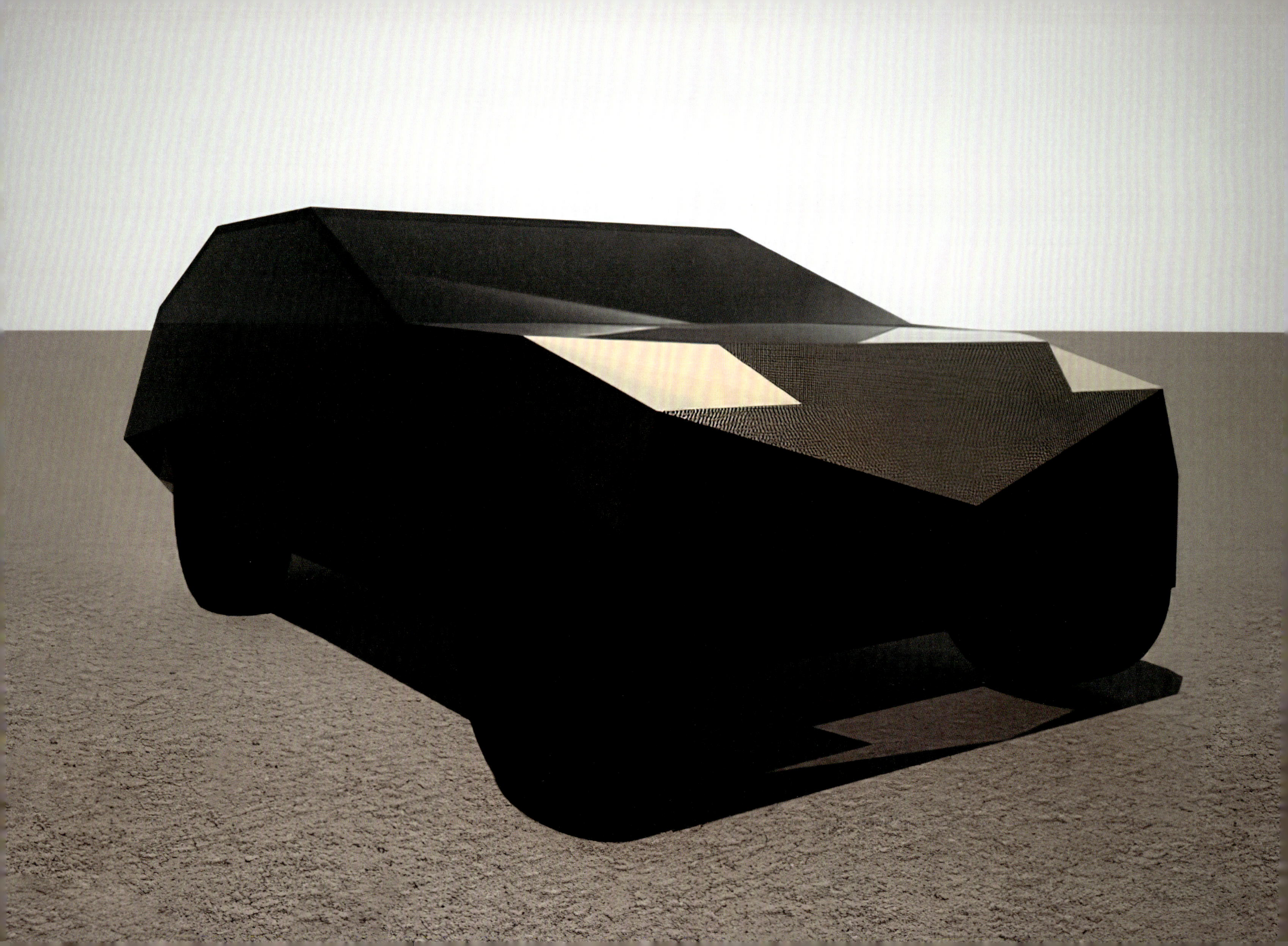

SUV EV CONCEPT CAR

2015

The SUV EV concept was developed to change consumer attitudes toward electric vehicles. Defined by a sharp-angled and aggressive silhouette reminiscent of stealth aircraft, the vehicle was designed without a single curve on its body. Its faceted design is defined by edges and angles, creating a truly beautiful crystalline structure with a geometric purity that is rare in the world of concept cars.

More than a redefinition of aesthetics, the design of the SUV EV concept represents the vehicle's advanced technologies with a form that provides a functionality well-suited to the needs of any EV. The overall triangular shape of the vehicle reduces air resistance and increases its range while setting it apart from contemporary EV designs. Flat facets and angled edges create a look of futurism, resulting in a legendary vehicle with one of the most visually iconic and arresting designs in recent history.

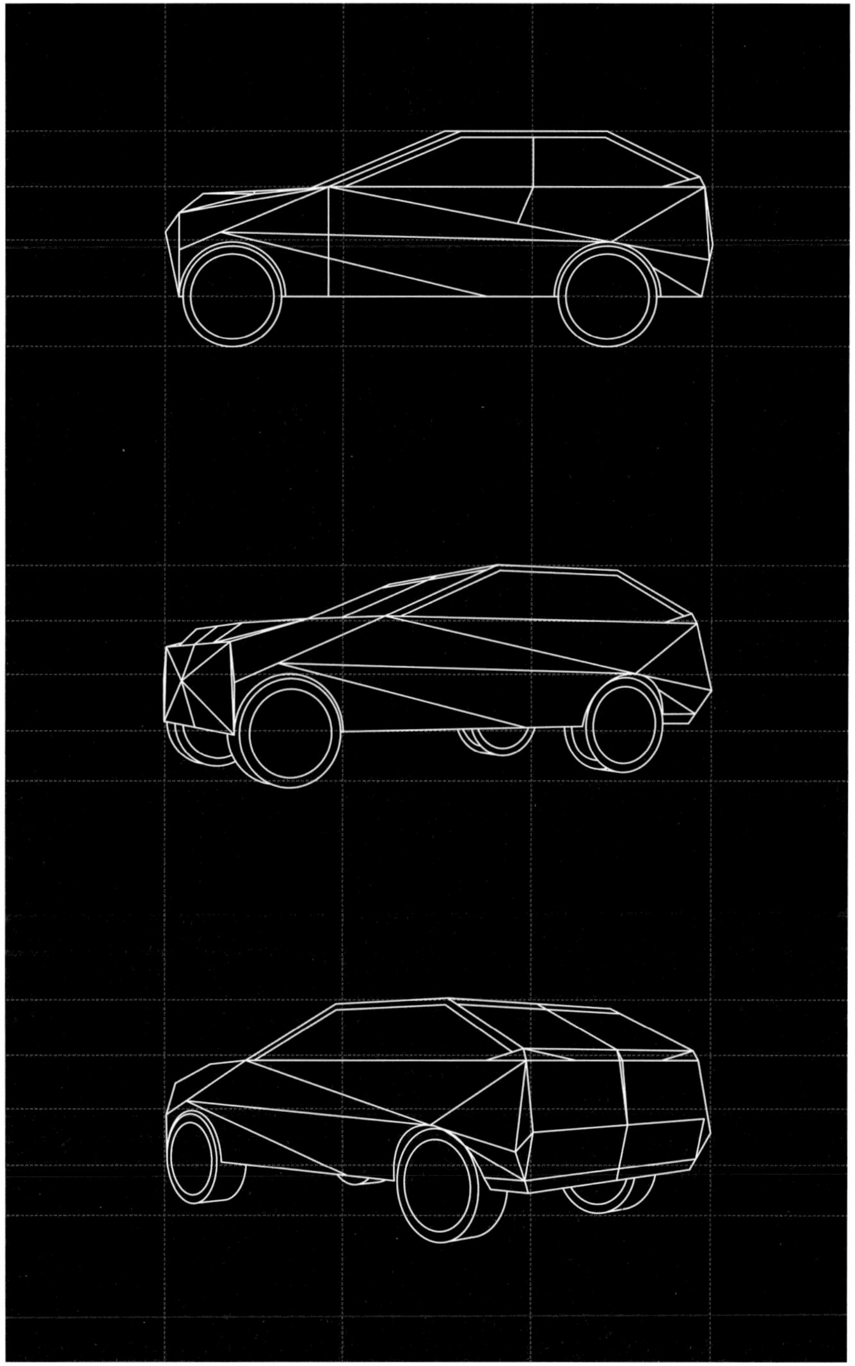

حلم
DREAM
PLACE
حلم

DREAMPLACE

2015

DreamPlace was a design concept developed for a proposed new global hub of technology, research and development, education, entertainment, and innovation located in Riyadh, Saudi Arabia.

The design drew inspiration from Saudi Arabia's storied Oil Well No. 7: the site of the first discovery of oil in the country in 1938. At that point, oil had yet to be discovered anywhere in the region, and drillers had already unsuccessfully searched in six other locations—they were just about to give up when No. 7 struck oil, thus making the well an iconic symbol of discovery, innovation, and prosperity in Saudi Arabia.

The stark, industrial, constructivist character of the project—inspired by the exposed steel trusses of that original oil well—was paired with contemporary interpretations of traditional Arabic ornamental geometries to create a design that was both completely modern and steeped in the rich heritage of the region.

Comprising shopping and entertainment areas, satellite campuses of numerous global universities, a central business district, and a vast R&D park to serve as the regional home for global technology firms, DreamPlace was designed to spark a new era of innovation, shared prosperity, and global connectedness in the region and beyond.

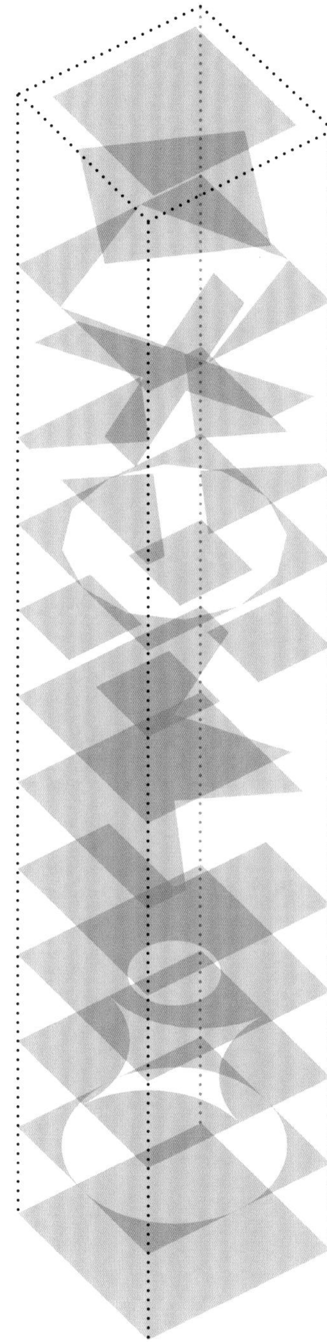

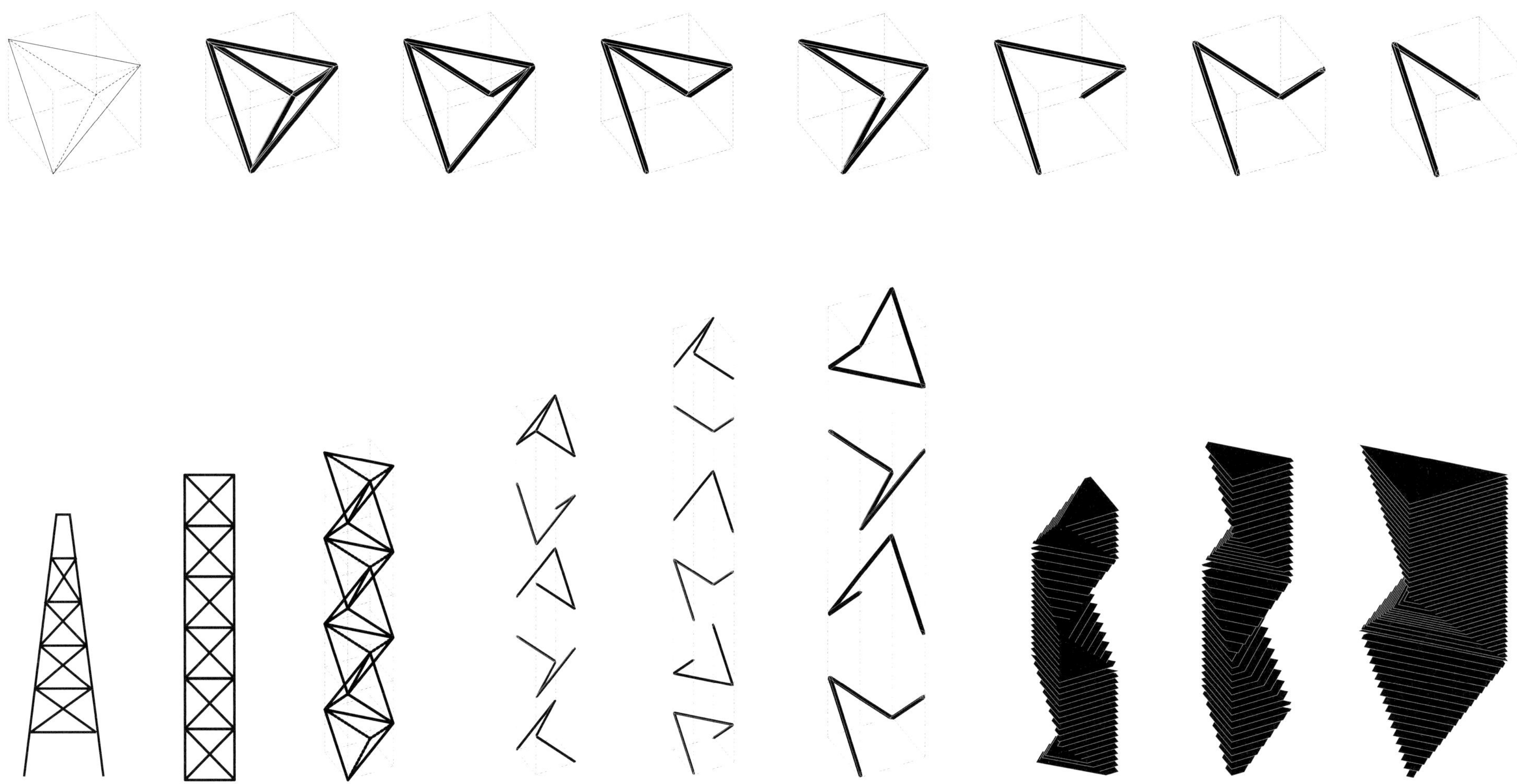

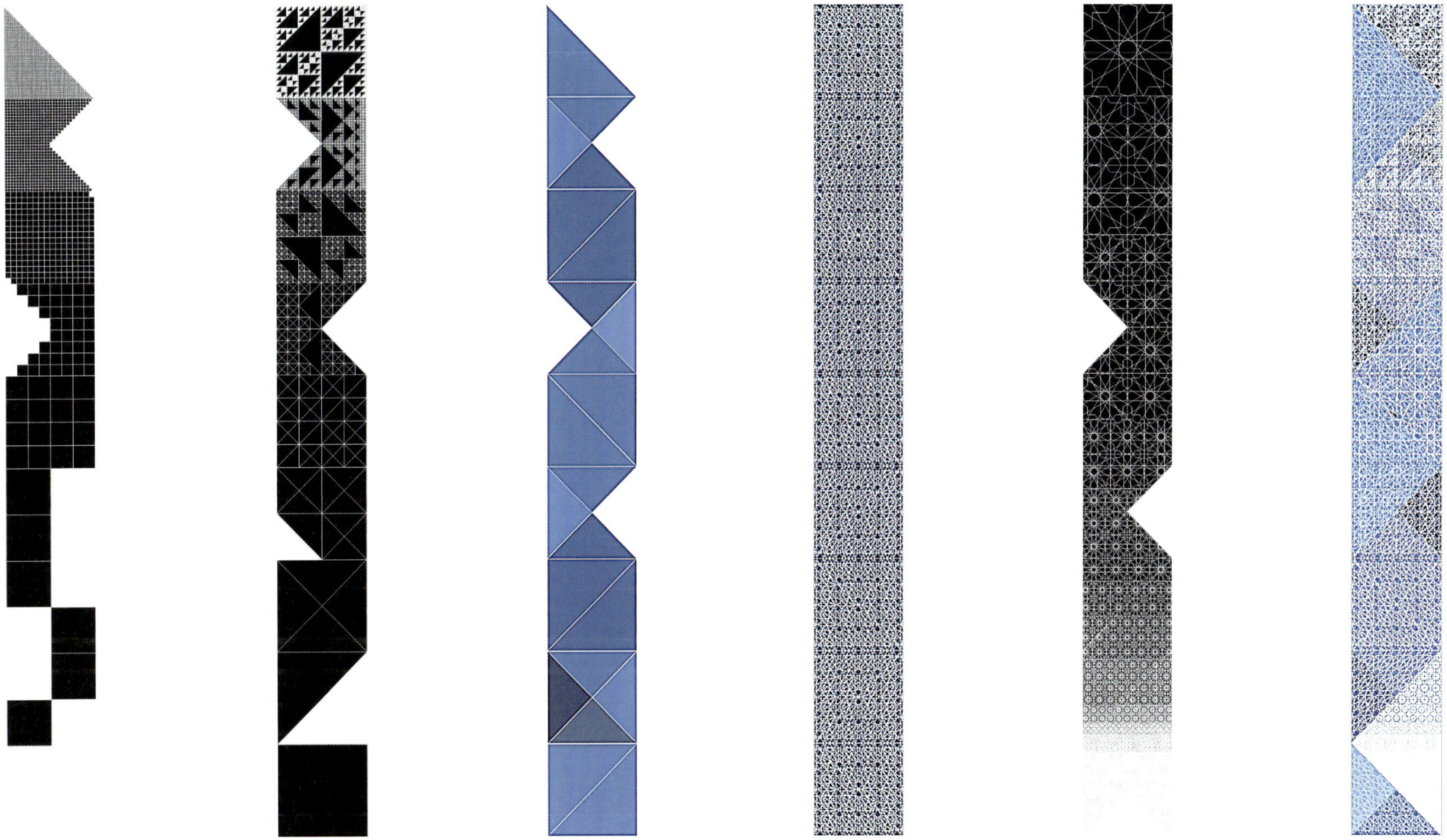

BVLGARI
Cartier
& Arpels

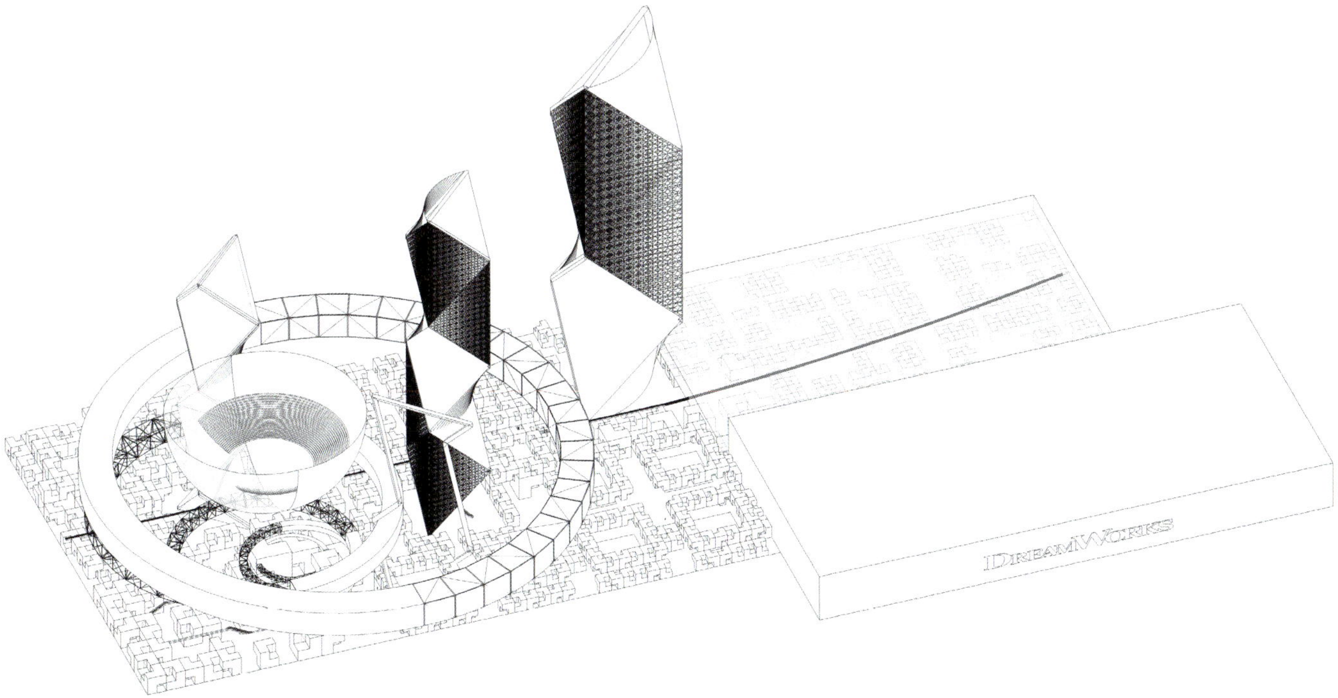
DreamWorks

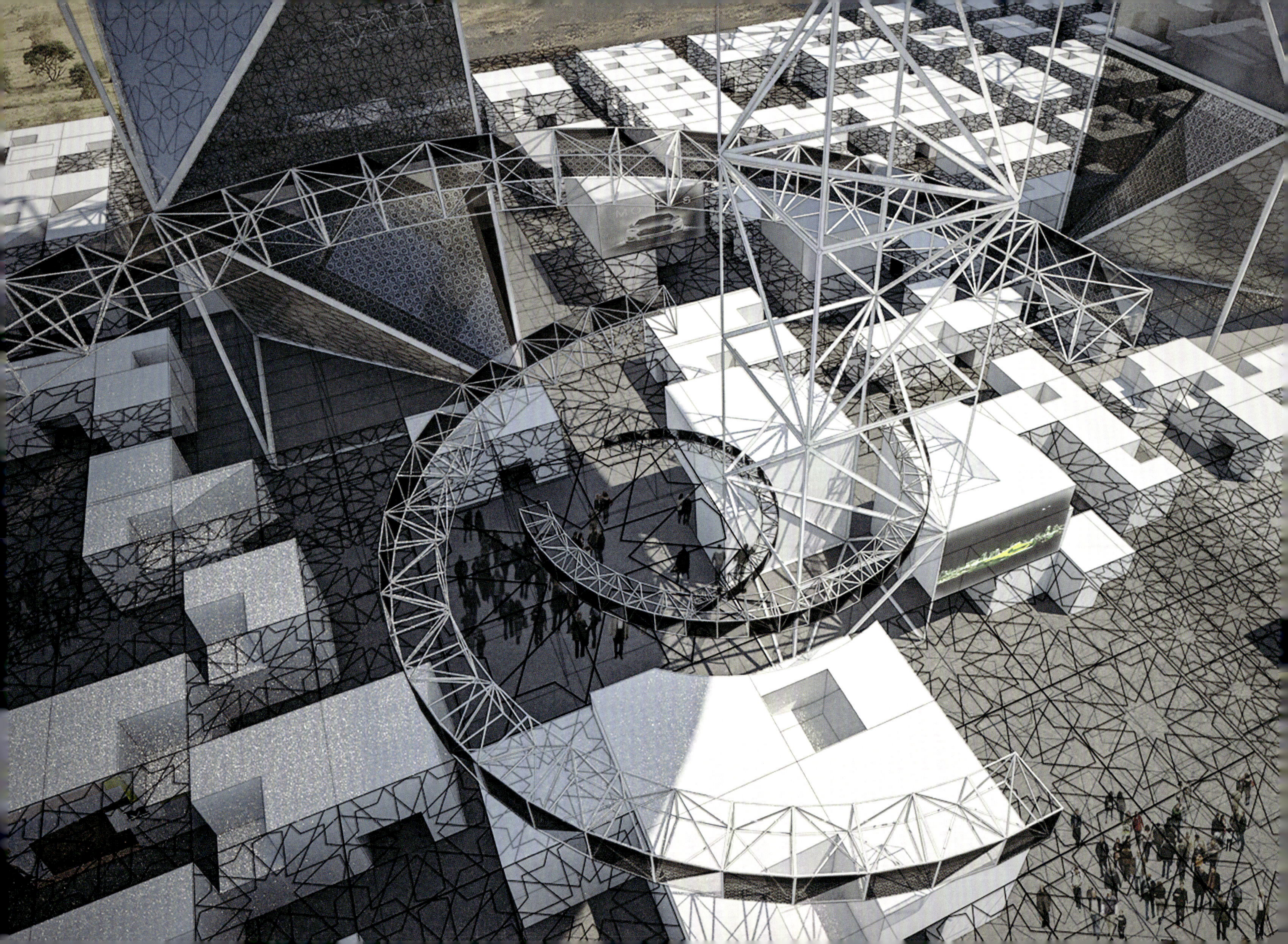

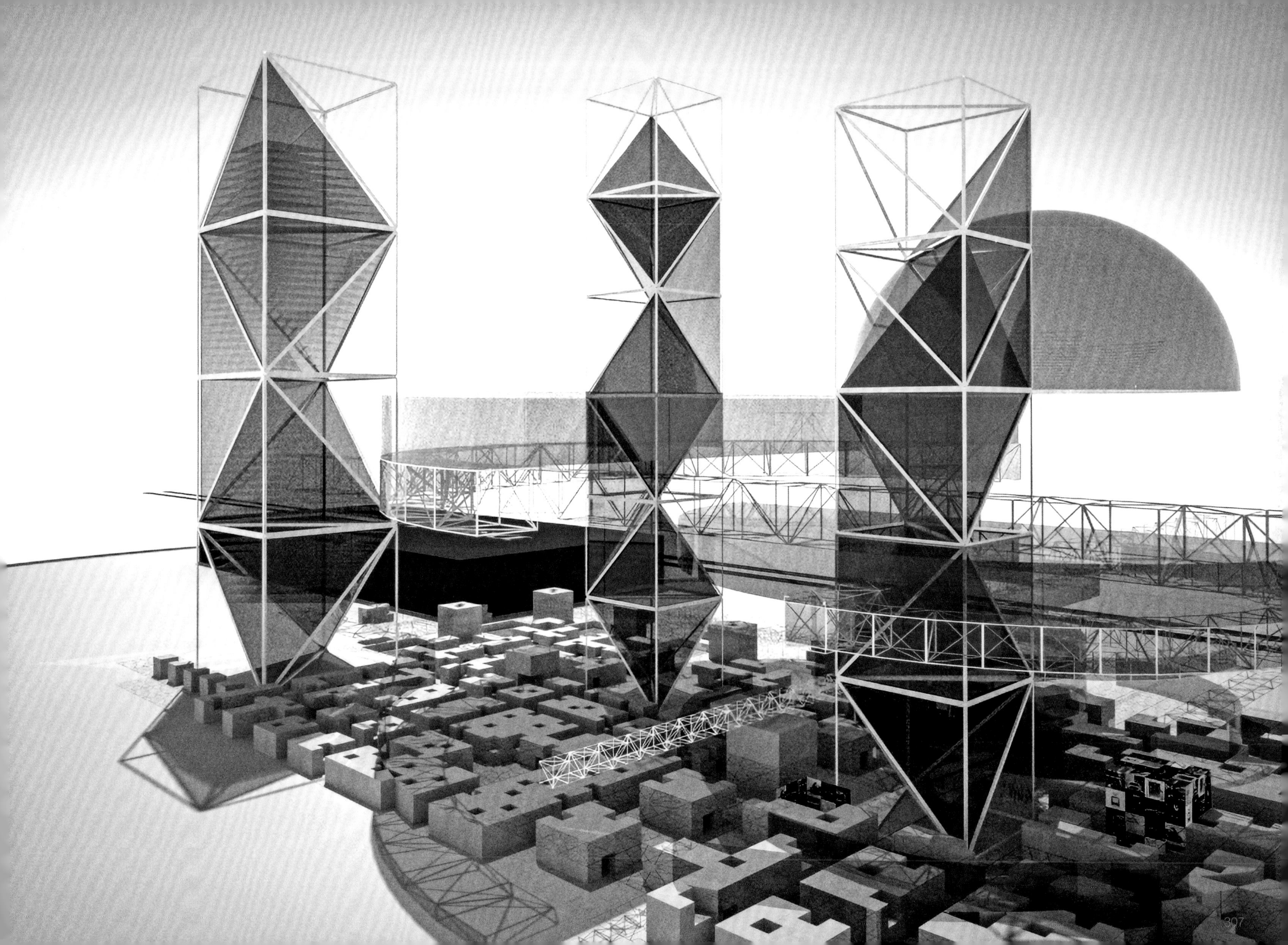

SPHERE ARABIA

2015

Sphere Arabia was developed as a concept for a global luxury brand that paid tribute to the traditions of art, design, and craftsmanship in the Arabian Peninsula region: a cultural patrimony that deserved to find a truly international audience.

A signature fragrance was created for the brand based on *oud*—the highly prized scent derived from the agar tree that can be found growing throughout the Middle East. The intricate and decorative fragrance bottle design incorporated historical motifs drawn from the regional traditions of art, architecture, and craft, layered over the distinctive but highly abstracted shape of a darkling beetle—the beautiful iridescent species native to (and indelibly associated with) the Arabian Peninsula.

Along with this signature scent, a range of concepts for fine jewelry, precision timepieces, writing instruments, and luxury home goods were also developed. Each design showcased the rich artistic traditions of the region, subtly evolving traditional motifs, patterns, and material palettes to create products that were unmistakably contemporary, yet profoundly rooted in a sense of place.

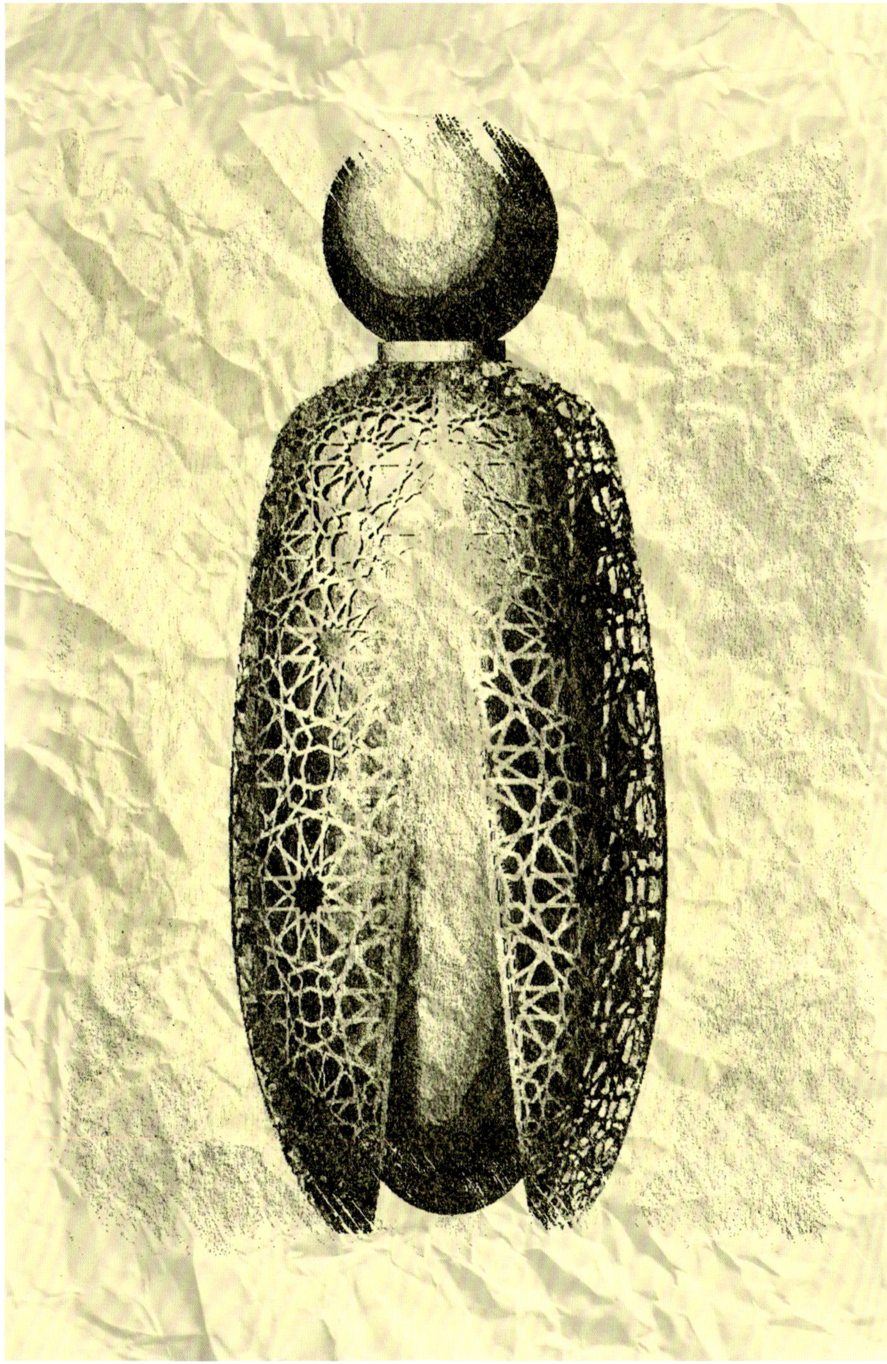

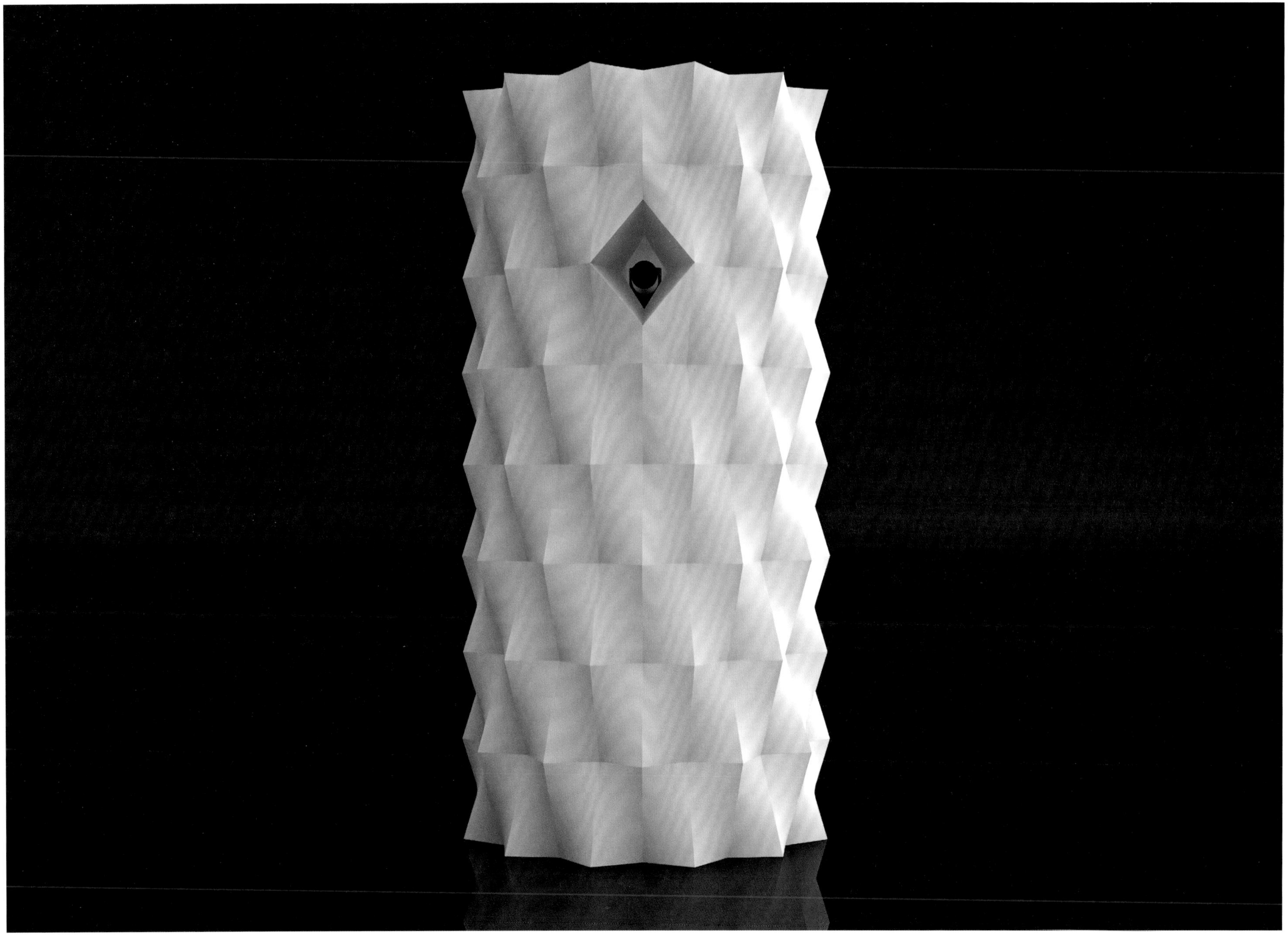

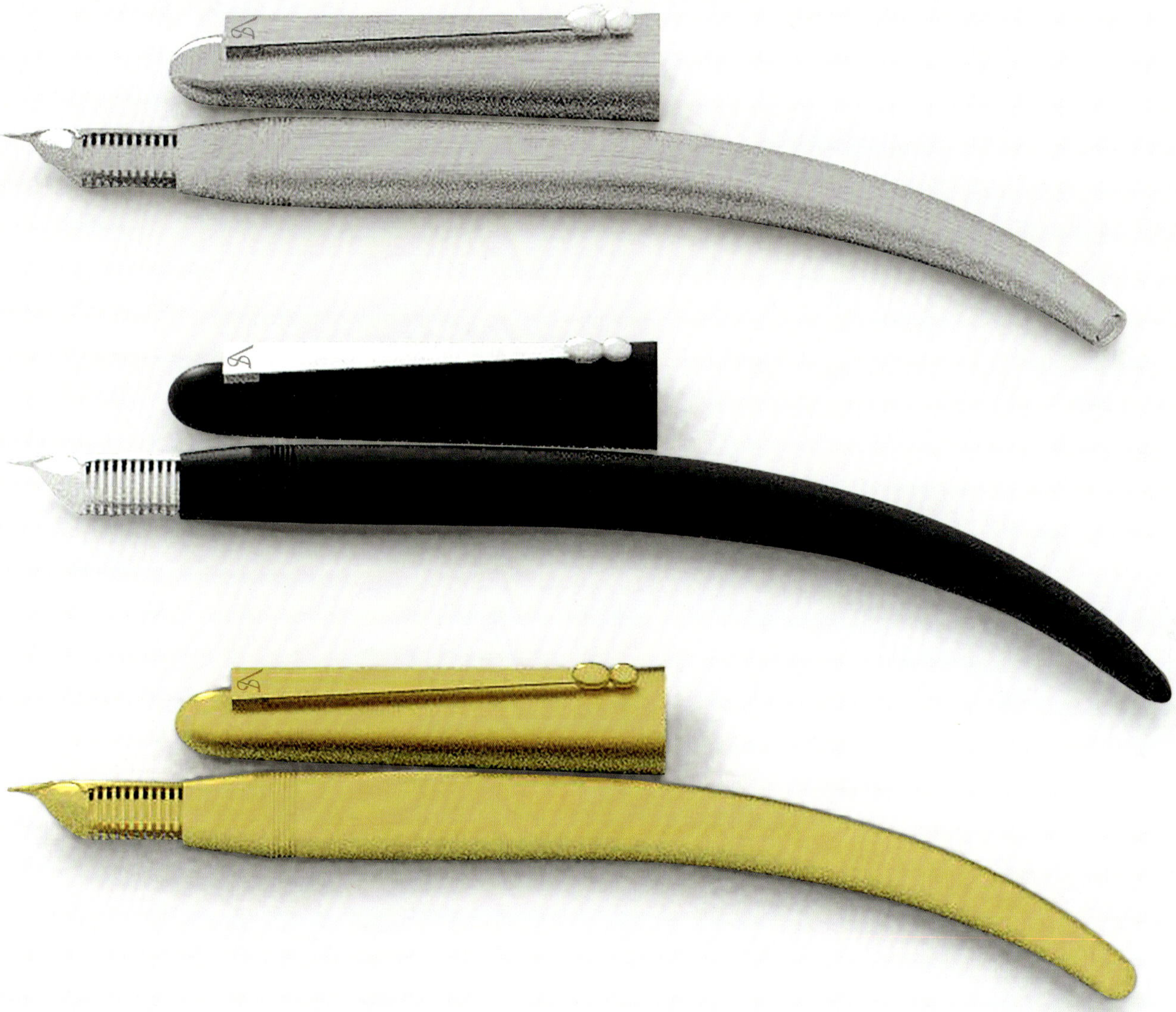

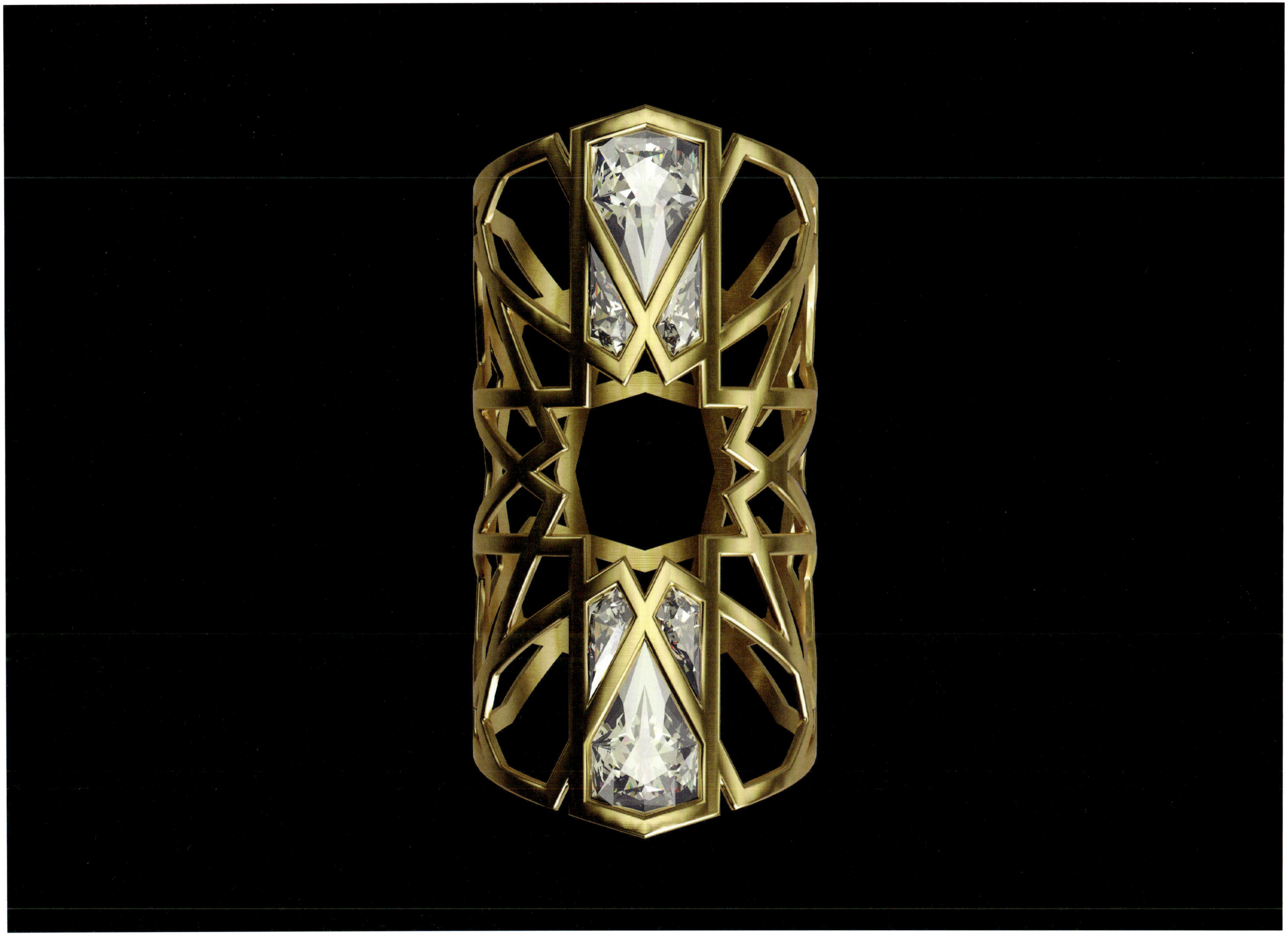

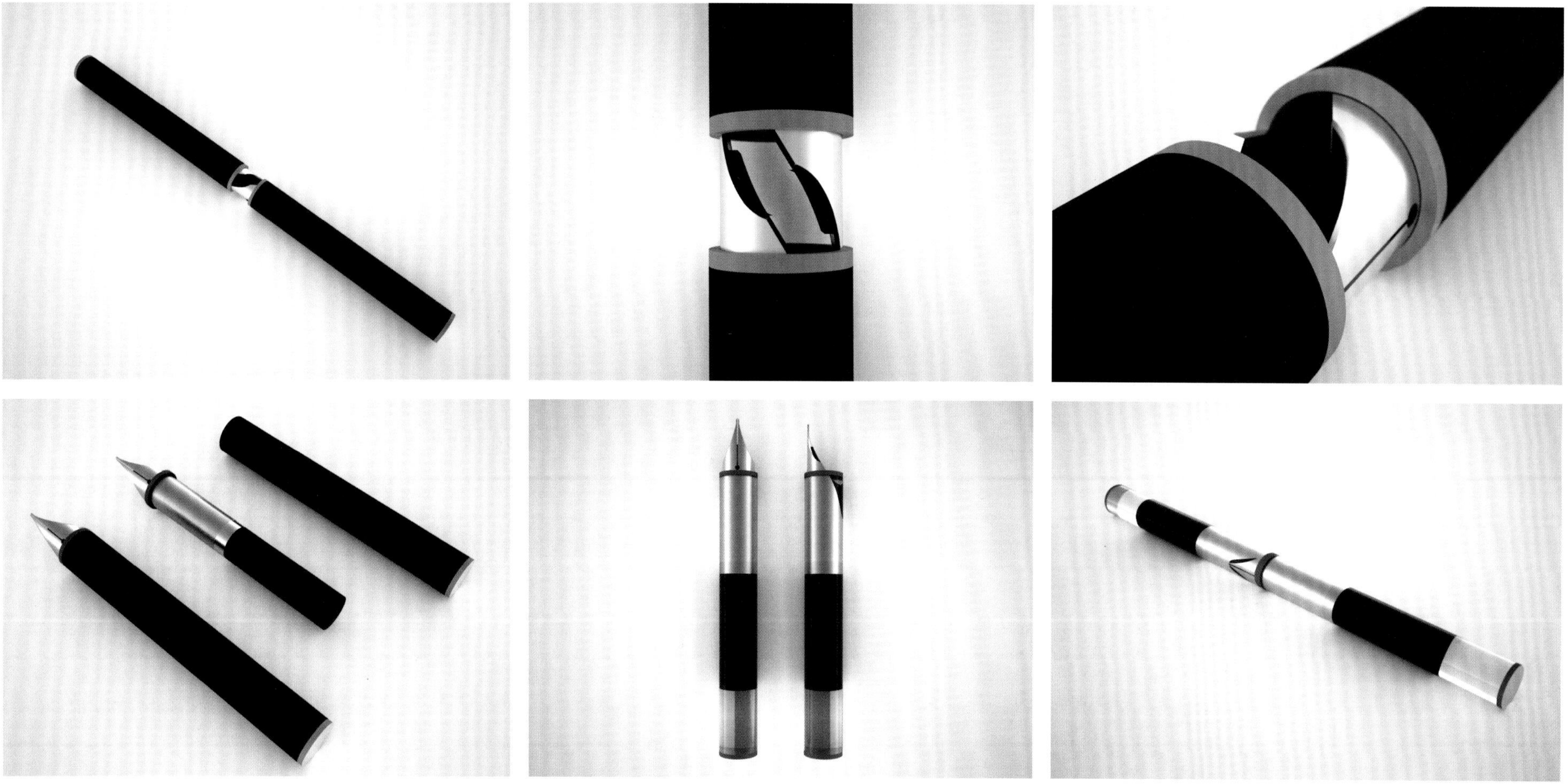

VISCONTI

2015

For Visconti, an Italian brand renowned for its handcrafted fountain pens, luxury timepieces, and elegant leather goods, a series of design concepts was developed for a new range of premium writing instruments as part of a overarching brand reinvigoration initiative.

A careful study of the ergonomics of the hand led to the development of new and more comfortable writing instrument form factors in a variety of unexpected and provocative shapes. In the signature piece, the female form was carefully engineered with curves strategically placed to fit snugly in the hand: expertly ergonomic, yet beautiful, whimsical, and irreverent. A series of accompanying communications displayed arresting ink blot graphics and contemporary takes on traditional Florentine motifs and designs: an homage to the roots of the brand that was still daring, fresh, and unmistakably modern.

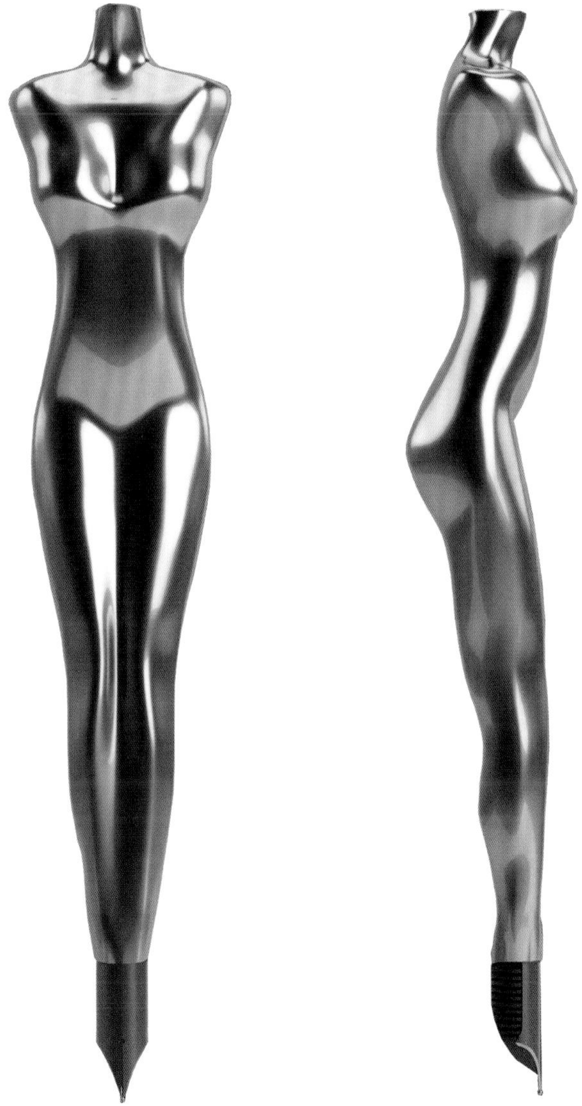

VISCONTI

VISCONTI
FIRENZE
VISCONTI
VISCONTI
VISCONTI
VISCONTI

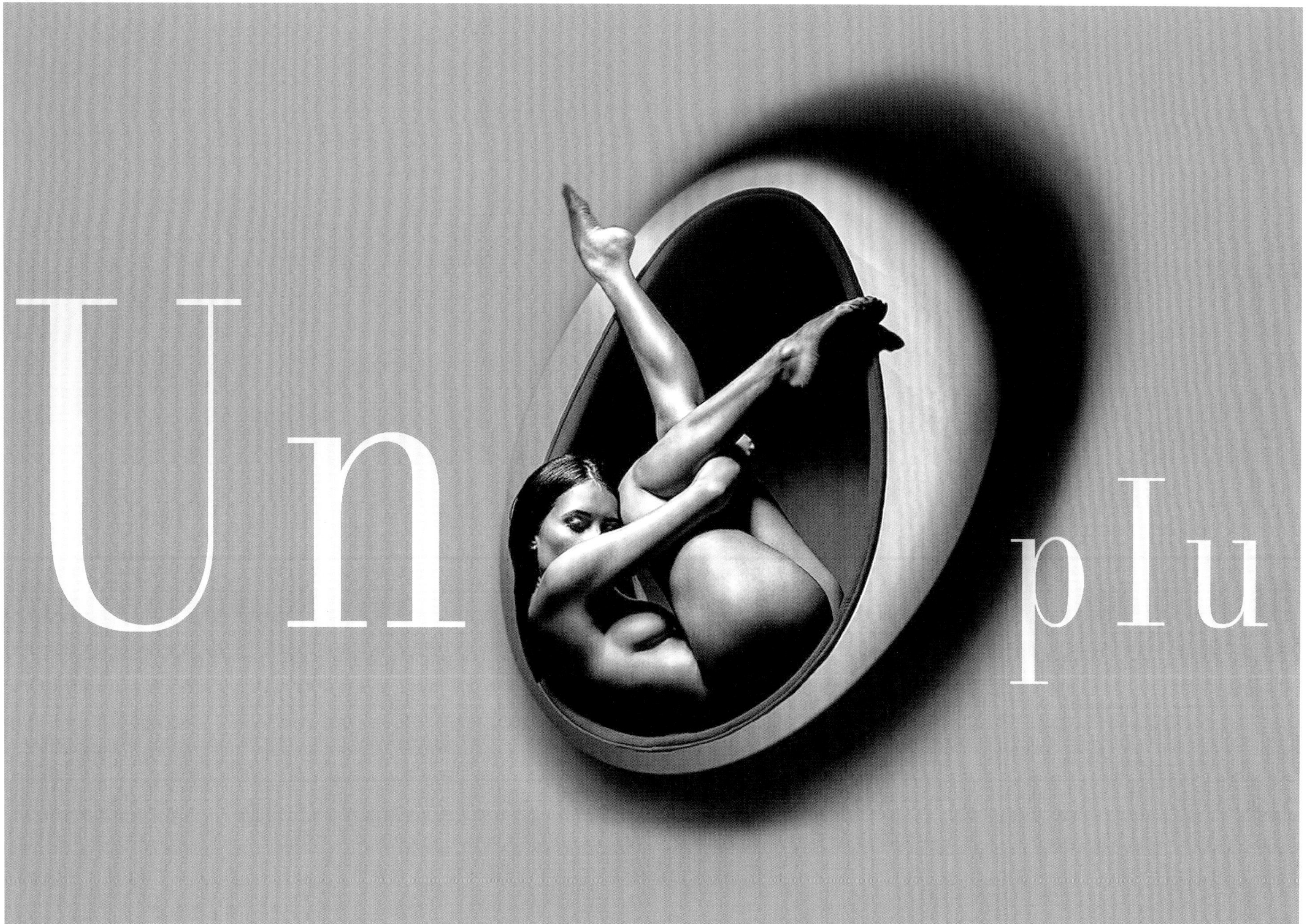
Un
plu

UNOPIÙ

2015

Unopiù, an Italian outdoor furniture brand, is known throughout Europe for its stylish, well-priced, high-quality pieces for the home and garden.

To introduce the brand into North America, a concept for a new brand mark was created by offsetting and dimensionalizing classic Roman serif letter forms to evoke the simplified form of a chair. A proposed launch campaign was also developed that evoked Italy's legendary sense of style, imbuing the brand with a distinctly Italian sense of wit, playfulness, and sensuality.

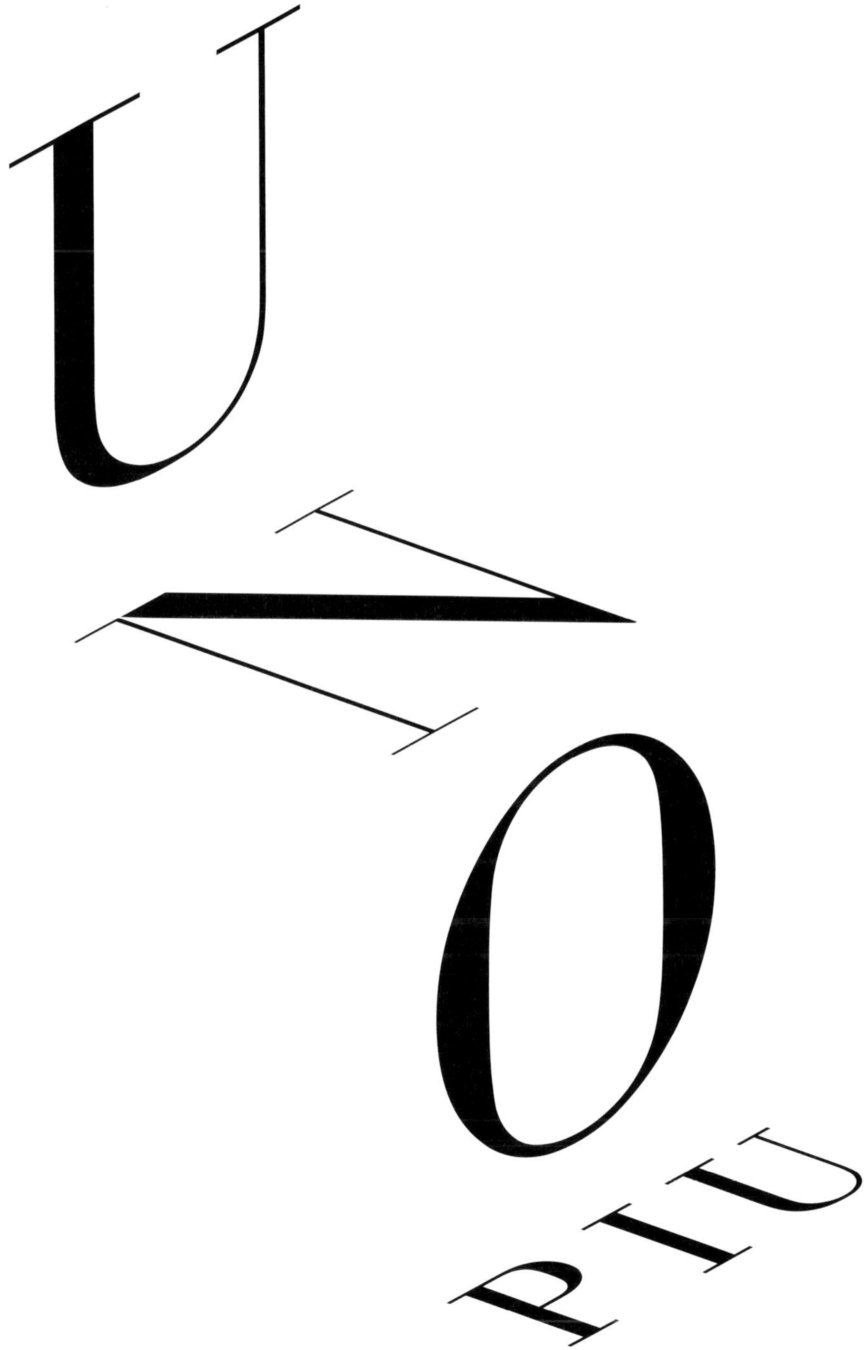

MARTHA GRAHAM COMPANY

2015

Inspired by the incredible beauty, strength, elegance, and artistry of the Martha Graham Company dancers, a short film was created to honor their exceptional talent and artistry, and to celebrate the legacy of one of the great pioneers of modern dance.

Composed of over two thousand still photographs, the individual images were methodically knit together into a stop-motion film that was a striking depiction of the dancers and their choreography. By lingering on static images of a dynamic art form, the spectacular finesse of the dancers was able to be appreciated on a far deeper level: minute gestures, tensing muscles, split-second movements, and explosions of power that might otherwise be invisible to the naked eye.

A new monogrammed brand mark was also developed that captured the timeless elegance of the dance company and paid homage to the Art Deco style that was prevalent in 1920s, when Martha Graham first founded her namesake company. The new brand mark was debuted alongside the film at a gala screening hosted by the Martha Graham Company in the winter of 2015.

DIANE VON FURSTENBERG

2015

To congratulate Diane von Furstenburg on the publication of her book *DVF: Journey of a Dress,* Peter presented the designer and fashion icon (and longtime neighbor) with a series of collages inspired by her legendary use of patterns and prints.

In the series, the body of the figure disappears and becomes supplanted by super-sized elements of various patterns that have appeared on the iconic DVF wrap dress over the years: a nod to the way in which the famous dress is, in the words of its creator, "truly like a second skin."

The photograph that is held up (the image that Andy Warhol based his portrait of Von Furstenberg on) obscures the face of the "everywoman" behind: a sly gesture that spoke to the universality of her designs, and a wink to her oft-quoted assertion that she has "never had anything to hide."

THE BOOES

2015

The Booes were conceived of as an original educational entertainment property that engages children while teaching them to care for the ocean, the planet, and each other. Inspired by the charming simplicity of classic paper dolls, the paper-thin Booes are enthusiastic stewards of the beautiful tropical island that they call home, and careful protectors of its many wildlife inhabitants.

Stories developed around the adventures of the Booes were fun, whimsical, and always contained an important central lesson that helped to teach kids to do their part to protect the environment and the many creatures in it. Planned activations for the Booes included a signature story book, animated television series, educational video game, and, of course, an array of beautiful and engaging collectible toys that would allow kids to immerse themselves in this bright and colorful world.

sharecare

SHARECARE

2015

For Sharecare, a revolutionary health care company started by the founder of WebMD, Jeff Arnold, a mobile app and online user experience were designed that changed the way health care is approached at the most foundational level. The app deployed technology in a way that upended traditional paradigms of medical treatment and encouraged users to take a proactive, preventative, holistic approach to wellness.

At the heart of the app was "Living in the Green," both a philosophy and a detailed analytics platform to help facilitate better living. Along with technology that monitored physical activity levels and other user vitals, a proprietary voice analysis tool that measured stress allowed users to take an "Emotional Selfie" to get a more complete picture of their health on a daily, or even hourly, basis. When a user's wellness profile slipped "out of the Green," clear steps were provided with a mind toward both short-term relief and longer-term, sustainable lifestyle changes.

To distill complex user data into actionable insights, an intuitive user interface was designed that connected cognitive behaviors with cognitive design, prompting a meaningful dialogue with consumers who wanted to empower themselves to live longer, better, happier, and healthier lives.

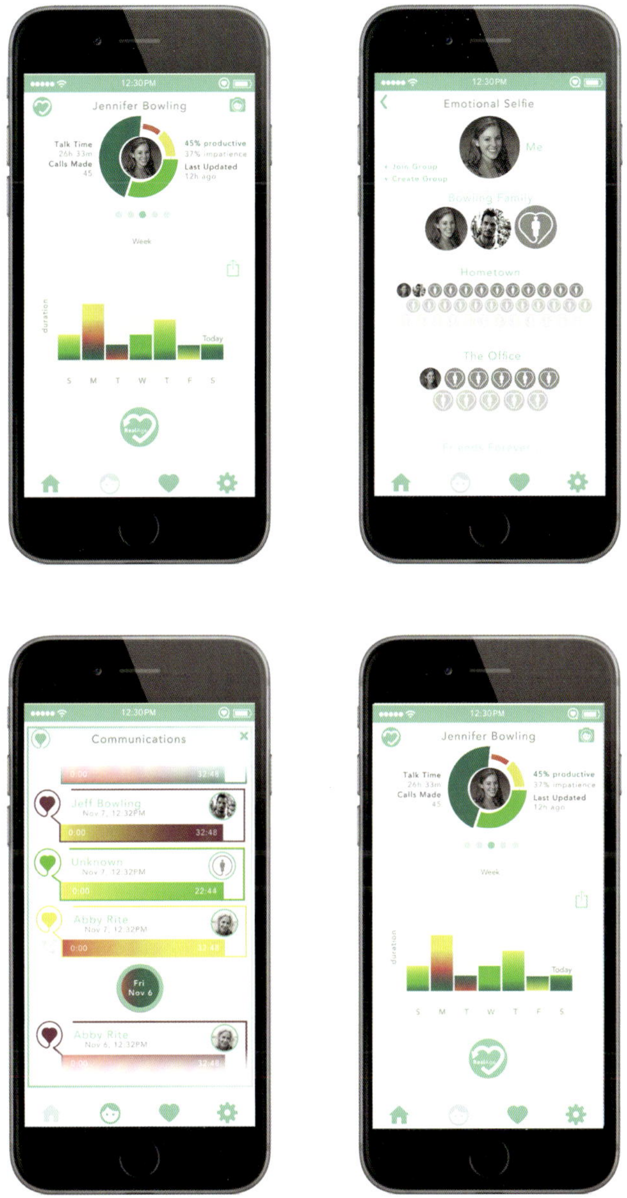

BALLY'S
OUR MEMBERS GET
SPECIAL RESTAURANT
UFC

UFC

2015

For UFC, the largest mixed martial arts promotion company in the world, a concept design was developed for a new, purpose-built UFC Arena. Designed around the unique shape, dimensions, and specifications of the eight-sided enclosure that UFC stages all of its matches in—known as the Octagon—the UFC Arena featured gently tiered seating, allowing every seat in the house to have an unobstructed, close-up view of the stage below. And unlike most other sporting or entertainment venues, the UFC Arena made a dramatic gesture toward inviting the outside world in by making one of its building faces a massive, transparent glass wall overlooking a public plaza below in the heart of the Las Vegas Strip. In this new ground-level social space, devoted fans and passersby alike were invited to join in the excitement of the world's most dynamic, intense, and thrilling spectator sport.

2016
MEMBER

NEW YORK CITY
FOOTBALL CLUB

adidas

NEW YORK CITY
BROOKLYN
MANHATTAN
BRONX
QUEENS
FOOTBALL CLUB
STATEN ISLAND

NYCFC

2015

When it launched in 2015 as the first Major League Soccer team in New York City, New York City Football Club needed to rapidly build up its fan base, instill local pride, and create a swell of interest in communities all around the city.

A fresh look and voice for NYCFC was designed to tap into the global passion for the sport, while also honoring unique local cultures of New York City. A new set of brand communications was developed around the club's greatest asset: its world-famous players who had relocated to New York to be a part of this exciting new era in professional soccer. NYCFC's roster of stars was featured in dramatic black-and-white photographs that elevated them to almost mythic proportions, showing them as the heroes they were to so many devoted fans all over the world.

At the same time, an exclusive range of new NYCFC merchandise was also developed—including commemorative coins, calendars, bracelets, T-shirts, scarves, caps, cleats, and sneakers—helping to elevate the founding members' sense of community and loyalty to New York's newest and most exciting major league sports team.

NYCityFC
FOUNDING MEMBER
2016

CITY IS SOCCER FOR NEW YORK.
BELONG
NYCFC
2016

NYCityFC
CALENDAR
2016
2016
January
February
March
April
May
June
July
August
September
October
November
December

STATEN ISLAND
BROOKLYN
QUEENS
MANHATTAN
BRONX

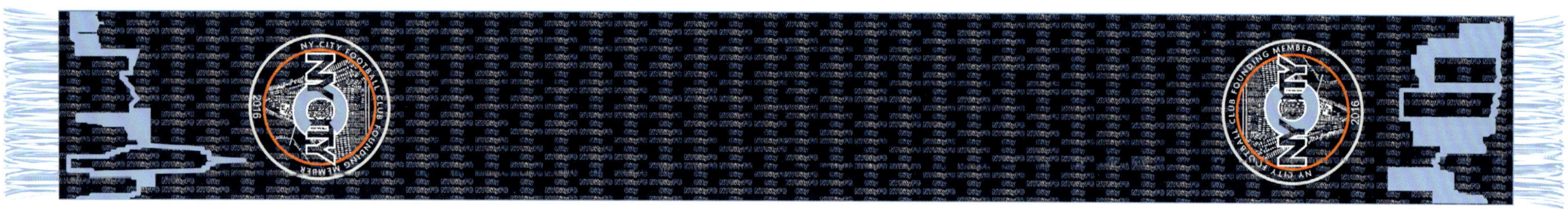
NY CITY FOOTBALL CLUB FOUNDING MEMBER
2016

NEW YORK CITY

adidas

7
VILLA
21
PIRLO
PIRLO
21
21
PIRLO
7
VILLA
NYCityFC
10
MIX
NYCityFC
NYCFC
7
DAVID
VILLA
NYCityFC

2016
MEMBER
NYCityFC
FOUNDING MEMBER
2015
NYCityFC

YOUR TICKETS TODAY
FOOTBALL CLUB

NYCFC CITYCAGE

2015

The NYCFC CityCage concept was a design proposal for a modular, pop-up soccer pitch that was developed to tap into the huge amount of athletic talent that existed in communities all around New York City, and create an authentic, grassroots buzz around NYCFC and its roster of world-class players. Current fans would be drawn deeper into the community, while new fans would organically want to check out this new phenomenon unfolding on the streets of their neighborhood.

The compact design of the CityCage made games even more intense and exciting for players and spectators alike: by miniaturizing a full-scale soccer pitch to suit a tight urban setting, the explosive energy of a soccer match was front and center, as were the extraordinary skill and sportsmanship of the amateur players who loved the game.

The proposed program tapped into the long-standing culture of street sports in New York City and the exploding popularity of soccer in North America, and was designed to help the game to become a mainstay in the cultural landscape of New York, ensuring that any community in any part of the city could get in on the action and share in the energy of New York's newest professional sports team.

Schneider
18 MEGAPIXEL

BLACKBERRY

2015

BlackBerry, the Canadian technology company that pioneered the smartphone in the early to mid-2000s, had built its reputation on providing the highest level of security with all of its products. Its devices and enterprise software were trusted by and depended upon by countless government agencies, corporations, and individuals for whom privacy was of the utmost importance.

In 2015, as the company developed its first Android-powered phone, a complete naming, brand identity architecture system, and global launch strategy were created to celebrate the company's heritage of privacy in a bold new way. The Priv was BlackBerry's most sophisticated product to ever come to market, and the name, which called up the two most important pillars of the BlackBerry experience—Privacy and Privilege—reminded users of the superlative quality of BlackBerry devices in terms of design, ease of use, and end-to-end encryption.

The accompanying launch campaign focused on the freedom that's gained through knowing your privacy is assured: that sense of peace that comes from knowing that personal, family, and business matters are perfectly safe and secure. The campaign also aimed to educate consumers on the many ways that the Priv allowed them to take control of their privacy, with software tools that made it easy to share as much, or as little, as they wanted with the world around them. This powerful message—executed across web, film, print, social media, and pop-up activations—reminded customers of the tremendous importance of privacy, and heralded a new era for this smartphone pioneer.

PRIV

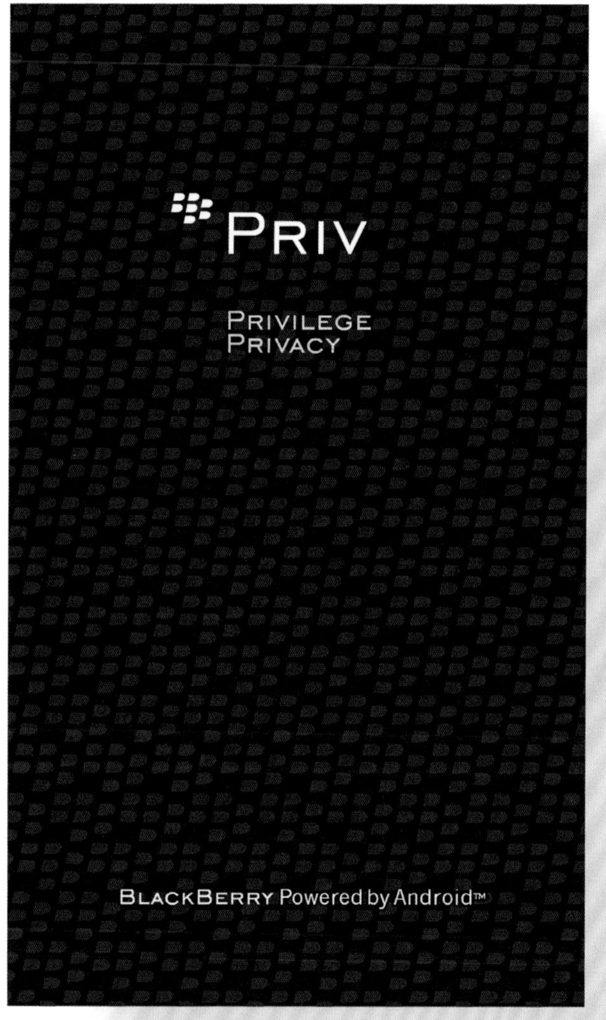

For
your
PRIV
ACY
PRIV

PRIV

entral line
Westbound platform 5
Bank
St. Paul's
Chancery Lane
Holborn
Tottenham Court Road
Oxford Circus
Bond Street
Marble Arch
Lancaster Gate
Queensway
Notting Hill Gate
Holland Park
Shepherd's Bush
White City
East Acton
North Acton
Hanger Lane
Perivale
Greenford
Northolt
South Ruislip
Ruislip Gardens
West Ruislip
West Acton
Ealing Broadway
DO
NOT
DISTURB
PRIV
TOUCH + TYPE
PRIV
AVAILABLE AT
SELFRIDGES
Historic Royal Palaces
Tower of London
PRIV
BAN
A 4706
MIND THE GAP

PRIVACY
PRIV

AMAN NEW YORK

2016

Aman is renowned for its exquisitely designed hotels and resorts located in some of the most naturally beautiful settings in the world. As its first North American urban property, the Aman New York, housed in the historic Crown Building on Fifth Avenue, was designed to set a new standard of luxury and exclusivity in New York City.

As an exclusive gift for prospective buyers of the Residences at Aman New York, a limited-edition art book was designed, featuring soulful, serene, and evocative photography that perfectly captured the essence of this world-renowned luxury brand, while also situating it in its new urban context. Inspired by the cubic volume that sits above the Crown Building and supports the namesake crown, the book—itself a perfect cube—was filled with over two thousand photographs that conveyed the singular sense of beauty, simplicity, and tranquility that only Aman can offer. An accompanying short film was also created that wove together these images of peace and repose with visualizations of the exquisite hotel and residences designed by celebrated Belgian architect Jean-Michel Gathy.

NEW YORK

SOLÉMIA

2017

The Solémia development in Miami brought together two powerful real estate dynasties, the Soffer and LeFrak families, to create a spectacular new multi-use master planned community unlike anything else.

In devising a name for this groundbreaking property, inspiration came from the names of its founders. The "Soffer" and "LeFrak" names were combined, yielding "Solé"—the Spanish word for sun—and attached on to that was the abbreviation of the city where it was to be located: "Mia." The result was a distinctive and proprietary name infused with the DNA of its founders that also felt familiar and timeless, as though it had always been there.

Soffer

50 years creating revolutionary lifestyle living

LeFrak

A legacy of building and enduring communities

Miami

America's most desired destination

VEON

2017

For VEON—a telecommunications company serving customers across Europe, Asia, and the Middle East—a complete reengineering of the brand's identity was developed to help launch a new chapter in the company's history.

This new brand image and attitude culminated in the launch of the VEON app: free to VEON customers worldwide, the VEON app allowed mobile users to talk, text, video chat, and send images to anyone, anywhere in the world, without being charged for any data usage. The app was launched with the "Be Truly Free" campaign, which highlighted both its practical and emotional benefits: complete freedom to communicate with the people who matter the most in your life, without any limitations.

Communications celebrated the power of connection and belonging, and were customized to each local market to be reflective of local pride and cultural customs, offering customers everywhere a respectful and authentic representation of life with VEON.

BE TRULY FREE

Chat Call Talk
Text Listen Surf
Watch Read
Connect Share
Relax Discover
Play Work
ImagineForFree
BE TRULY FREE
AVAILABLE AT APP STORE
AND GOOGLE PLAY

be truly free
CHAT AND CALL FOR FREE ON VEON
AVAILABLE AT APP STORE AND GOOGLE PLAY

be free
CHAT AND CALL FOR FREE ON VEON

BE TRULY FREE
AVAILABLE AT APP STORE
AND GOOGLE PLAY
CHAT AND CALL FOR
FREE ON VEON
LOVE

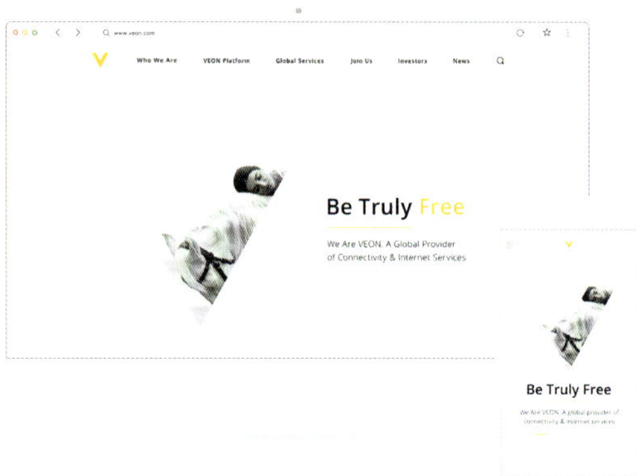
Who We Are
VEON Platform
Global Services
Join Us
Investors
News
Be Truly Free
We Are VEON. A Global Provider
of Connectivity & Internet Services
Be Truly Free

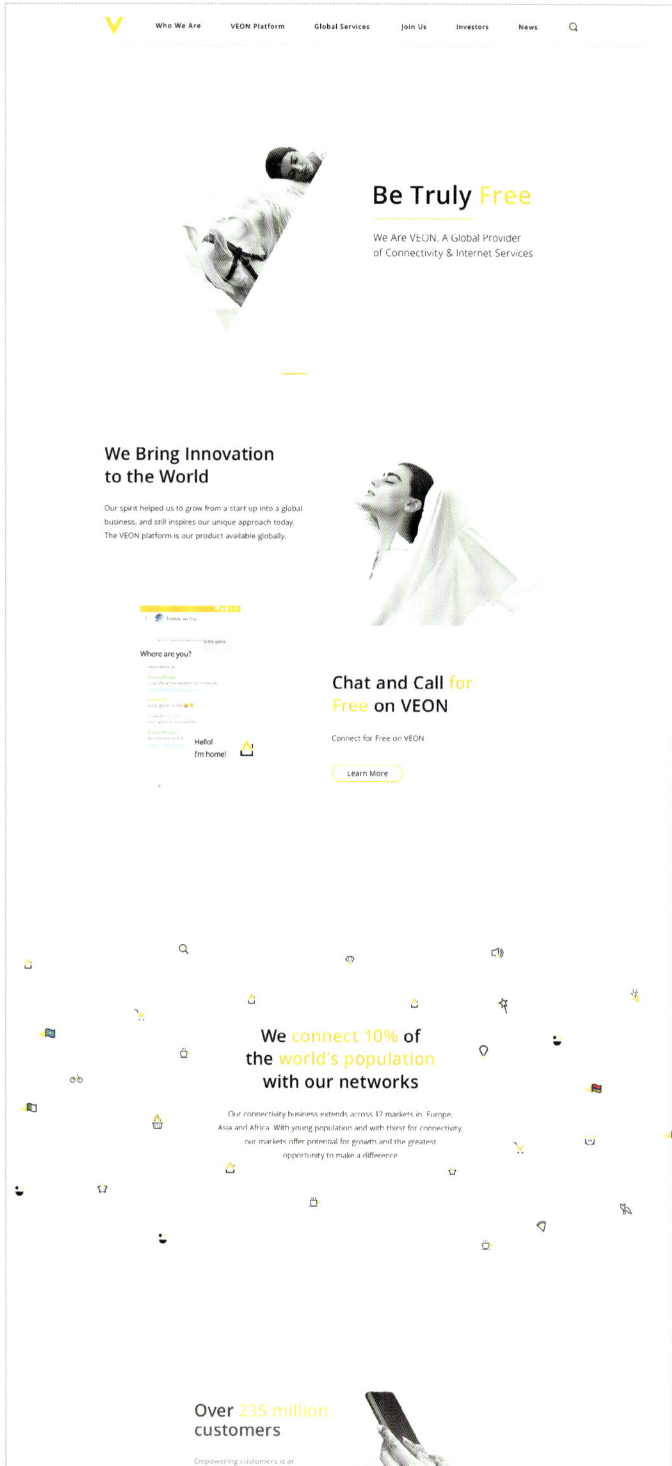
Who We Are
VEON Platform
Global Services
Join Us
Investors
News
Be Truly Free
We Are VEON. A Global Provider
of Connectivity & Internet Services
We Bring Innovation
to the World
Our spirit helped us to grow from a start up into a global
business, and still inspires our unique approach today.
The VEON platform is our product available globally.
Where are you?
Hello!
I'm home!
Chat and Call for
Free on VEON
Connect for Free on VEON
Learn More
We connect 10% of
the world's population
with our networks
Our connectivity business extends across 12 markets in Europe,
Asia and Africa. With young population and with thirst for connectivity,
our markets offer potential for growth and the greatest
opportunity to make a difference.
Over 235 million
customers

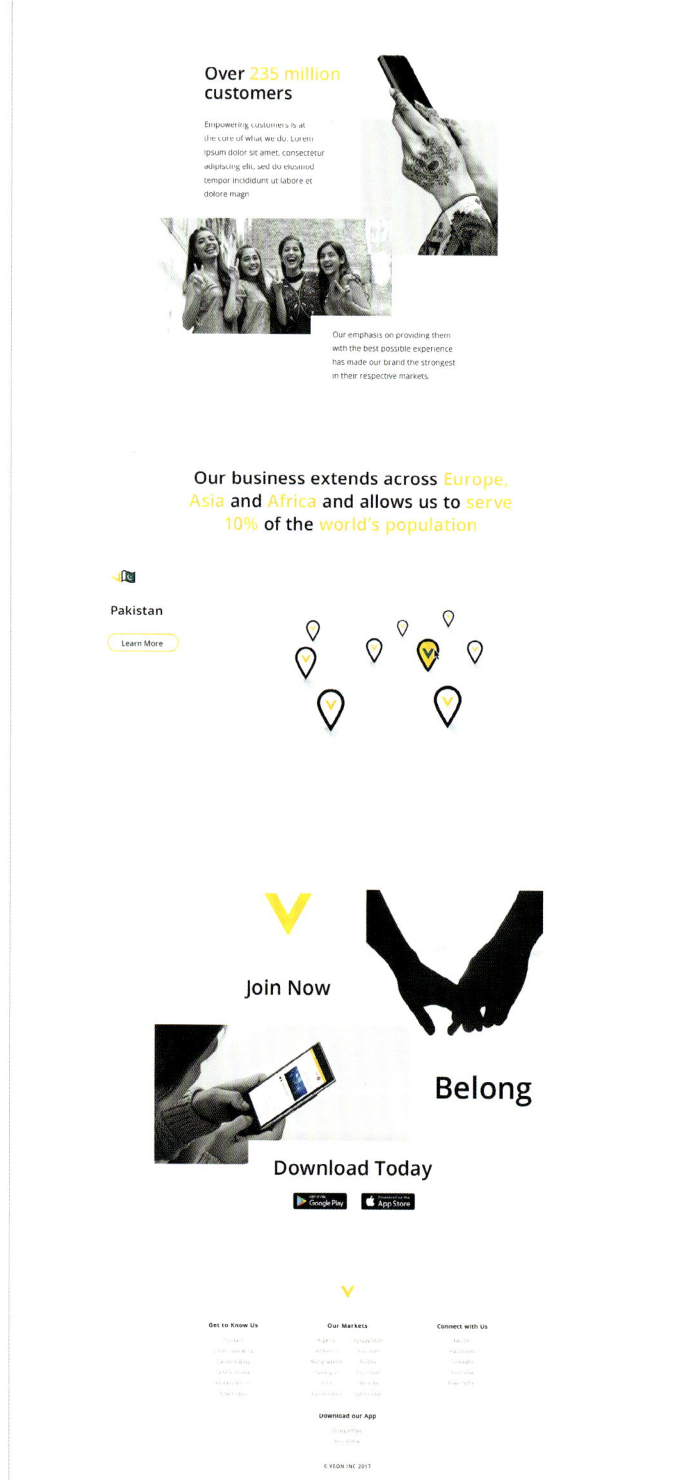
Over 235 million
customers
Empowering customers is at
the core of what we do. Lorem
ipsum dolor sit amet, consectetur
adipiscing elit, sed do eiusmod
tempor incididunt ut labore et
dolore magn
Our emphasis on providing them
with the best possible experience
has made our brand the strongest
in their respective markets.
Our business extends across Europe,
Asia and Africa and allows us to serve
10% of the world's population
Pakistan
Learn More
Join Now
Belong
Download Today
Google Play
App Store
Get to Know Us
Our Markets
Connect with Us
Download our App
© VEON INC 2017

CUSTOMER LOVE
Our Vision & Strategy
Learn More
Our Markets
Pakistan
Manage Your Mobile Plan
Corporate Governance
Learn More
Jazz
Pakistan
Our business extends across Europe, Asia and Africa and allows us to serve 10% of the world's population
Italy
Georgia
Algeria
Tajikistan
Pakistan
Ukraine
Kazakhstan
Russia
Bangladesh
Armenia
Uzbekistan
Kyrgyzstan
Take Control on VEON
Who We Are
VEON Platform
Global Services
Join Us
Investors
News
Be Truly Free
We Are VEON. A Global Provider of Connectivity & Internet Services
Join Us
Do what you love and join our team
Get Started
We Bring Innovation to the World
Chat and Call for Free on VEON
Connect for free on VEON
Learn More
Where are you?
Hello! I'm home!
Find Your Dream Job Now
LOCATIONS
TEAMS
Amsterdam
Lviv
Islamabad
connect 10% of
population
works

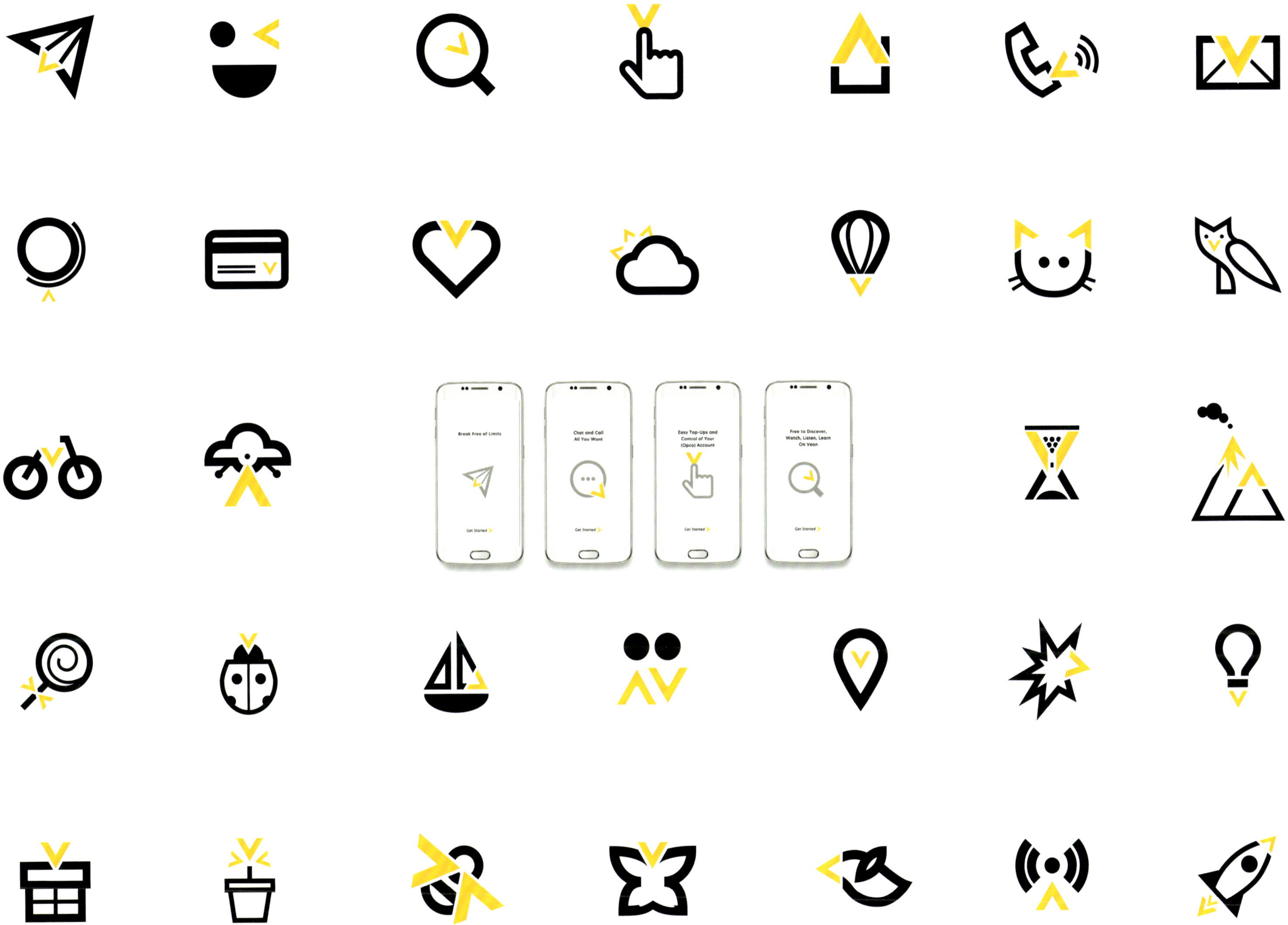

Break Free of Limits
Get Started
Chat and Call All You Want
Get Started
Easy Top-Ups and Control of Your (Opco) Account
Get Started
Free to Discover, Watch, Listen, Learn On Veon
Get Started

BE TRULY FREE

A SPIKE LEE JOINT

VEON BE TRULY FREE

2017

For VEON's presence at the Mobile World Congress 2017 in Barcelona, a short film, *Be Truly Free,* was created to celebrate the true meaning of freedom.

Directed by Oscar-winning filmmaker Spike Lee, this striking and evocative film featured video portraits of people from all cultures and walks of life, who embodied the spirit of freedom in all its many forms.

These deeply personal and candid scenes, captured by Lee's intimate and humanizing lens, were set to a powerful manifesto on the universal need for freedom narrated by the famed director: a stirring vision of the great mosaic of humanity and a powerful evocation of a better, freer, and more equal world.

Free

AdobeStock

LIFECELL

2018

A concept created for Lifecell—a subsidiary of Turkcell, a major telecommunications player in Turkey, Eastern Europe, and the Caucasus—LifeBots were envisioned as whimsical and engaging new guides that could help to make Lifecell's suite of technological innovations more accessible and approachable.

A perfect mix of nature and technology, each LifeBot had its own distinct character, backstory, and particular set of skills that aligned with and helped to illuminate Lifecell's core offerings. Relatable to children and adults alike, Lifebots were charming, adorable, and engaging representatives of the Lifecell brand that ignited imagination and helped to evoke a vision of the future where technology is friendly, accessible, personal, and, above all else, genuinely helpful.

LIFECELL

LIFECELL

LIFECELL

LIFECELL

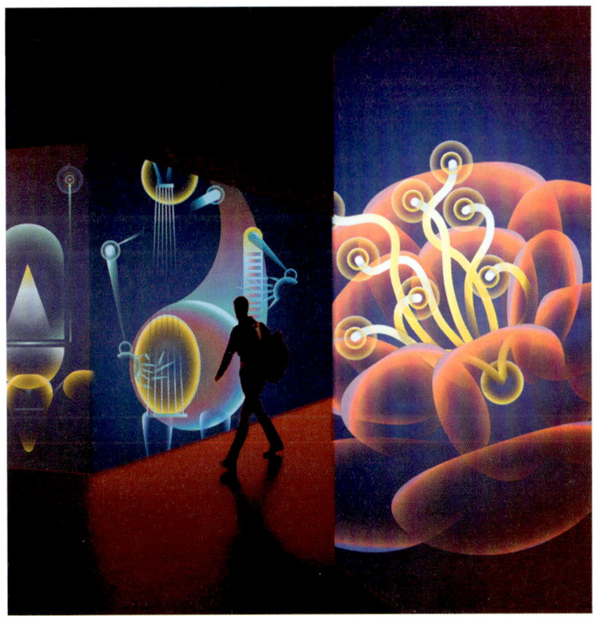

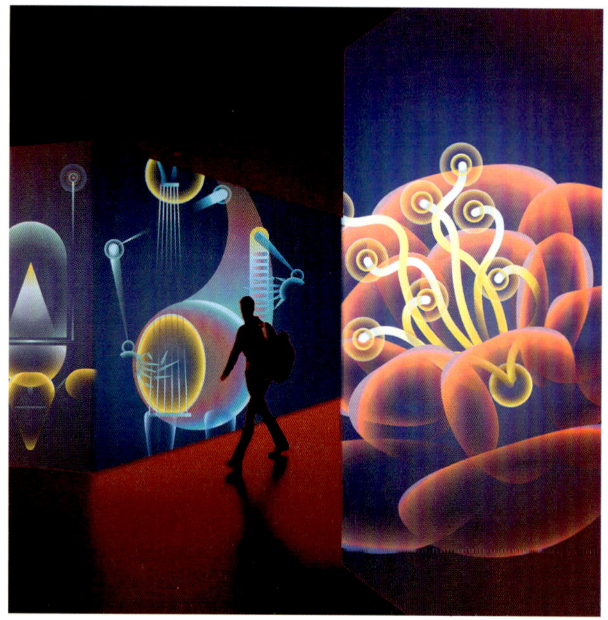

Good Design is
Good Business
Be the Future.
Join the 2019 Turkcell Design-Led Innovation Academy Program
Values and Vision of Excellence
Human Centric
Inclusive Design Thinking
Be the Future.
Join the 2019 Turkcell Design-Led Innovation Academy Program
Values and Vision of Excellence
Led
Innovation
To design is much more than simply to assemble, to order, or even to edit: it is to add value and meaning, to illuminate, to simplify, to clarify, to modify, to dignify, to dramatize, to persuade, perhaps even to amuse. To design is to grow and to transform Turkcell.
Design
Led
Innovation
Design Creates Culture.
Culture Shapes Values.
Values Determine the Future.
Design is How it Works
(Not What it Looks Like)
Good Design is Honest

TDLI

2018

To help Turkish telecommunications company Turkcell foster a company-wide culture of innovation, disruptive thinking, and design excellence, the Turkcell Design-Led Innovation Academy (TDLI) was created to provide a comprehensive Design Thinking curriculum to Turkcell's thousands of employees worldwide. At the heart of this new curriculum was the DO1440 initiative: a set of proactive, action-oriented new strategies for enhancing the culture of innovation at Turkcell that encouraged employees to make the most of each and every one of the 1,440 minutes in a day with fearless, creative, outside-the-box thinking.

To help promote the new TDLI initiative internally to Turkcell employees, a series of eye-catching posters was designed and installed throughout Turkcell's Istanbul headquarters. Featuring bold statements about the power of Design Thinking, these mantras were brought to life through elegant, minimalist, primary-color graphic compositions. Using only red, green, blue, and yellow circles, these posters still managed to capture the essence of the aphorisms they accompanied, and were themselves a testament to the power of good design to communicate complex information simply, clearly, and beautifully.

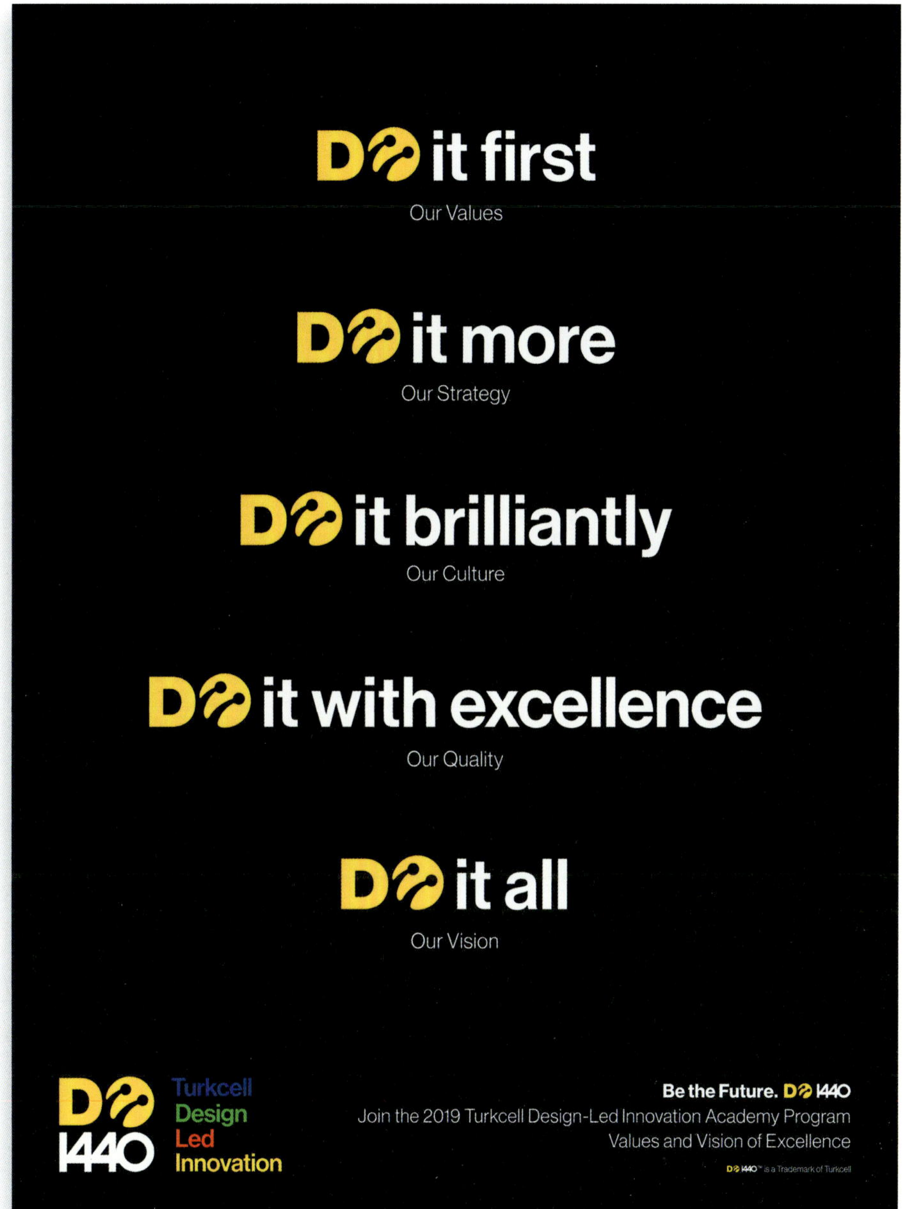

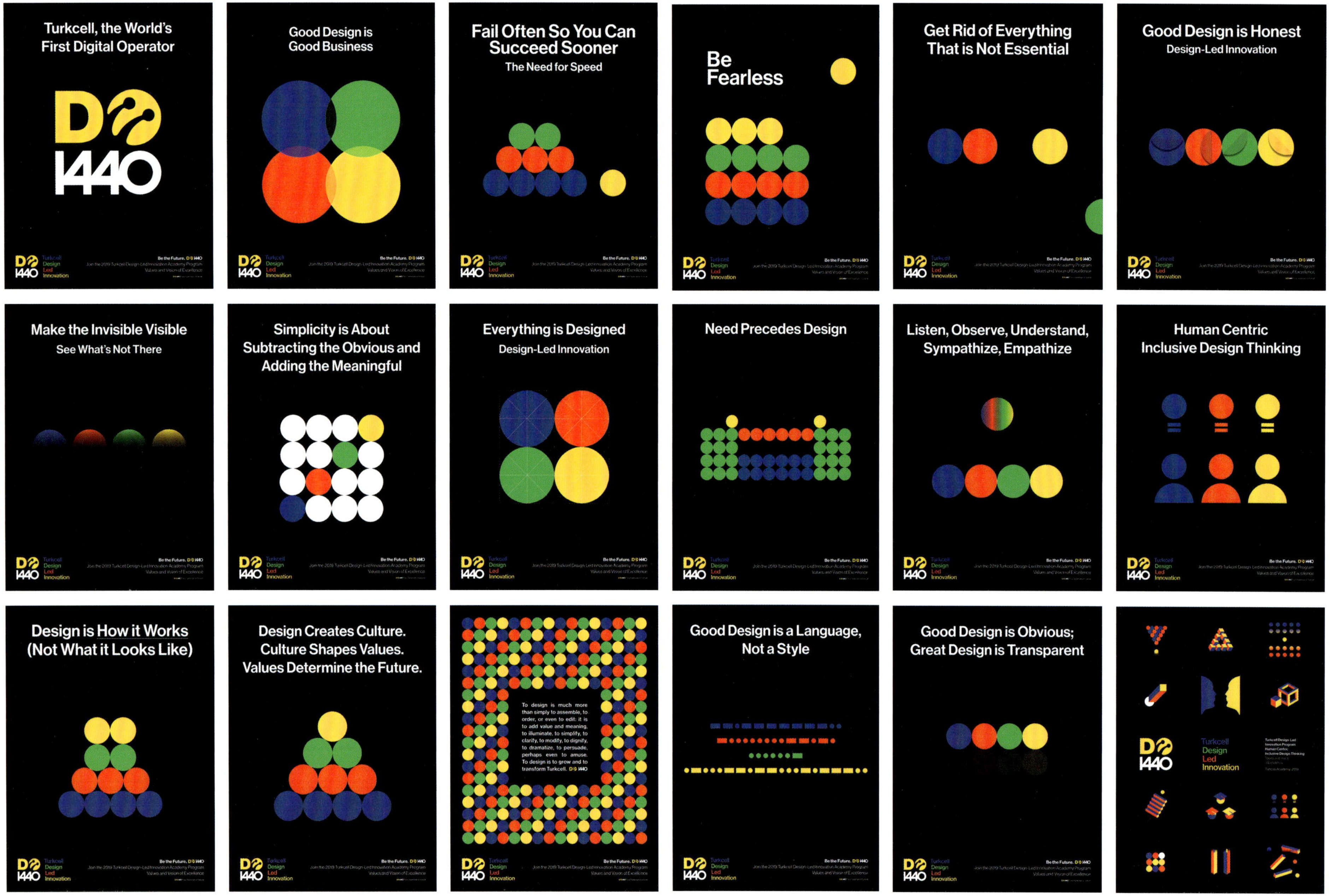
Turkcell, the World's First Digital Operator
Good Design is Good Business
Fail Often So You Can Succeed Sooner
The Need for Speed
Be Fearless
Get Rid of Everything That is Not Essential
Good Design is Honest
Design-Led Innovation
Make the Invisible Visible
See What's Not There
Simplicity is About Subtracting the Obvious and Adding the Meaningful
Everything is Designed
Design-Led Innovation
Need Precedes Design
Listen, Observe, Understand, Sympathize, Empathize
Human Centric Inclusive Design Thinking
Design is How it Works (Not What it Looks Like)
Design Creates Culture. Culture Shapes Values. Values Determine the Future.
To design is much more than simply to assemble, to order, or even to edit: it is to add value and meaning, to illuminate, to simplify, to clarify, to modify, to dignify, to dramatize, to persuade, perhaps even to amuse. To design is to grow and to transform Turkcell.
Good Design is a Language, Not a Style
Good Design is Obvious; Great Design is Transparent
Turkcell Design Led Innovation

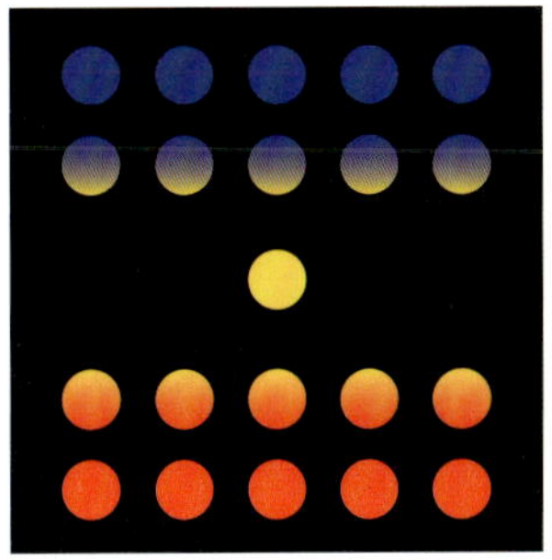

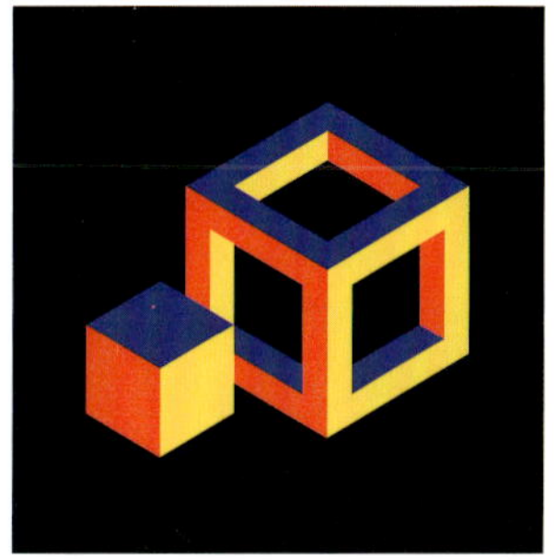
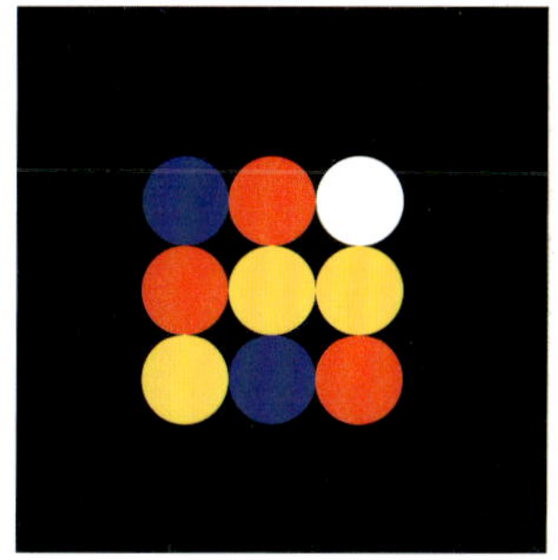

Turkcell
Design
Led
Innovation

Turkcell Design-Led
Innovation Program
Human Centric
Inclusive Design Thinking
Values and Vision
of Excellence

Turkcell Academy 2019

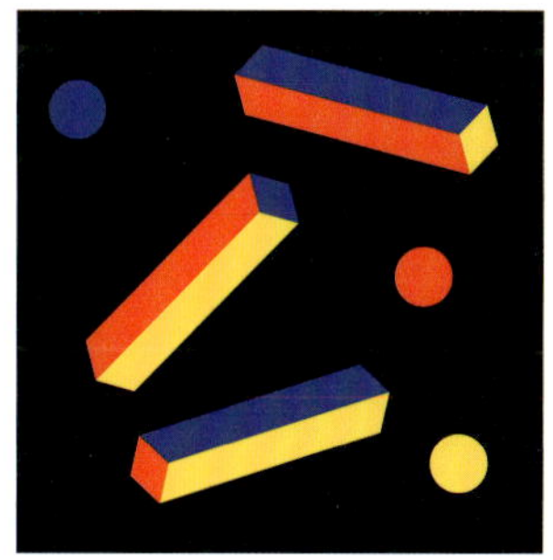

TURKCELL PRESENTS
THE LEGENDS OF CREATIVITY & DESIGN
LANCE ★ WYMAN
GEORGE ★ LOIS
TOM ★ GEISMAR
IN SYNC ★ IN DESIGN ★ ISTANBUL ★ SPRING 2019
PANEL DISCUSSION AND INDIVIDUAL LECTURES HOSTED BY SAGI HAVIV, PETER ARNELL & TOM HARDY
APRIL 22ND CALL FOR TICKETS +90 532 532 0000
D 1440 Turkcell Design Led Innovation
TURKCELL

THE LEGENDS OF CREATIVITY & DESIGN

2018

As part of the Turkcell Design-Led Innovation Academy's new curriculum, plans were made for a lecture series that would bring together some of the world's most prolific design practitioners for panel discussions and solo talks, providing Turkcell's employees the opportunity to learn firsthand from the greatest design minds across a wide range of disciplines.

An early TDLI lecture series proposed convening three of the all-time greats of graphic design: the legendary designers George Lois, Tom Geismar, and Lance Wyman. Lois, best known for his arresting and often subversive *Esquire* magazine covers throughout the 1960s and 70s, was to be joined Geismar, a master of corporate branding who was responsible for iconic logos like Chase Manhattan Bank, Xerox, PBS, and Mobil, and Wyman, the creator of the iconic design campaign for the 1968 Mexico City Olympics. An unprecedented meeting of the minds, the Legends of Creativity & Design, as the event was billed, aimed to provide an unprecedented look into the design philosophy and creative process of three exceptional artists.

To promote the event, a series of posters were designed that gave the meeting of these three legendary designers the excitement and gravitas of an elite sporting event—reminiscent of the newspaper ads and billboards that promoted Muhammad Ali's legendary matchups at the height of his career.

Calls
Discover
Games
100%
4754
Phone Number
Annie Osburn (Group Admin)
Andrew Whitlow (Group Admin)
Michael S. Stovall
Shared Media
GIF
Hold for video, tap for photo
Looking forward to see you tonight
Yeah you too.
Documents
Gallery
Contact
Location
Buzz
Send Money
Follow Me
New Services
Moovit Toplu Taşıma
BiP İçerik
Trending Services
BiP Translate
Catego
bip
go. be. do...
Get BiP'n
Messages
Edit
Chats
Services
Let's Get BiP'N!
To start a new chat, tap + icon and select a contact
Messages
Calls
Done
Deselect All
esterday
Annie Osburn
Tap here for contact info
Incoming Call
19 sec 10:53
Settings
Chat Settings

BIP

2019

BiP is a secure, free-of-charge communications platform that features instant messaging and video chat capabilities, along with access to hundreds of dynamic services, from weather to exchange rates. As part of a coordinated international expansion strategy, a complete redesign of the app UX and UI was undertaken to help make BiP the world's best, most functional, and most delightful communication tool.

This comprehensive overhaul touched every aspect of the app, providing an enhanced user experience through dozens of new functionalities like custom themes, auto-translate, free HD video and voice calling, automatic gifts and giveaways, and more. A new light, bright, and welcoming visual identity helped to enhance BiP's appeal, usability, and accessibility, while a suite of custom emojis and stickers gave users yet another way to express themselves and share every moment.

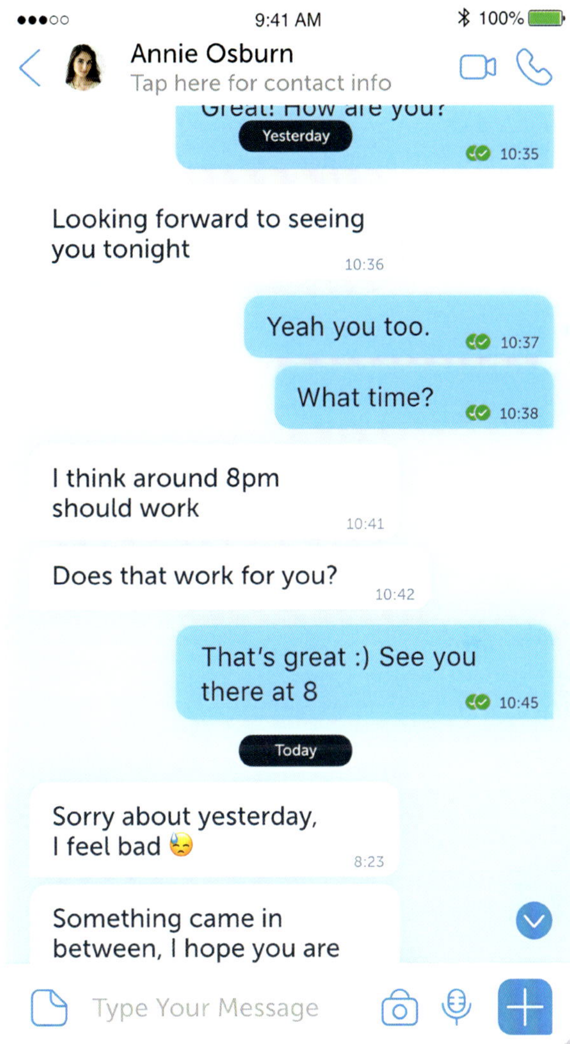

192
Countries in use
23M
App Downloads
5M
Monthly Active Users
80M
Messages Daily
100%
Ensured Security
what's new
what's better
what's now
what's bip
what's bip'n
bip
So much more than messaging
With all that BiP has to offer, how can we not show proudly and boldly to the world that we are already better than what they are using today?
Do More
do
Discover channels & services, play games and more
Be Connected
be
Send photos, documents, videos and share yourself
Let's Go Bip'n
go
Enjoy chat, voice, and video calls with your friends, family, and groups
bip
think bip

1440 FONT

2019

Developed for BiP, the 1440 font was a signature typeface that was designed to be used within the instant messaging and video chat app, as well as in communications, campaigns, collateral materials, and numerous other brand activations.

The sleek, minimalist, sans serif typeface perfectly captured BiP's dynamic and modern character, while also projecting a friendly, welcoming, and distinctly approachable personality.

1440

ROUNDED

a b c d

e f g h

i j k l

m n o p

q r s t

u v w x

y z

YAANI MAIL

2019

Developed for Turkcell, YaaniMail was conceived as an entirely new kind of email experience. Safe, secure, and beautifully designed, the mobile and desktop app UX and UI paired a clean, bright, and friendly design language with streamlined functionality and intuitive features like offline work mode, auto spam detection, and smart filters. YaaniMail also offered advanced security features, with all customer data stored securely on Turkcell's centralized company-run data centers, making YaaniMail the ideal all-in-one solution for personal and business use.

A brand identity system was also developed to accompany the launch of the product: a stylized logo mark inspired by the @ symbol and punctuated by a signature yellow bullseye, and a clean, modern sans serif word mark in bright Yaani blue.

YaaniMail

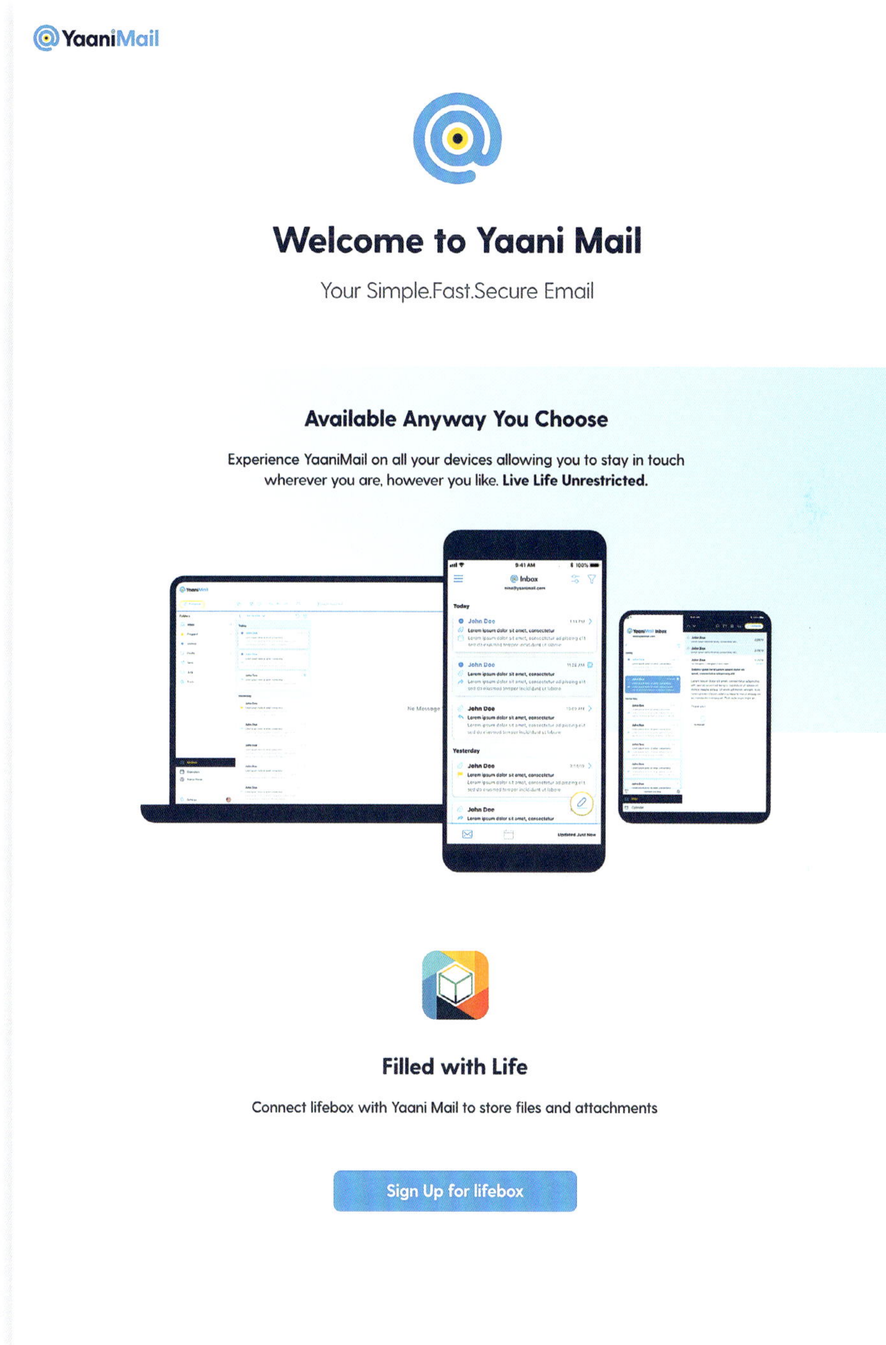

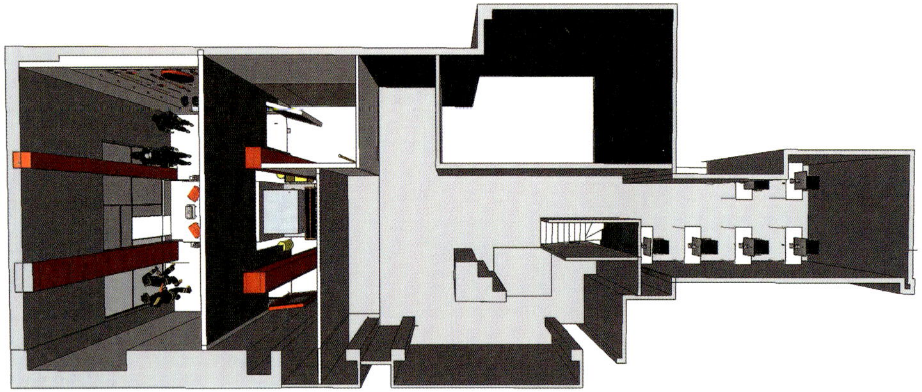

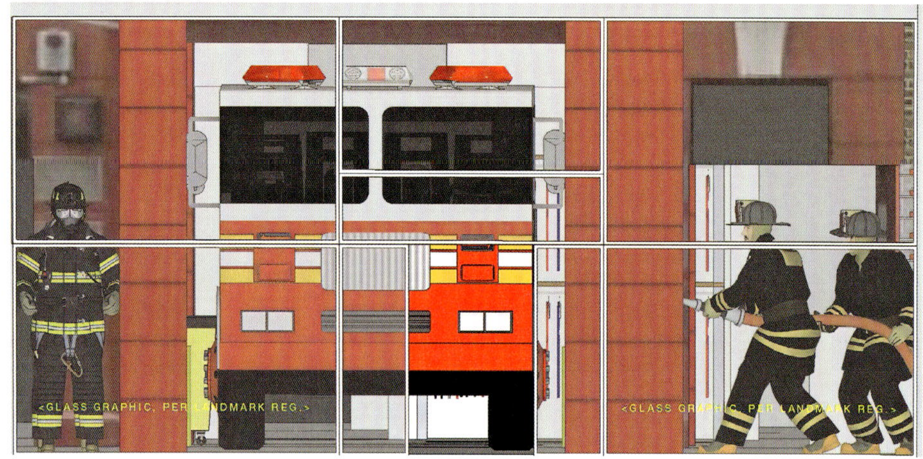
<GLASS GRAPHIC, PER LANDMARK REG.>

55 ENGINE 55

NEW YORK'S BRAVEST
F.D.
N.Y.
55

FDNY FIREZONE

2020

For the FDNY Fire Zone—a state-of-the-art fire-safety learning center and flagship retail space located at New York City's Rockefeller Center—a comprehensive new exhibition design concept was developed to help bring the life-saving work of the FDNY to life.

With a particular emphasis on engaging and inspiring children, the space featured a wide range of interactive elements that allowed younger visitors to step into the shoes of their heroes and truly get a feel for what it means to be a part of New York's Bravest.

At the heart of the Fire Zone experience was an innovative fire simulator that utilized special effects and multimedia to help teach visitors about fire safety and what to do in case of a fire emergency. Visitors left the Fire Zone with the knowledge necessary to eliminate fire hazards in their own homes and to create escape plans for their families.

New merchandising and customer experience strategies were also developed for the flagship retail space where official FDNY-branded merchandise was sold, with one hundred percent of proceeds being cycled back into funding for the FDNY Foundation's life-saving work.

DIGICEL

2020

For Digicel, a global telecommunications company operating in thirty-two markets across the Caribbean, Central America, and the South Pacific, a total brand reengineering effort was undertaken to help the company transform from a mobile carrier to a true Digital Operator.

To reintroduce Digicel to its many customers, new brand communications were developed, featuring candid portraits of Digicel customers doing more with the power of communication at their fingertips—working more efficiently, connecting more meaningfully with friends and family, and simply living happier, better, more connected lives.

Digicel's markets are home to some of the most vibrant cultures in the world, so it only made sense that Digicel should seek to honor the customers it serves by reflecting their regional pride cues back to them. Inspired by the color, energy, and vitality of daily life in these diverse communities, Digicel's new spirit was bright, joyful, and always put the customer front and center. A new global brand mark was also developed to help bring this new energy to life: dubbed the D-Smile, this beaming icon was designed to be a warm, welcoming, and approachable beacon to customers, which would always help to bring a smile to their faces.

In addition to the new overarching brand framework, a complete UX and UI overhaul of Digicel's family of apps and digital services was undertaken to streamline use, offer an enhanced customer service experience, and unify what had previously been a disparate collection of apps and platforms. Comprising D'Music for music streaming, PlayGo for streaming entertainment, SportsMax for premium live sports content, BiP for talk, messaging, and video chat, Loop for all the latest news, GoLoud for podcasts, Billo for cloud storage, and myDigicel for managing every Digicel service, this complete suite of apps and services offered customers everything they needed for their digital life, all within a cohesive, unified, and accessible Digicel ecosystem.

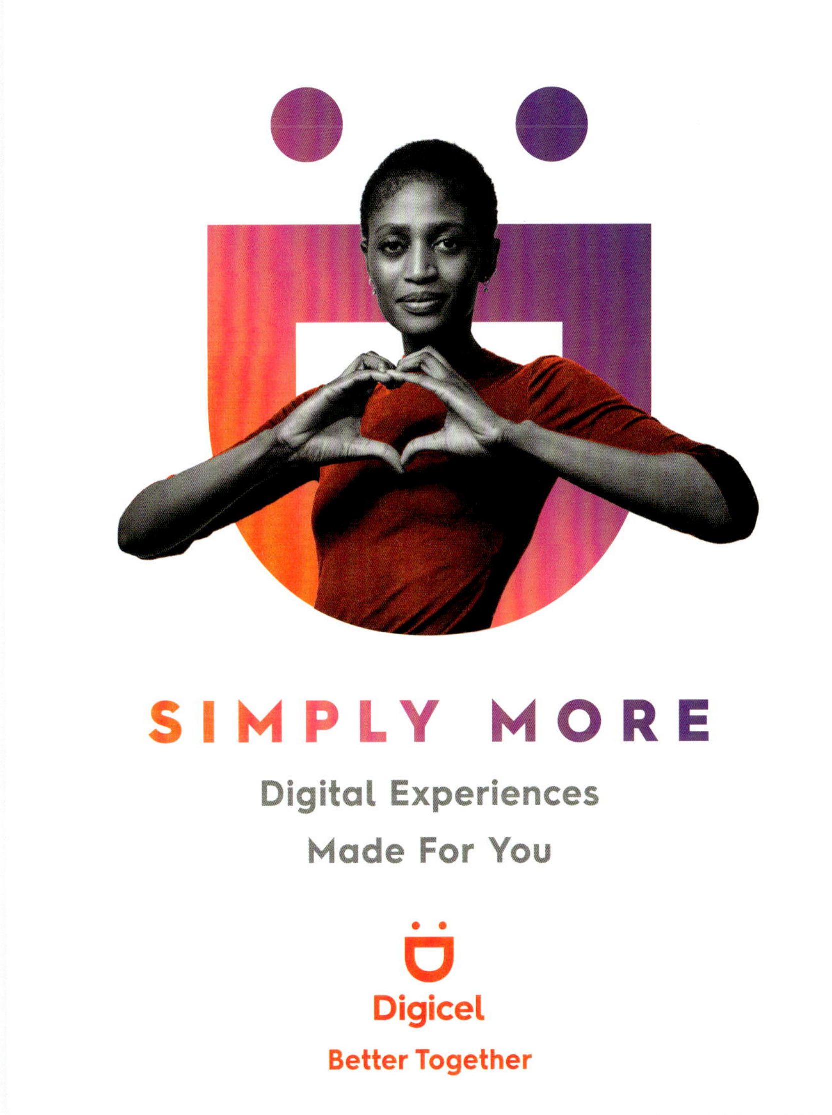

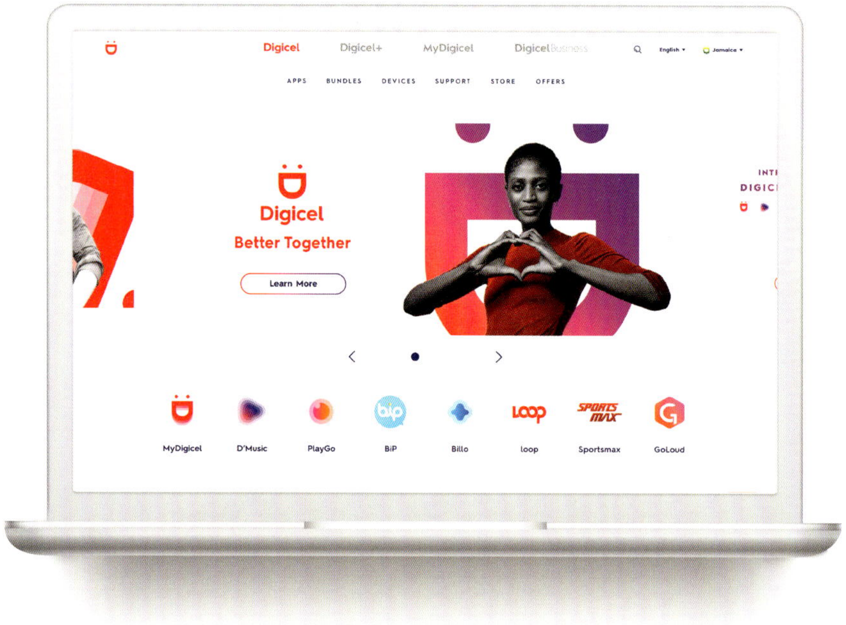
Digicel
Digicel+
MyDigicel
APPS
BUNDLES
DEVICES
SUPPORT
STORE
OFFERS
Digicel
Better Together
Learn More
MyDigicel
D'Music
PlayGo
BiP
Billo
loop
Sportsmax
GoLoud

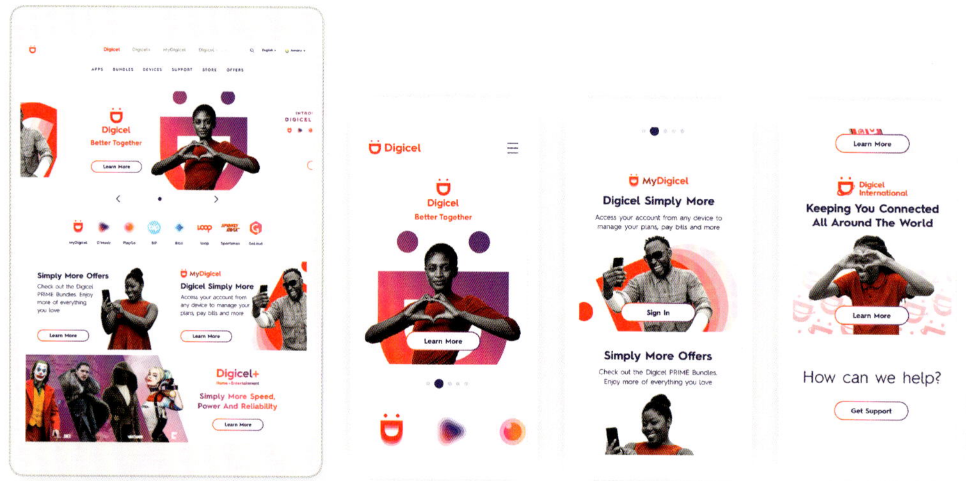
Digicel
Better Together
Learn More
Simply More Offers
MyDigicel
Digicel Simply More
Digicel+
Simply More Speed, Power And Reliability
Digicel Simply More
Access your account from any device to manage your plans, pay bills and more
Sign In
Simply More Offers
Check out the Digicel PRIME Bundles. Enjoy more of everything you love
Learn More
Digicel International
Keeping You Connected All Around The World
How can we help?
Get Support

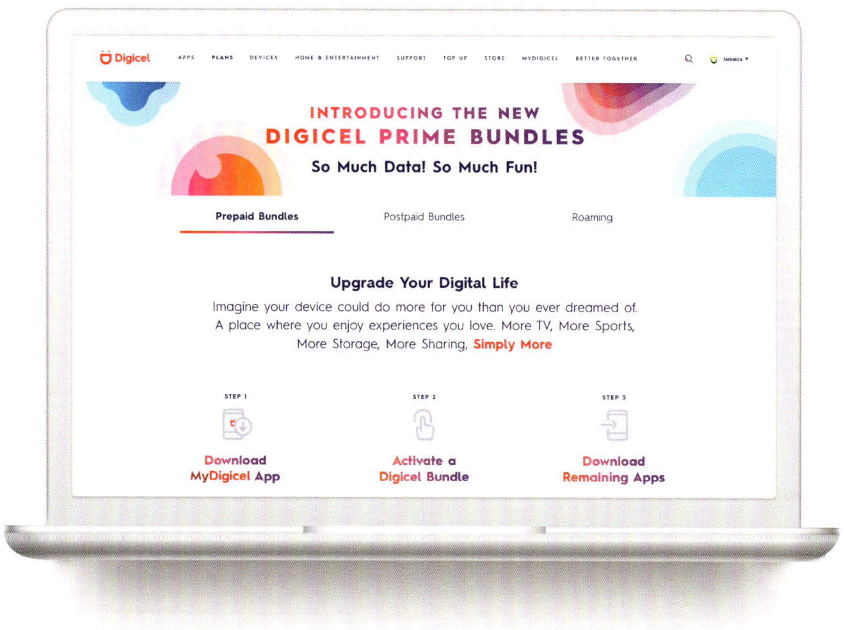
Digicel
APPS
PLANS
DEVICES
HOME & ENTERTAINMENT
SUPPORT
TOP UP
STORE
MYDIGICEL
BETTER TOGETHER
Jamaica
INTRODUCING THE NEW
DIGICEL PRIME BUNDLES
So Much Data! So Much Fun!
Prepaid Bundles
Postpaid Bundles
Roaming
Upgrade Your Digital Life
Imagine your device could do more for you than you ever dreamed of.
A place where you enjoy experiences you love. More TV, More Sports,
More Storage, More Sharing, Simply More
STEP 1
STEP 2
STEP 3
Download MyDigicel App
Activate a Digicel Bundle
Download Remaining Apps

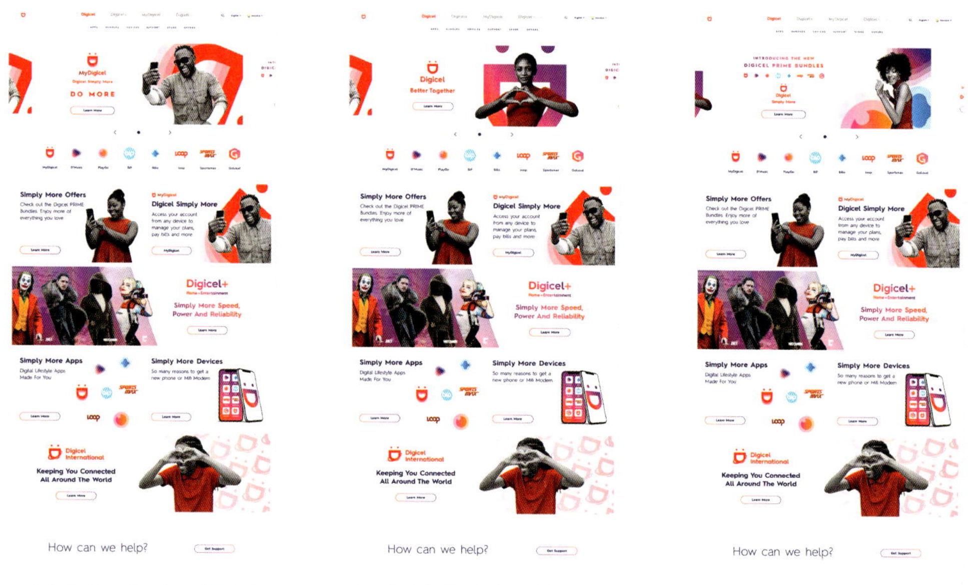
Simply More Offers
Digicel Simply More
Digicel+
Simply More Speed, Power And Reliability
Simply More Apps
Simply More Devices
Keeping You Connected All Around The World
How can we help?

Digicel
Better Together
WATCH 14 HOURS OF
TELEVISION ON PLAYGO
Introducing Our New
Digicel Prime Bundles
7GB of PlayGo for 7 Days
Ask In-Store Today

Digicel
Better Together
LISTEN TO 100 HOURS
OF RADIO ON GOLOUD
Introducing Our New
Digicel Prime Bundles
7GB of GoLoud for 7 Days
Ask In-Store Today

Digicel
Better Together
LISTEN TO 100 HOURS
OF SONGS WITH D'MUSIC
Introducing Our New
Digicel Prime Bundles
7GB of D'Music for 7 Days
Ask In-Store Today

INTRODUCING THE NEW
DIGICEL PRIME BUNDLES
Digicel
Better Together

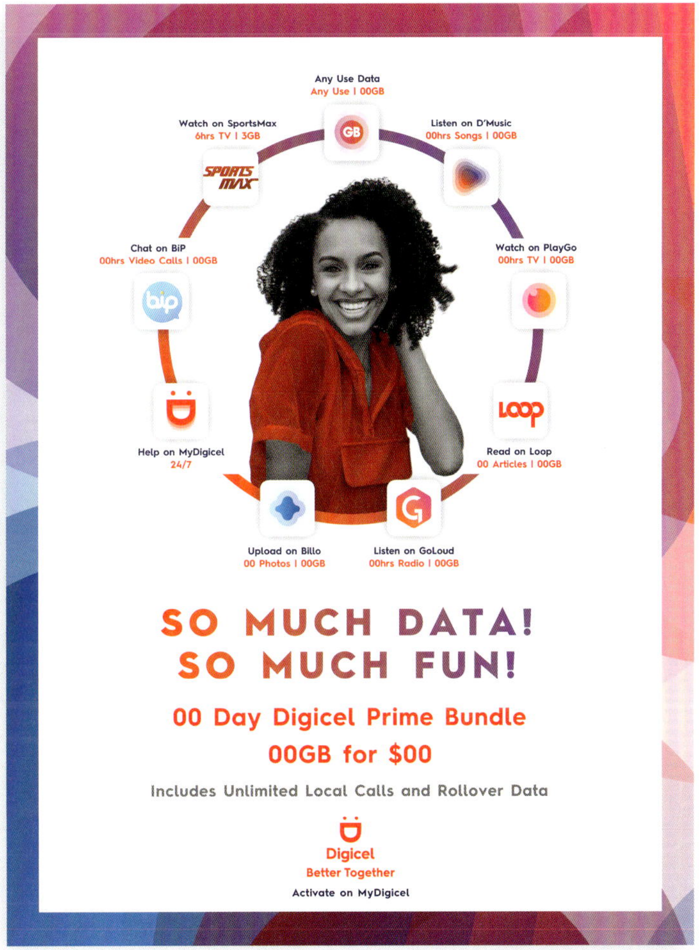
Any Use Data
Any Use I 00GB
Watch on SportsMax
6hrs TV I 3GB
Listen on D'Music
00hrs Songs I 00GB
Chat on BiP
00hrs Video Calls I 00GB
Watch on PlayGo
00hrs TV I 00GB
Help on MyDigicel
24/7
Read on Loop
00 Articles I 00GB
Upload on Billo
00 Photos I 00GB
Listen on GoLoud
00hrs Radio I 00GB
SO MUCH DATA!
SO MUCH FUN!
00 Day Digicel Prime Bundle
00GB for $00
Includes Unlimited Local Calls and Rollover Data
Digicel
Better Together
Activate on MyDigicel

bip

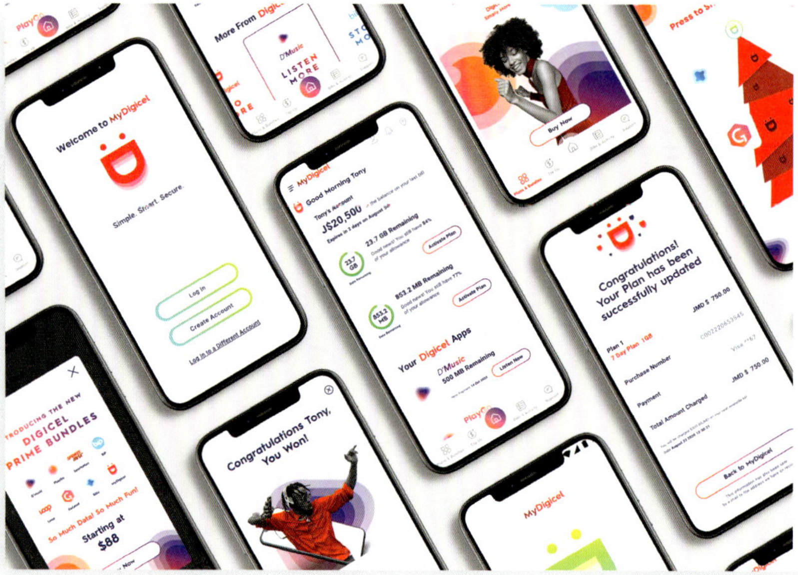
Welcome to MyDigicel
Simple. Smart. Secure.
Log In
Create Account
Good Morning Tony
Your Digicel Apps
Congratulations! Your Plan has been successfully updated
Congratulations Tony, You Won!
Introducing the new Digicel Prime Bundles
So Much Data! So Much Fun!
Starting at $88

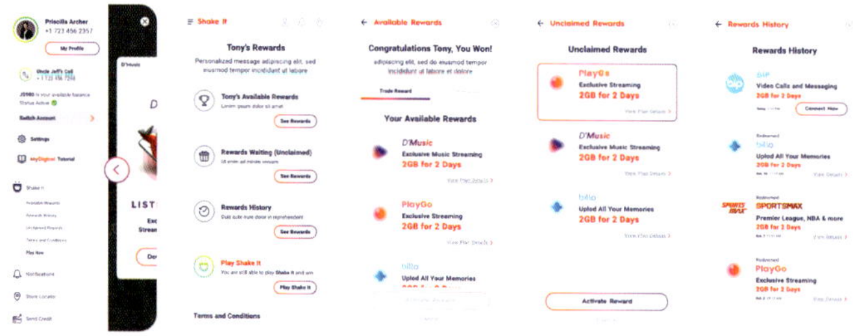

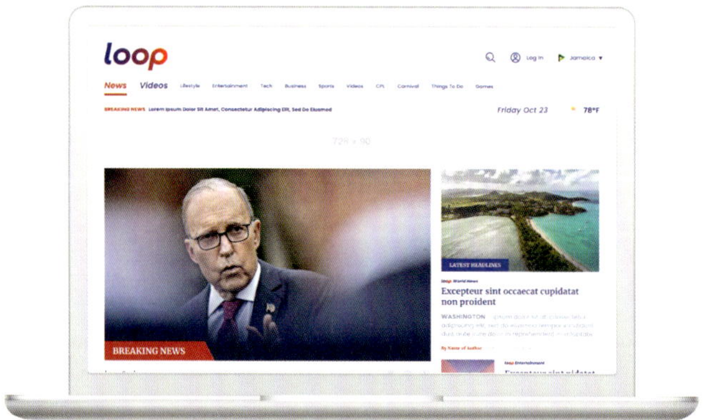
loop
News
Videos
BREAKING NEWS
Excepteur sint occaecat cupidatat non proident

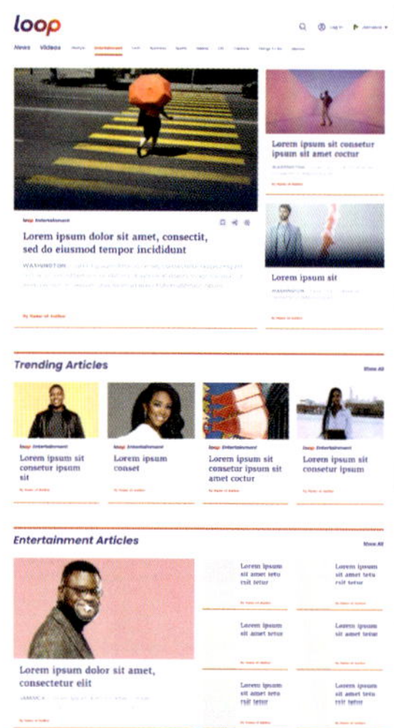
loop
Trending Articles
Entertainment Articles

Entertainment Articles
Sponsored Stories
loop

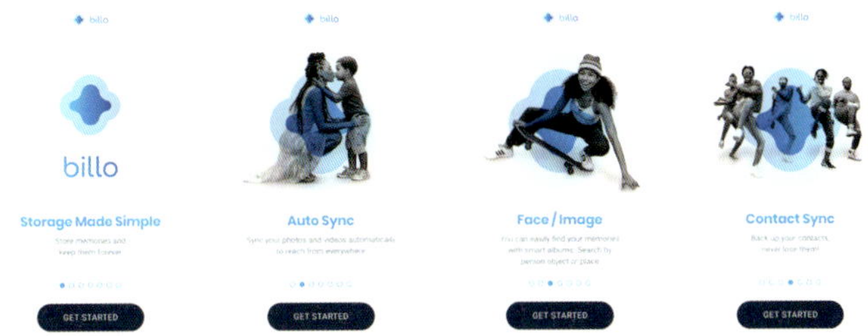
billo
Storage Made Simple
GET STARTED
Auto Sync
GET STARTED
Face / Image
GET STARTED
Contact Sync
GET STARTED

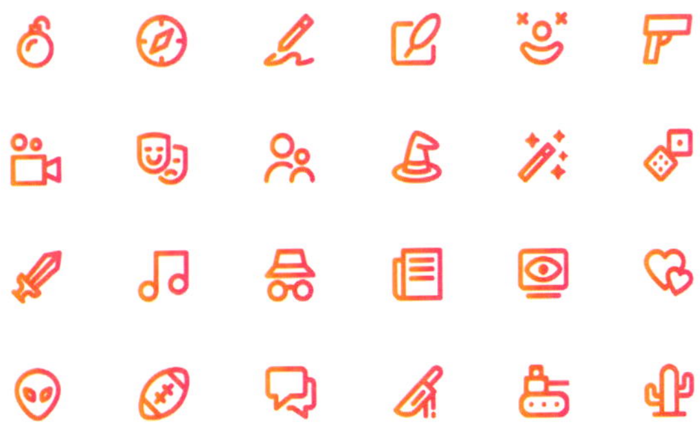

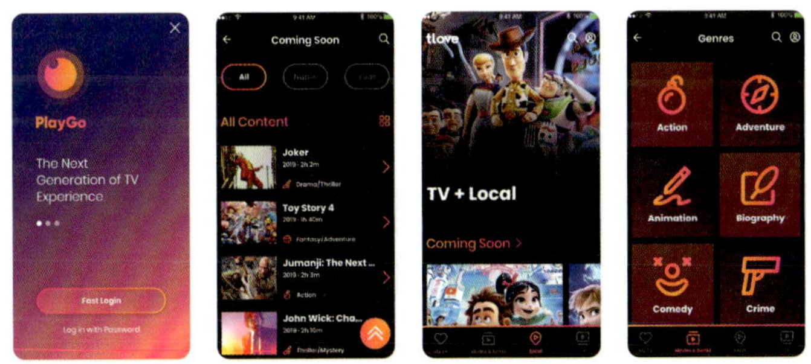
PlayGo
The Next Generation of TV Experience
Fast Login
Coming Soon
All Content
Joker
Toy Story 4
Jumanji: The Next ...
John Wick: Cha...
TV + Local
Coming Soon
Genres
Action
Adventure
Animation
Biography
Comedy
Crime

D'Music
FEEL IT!
CARIBBEAN'S
SICKEST
MUSIC APP

SOCA
DANCE HALL
HOTTEST LOCAL ARTISTS

THE SOUND OF YOUR LIFE
D'Music
Simply More Music
Download Today

YOUR ACCOUNT
YOUR PROFILE
Manage your Favorites and the content you Follow
YOUR SETTINGS
General Settings
Reminders
Choose Your Theme
HELP
Contact Us
Support Request
Give Feedback
LIVE CHANN
View All
DON'T MISS OUT
HOME
LIVE TV
VIDEOS
STORIES
EPL
NBA
Add Team
YOUR TEAMS
Manchester United
LA Lakers
Find and
FAVORITE
Find Sport, Leagues or Teams
Campions League
Premier Leagu
Powered by Digicel
WELCOME TO
SPORTSMAX
Skip
et acmus sire dignissimos
View All
CHANNELS
ADD LEAGUE OR TEAM
Find and follow your
FAVORITE leagues or teams

SPORTSMAX

2020

For SportsMax, the Caribbean's most popular sports streaming platform, a total brand reinvention effort was created, encompassing a new brand image and identity, along with a completely revamped UX and UI for SportMax's signature app.

Conceived of as the definitive source for sports, the new SportsMax app featured a massive selection of live-streaming sporting events from around the world, plus a vast video archive, up-to-the-second news highlights, and an online social hub for SportsMax's global community.

The reimagined app, layered in next-generation features and functionality, transformed SportsMax from a simple streaming app to a fully-fledged virtual destination that truly immersed fans in the action and excitement of their favorite sports.

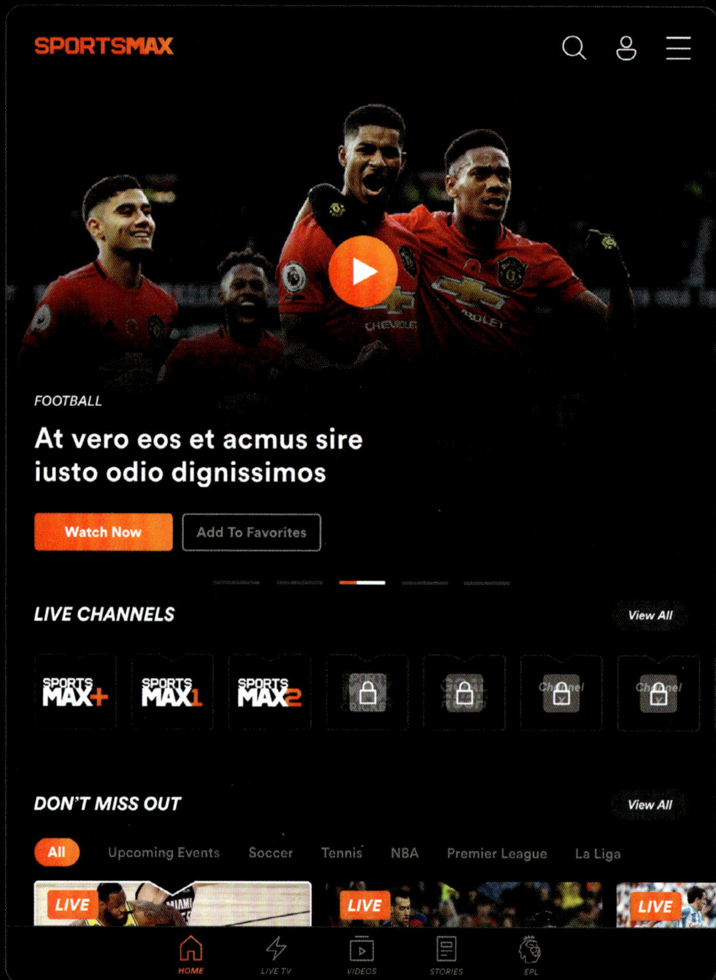
SPORTSMAX
FOOTBALL
At vero eos et acmus sire iusto odio dignissimos
Watch Now
Add To Favorites
LIVE CHANNELS
View All
SPORTS MAX+
SPORTS MAX1
SPORTS MAX2
DON'T MISS OUT
View All
All
Upcoming Events
Soccer
Tennis
NBA
Premier League
La Liga
LIVE
LIVE
LIVE
HOME
LIVE TV
VIDEOS
STORIES
EPL
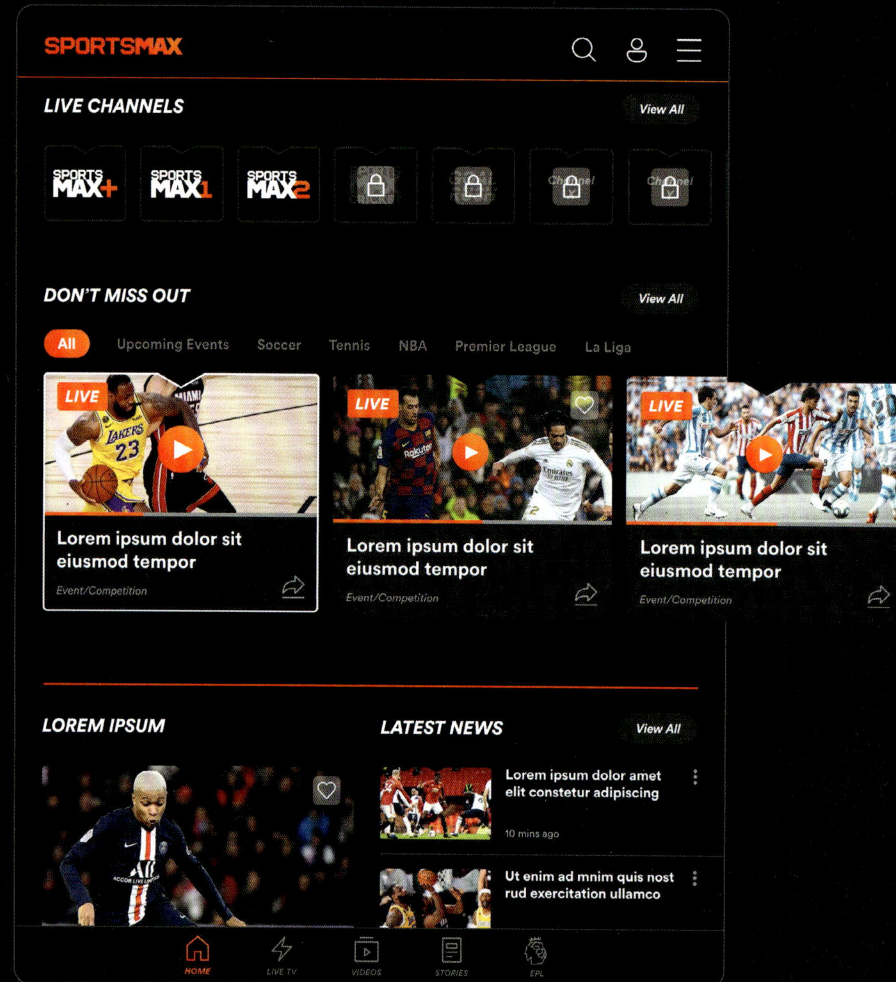
SPORTSMAX
LIVE CHANNELS
View All
SPORTS MAX+
SPORTS MAX1
SPORTS MAX2
DON'T MISS OUT
View All
All
Upcoming Events
Soccer
Tennis
NBA
Premier League
La Liga
LIVE
Lorem ipsum dolor sit eiusmod tempor
Event/Competition
LIVE
Lorem ipsum dolor sit eiusmod tempor
Event/Competition
LIVE
Lorem ipsum dolor sit eiusmod tempor
Event/Competition
LOREM IPSUM
LATEST NEWS
View All
Lorem ipsum dolor amet elit constetur adipiscing
10 mins ago
Ut enim ad mnim quis nost rud exercitation ullamco
HOME
LIVE TV
VIDEOS
STORIES
EPL

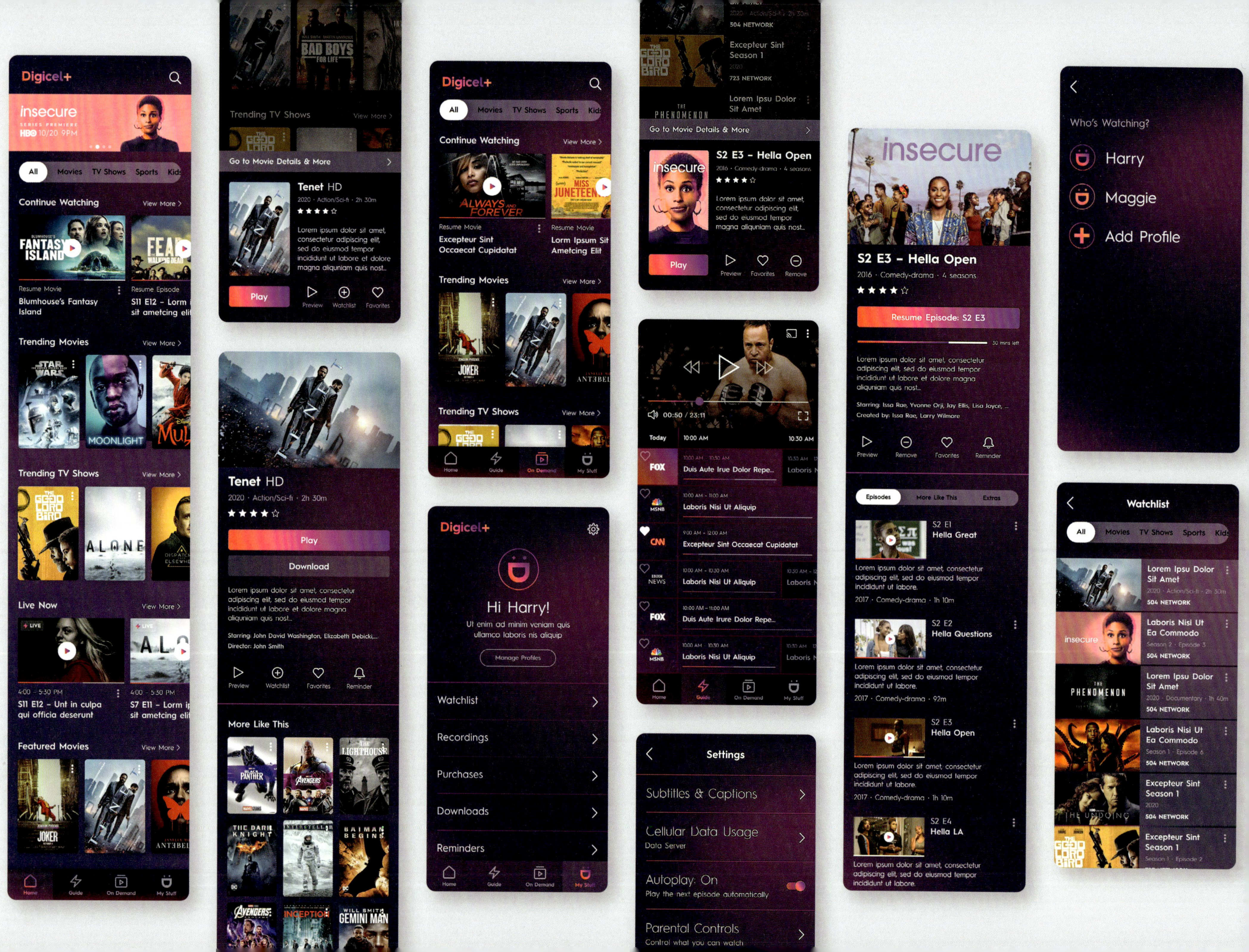
Digicel+
insecure
All
Movies
TV Shows
Sports
Continue Watching
View More
Resume Movie
Blumhouse's Fantasy Island
Resume Episode
Trending Movies
Trending TV Shows
Live Now
Featured Movies
Home
Guide
On Demand
My Stuff
Go to Movie Details & More
Tenet HD
2020 · Action/Sci-fi · 2h 30m
Play
Preview
Watchlist
Favorites
Download
Starring: John David Washington, Elizabeth Debicki...
Director: John Smith
Reminder
More Like This
Excepteur Sint Occaecat Cupidatat
Lorm Ipsum Sit Ametcing Elit
Hi Harry!
Ut enim ad minim veniam quis ullamco laboris nis aliquip
Manage Profiles
Watchlist
Recordings
Purchases
Downloads
Reminders
S2 E3 – Hella Open
Remove
Today
10:00 AM
10:30 AM
Laboris Nisi Ut Aliquip
Excepteur Sint Occaecat Cupidatat
Settings
Subtitles & Captions
Cellular Data Usage
Data Server
Autoplay: On
Play the next episode automatically
Parental Controls
Control what you can watch
2016 · Comedy-drama · 4 seasons
Resume Episode: S2 E3
Starring: Issa Rae, Yvonne Orji, Jay Ellis, Lisa Joyce, ...
Created by: Issa Rae, Larry Wilmore
Episodes
More Like This
Extras
S2 E1
Hella Great
S2 E2
Hella Questions
S2 E3
Hella Open
S2 E4
Hella LA
Who's Watching?
Harry
Maggie
Add Profile
Watchlist
Lorem Ipsu Dolor Sit Amet
504 NETWORK
Laboris Nisi Ut Ea Commodo
Excepteur Sint Season 1

DIGICEL+

2020

Digicel+ was developed as a new unified platform for all of Digicel's home and entertainment offerings. As part of this multipronged effort, a new naming strategy, brand identity, and launch communications were developed, along with totally reengineered UX and UI for the namesake Digicel+ app.

The Digicel+ app was designed to offer a seamless multiscreen viewing experience: whether watching on a TV, laptop, tablet, or phone, the user experience was designed to be easy, intuitive, and delightful. The beautiful and elegant design of the app was matched by enhanced functionality that made it easier than ever for customers to access a world of entertainment, customized and personalized just for them.

The design of the Digicel+ brand mark was a clever and inviting adaptation of Digicel's global brand icon, the D-Smile. With a plus symbol substituted for one of its "eyes," the colorful Digicel+ logo seemed to give a little wink: a lighthearted way of connecting with customers and assuring them that Digicel+ has got them covered for all of their home and entertainment needs.

Launch communications for the new Digicel+ app featured images of the countless movies, TV shows, and live sporting events available to enjoy, all backed by the overarching Digicel+ brand promise of "Simply More Power and Reliability."

DB

DB

DB

DIGICEL BUSINESS

2020

Digicel Business offers custom solutions tailor-made to meet the specific needs of businesses: whether large or small, and across countless industries, Digicel Business provides everything needed to keep a business up and running, efficiently and seamlessly.

As part of the overarching Digicel brand reboot initiative, a new visual identity and online presence were developed for Digicel Business. Featuring an inviting graphic style and a bright and cheerful color palette, this new look and feel lent a friendlier face to Digicel Business, while intuitive, universally-comprehensible icons and illustrations helped to make the many products, services, and solutions offered by Digicel more accessible, engaging, and, most of all, useful.

QatarAmerica
INSTITUTE FOR CULTURE

QAIC

2020

Founded in 2017, the Qatar America Institute for Culture is an independent non-profit organization that cultivates artistic expression and facilitates a cultural dialogue between the United States, Qatar, and the larger Arab and Islamic worlds. Through art exhibitions, educational programs, scholarly research, charitable endeavors like the Qatar America Network for Autism initiative, and a range of other cross-cultural partnerships, QAIC provides interactive experiences in an inclusive environment to celebrate and appreciate art and culture.

To help this organization maximize its reach, elevate its stature, and further broaden its mission, a comprehensive new brand image concept was developed for execution across multiple platforms and touchpoints.

The new logo mark—a bold serif Q abutting an abstracted A letterform—was designed to be an iconic symbol that conveys the spirit of openness, optimism, and unity that is at the heart of QAIC's mission and mandate. This new brand mark represented a truly harmonious cultural hybrid of Qatar and America and served as an invitation for people from all nations to come together in the name of collaboration, exchange, and steady progress toward a brighter future.

In addition to this new logo, a redesigned web presence was proposed, alongside concepts for social media campaigns, print and digital collateral, and a range of QAIC-branded merchandise.

TB12

TB12

2020

TB12 is a holistic wellness and performance brand founded by American football legend and seven-time Super Bowl champion Tom Brady, in partnership with his longtime trainer Alex Guerrero. Rooted in the personal performance regimen of one of the greatest athletes of all time, TB12 represents the culmination of Brady and Guerrero's decades-long collaboration to redefine human performance on holistic and functional terms. TB12's mission is to provide a complete suite of tools to empower people to live well and perform at their very best, with a particular focus on longevity: helping people to do what they love, better and for longer. TB12 provides customers with everything they need to realize their wellness aims and push the limits of their potential.

Working closely with Tom, the existing brand identity and mandate were clarified, strengthened, and streamlined, while a variety of bold new products and brand experiences were also developed—including a range of daily supplements and high-protein snacks grouped by their more essential wellness functions: Protect, Perform, and Recover.

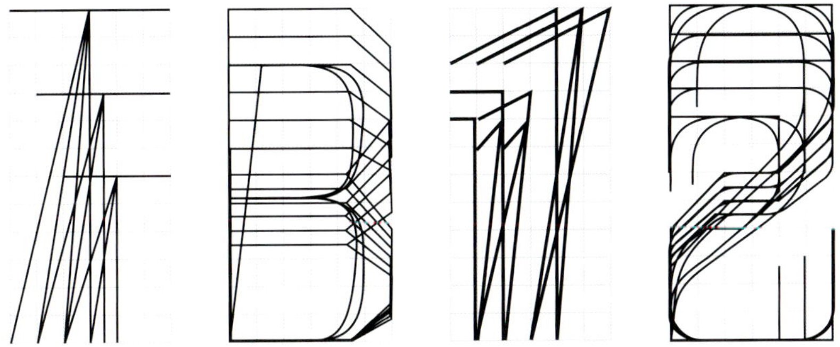

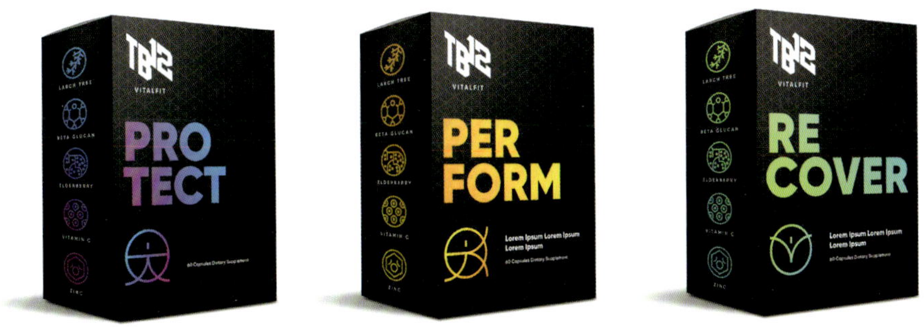

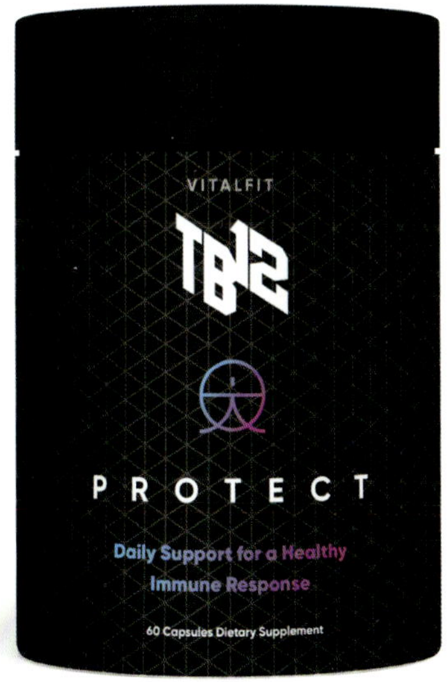

KOBALT TOTES

2020

Informed by the dimensionalized geometric language of the hexagon, this flagship collection of Kobalt's next-generation storage platform embodies the brand's ethos of strength, economy, and modularity.

Boasting best-in-class durability and features, the three totes and the sportsman trunk feature a 360 degree sealing gasket for weather-resistance, an ergonomic fatigue-mitigating handle assembly, and convenient transparent lid.

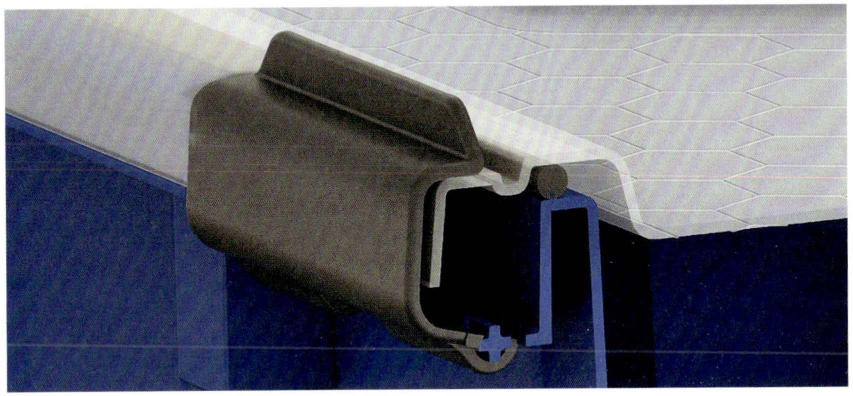

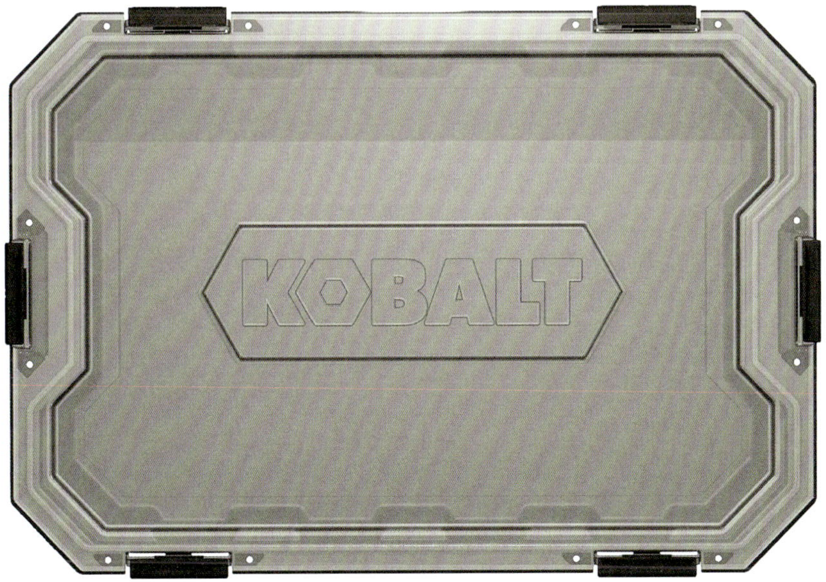

ar

ALLEN + ROTH

2020

Known for its well-designed home furnishings, accessories, and fixtures, Lowe's Allen + Roth brand has always embodied a sophisticated but comfortable aesthetic that is equally welcome in both modern and traditional styles—and everything in-between. Offering pieces in a wide range of styles at an accessible price point, Allen + Roth offers something for every type of customer, and every type of home.

To help the brand further broaden its appeal and stand out among its many competitors, a brand rejuvenation effort was rolled out across all consumer touchpoints. A gentle evolution rather than a dramatic reinvention, the refreshed Allen + Roth presented a more refined and upmarket air, and spoke to a greater diversity of tastes and lifestyles. The Allen + Roth word mark was subtly updated with a more elegant serif typeface and a rich, warm gray tone—the brand's new signature color—while an accompanying monogram logo mark was also introduced to provide the Allen + Roth brand with an instantly recognizable symbol that could be emblazoned across products, packaging, in-store displays, brand communications, social media, and more.

allen + roth

KOBALT SCREWDRIVER

2020

The development of next-generation Kobalt products was driven by the ethos to differentiate and improve—to both embody the unique visual identity architecture of the Kobalt brand, while also exercising a critical focus on improving product experience and performance for the end user.

With the Kobalt hexagon at its core, the design language was formed as a synthesis of brand and function. The handle geometry offers versatility in function, improved ergonomics, and visual brand differentiation. The resulting proprietary design utilizes faceted surfacing to maximize surface area for improved leverage, grip, and overall purchase with the tool.

The Helical Torque Zones are located at the base and upper end of the grip geometry. These zones improve handle purchase and leverage during rotation of the tool—increasing torque transfer without compromising comfort. Anchor Zones are located in the center of the grip geometry. These zones provide additional purchase area for fingers, which assists in stabilizing and driving the handle during operation.

A color-coded ring was incorporated into the handle design to assist in quick identification of the various driver tips.

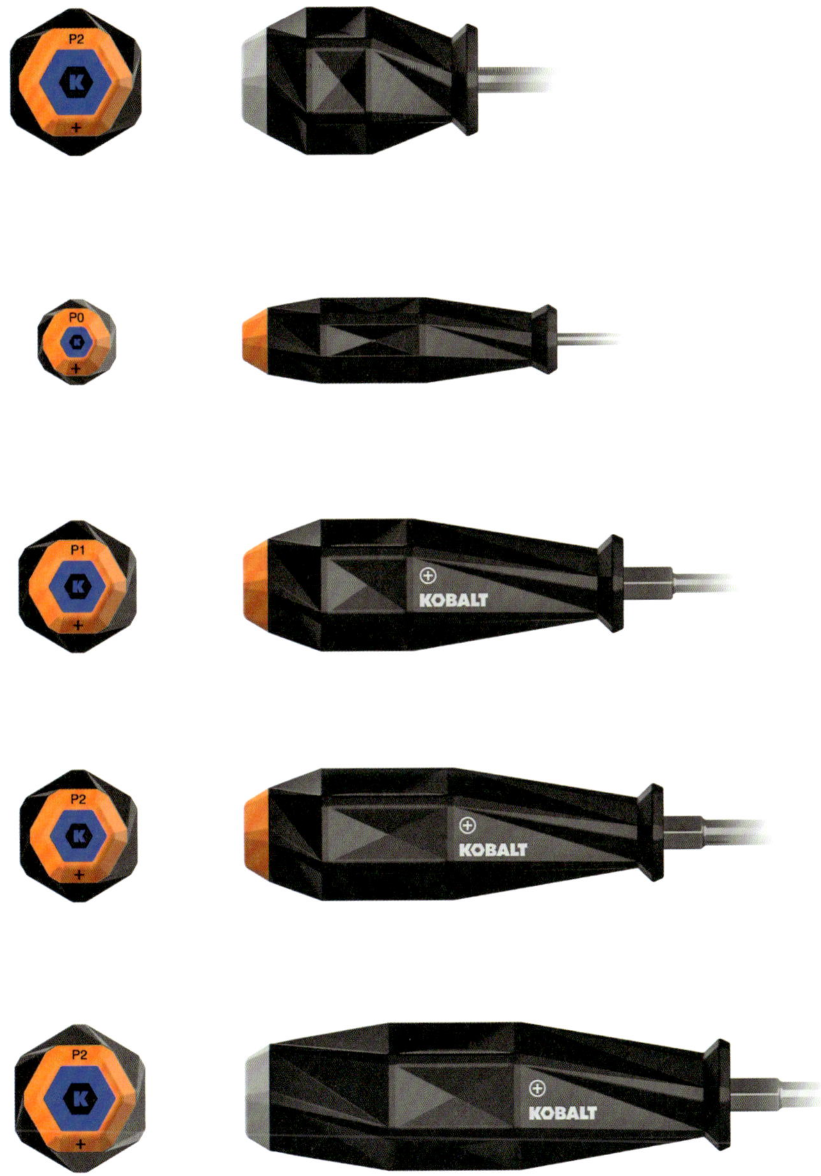

ACKNOWLEDGMENTS

The process of putting together this book has demanded that I spend time reconstructing memories and moments over the past forty years and recognizing the many talented and gifted people who have helped to make this whole extraordinary journey possible. Great projects always involve a team of collaborators, and I wish to share my thanks with the many who have been instrumental in bringing the work and this book to life.

First, I would like to thank Nadine Barth for her tireless effort in helping to publish this book, and who challenged me for years to share this work, as well as her colleagues Claire Cichy, Stefanie Kruszyk, Sonja Altmeppen, Aaron Bogart and all the wonderful people at Hatje Cantz.

I also thank Liyuan Tong, Joanne Min, Yuko Yamazaki, Vincenzo Piscitelli, Tristan Thom, and Dan Zuzunaga for their creativity, dedicated work, and endless contributions to this book.

Thank you to Frank Gehry for your bracing honesty, mentorship, and abiding friendship.

Thank you to Peter Lindbergh for allowing me to be in the presence of your greatness and for always demonstrating such humility, grace, and love to all those around you—you are sorely missed.

Thank you to Michel Marriot for your brilliant, kind, powerful words, and for your decades of friendship.

Thank you to Ted Bickford, my original partner in the Arnell-Bickford Agency, Sara Arnell and our first employees, Jack Ader, Barbara Lederberg, Cathy Hantz, and Lisa Stevens, with whom, and because of whom, it all began.

And thank you to my closest of friends, whose camaraderie, generosity, and encouragement have been invaluable and deeply appreciated: Tom Brady, Tie Domi, Robert Nardelli, Muktesh Pant, Alexey Reznikovich, Jeff Soffer, Richard Stark, and Andrew Rosen.

I've had the good fortune to have been inspired by extraordinary people who have shared opportunities and their friendship with me throughout the years, including: Alberto Alessi, Richard Avedon, Buzz Aldrin, Muhammad Ali, Loni Ali, Chris Bishop, Naomi Campbell, Tony Carbonetti, Bill Clinton, Celine Dion, Carla Fendi, Diane von Furstenberg, Michael Graves, Paul Gredinger, Tommy Hilfiger, David Hockney, Jimmy Iovine, Michael Jackson, Jadakiss, Quincy Jones Donna Karan, Jeffrey Katzenberg, John F. Kennedy Jr., Christian Lacroix, Karl Lagerfeld, Phyllis Lambert, Spike Lee, Jay Leno, Jet Li, George Lois, Peter Lopez, Jim Lovell, Lorne Michaels, Gian Franco Monicelli, Tommy Mottola, Yao Ming, Gwyneth Paltrow, Jean-Michel Pinault, Paul Rand, Aldo Rossi, Austin Rosen, Maury Satin, Howard-Yana Shapiro, Eunice Kennedy Shriver, Maria Shriver, Timothy Shriver, Steven Spielberg, Sylvester Stallone, James Stirling, Martha Stewart, Lynn Tierney, Tina Turner and Erwin Bach, André Leon Talley, Steven Tyler, Mark Wahlberg, John Carl Warnecke, Margot Warnecke, Lance Wyman, and Bernie Yurman to name only a few.

Thank you to my many collaborators who have honored me with their trust, support, and unwavering faith: Fred Agrusa, Jason Binn, Patti Cohen, Edwin Chan, Carrie Donovan, Miles O'Brien, Ron Feemster, Raymond Foye, Christian Haub, Dr. Fred Kass, Steve Kettmann, Rosemary McGrotha, Dawn Mello, Lisa Peake, Sid Rapoport, Sonia Ryzy-Ryski, Norman Schulman, Ann Thomopoluos, Tony To, John Varnarthos, Carol Vogel, Lawrence Weschler, and Scott Wittman.

Thank you to the clients, partners, and mentors who took a leap of faith and invited me to join them and bring their work to life: Bill Boltz, Jean-Yves Charlier, Massimo d'Amore, Mickey Drexler, Paul Fireman, Bob Gamgort, Richard Golub, Christiaan Houtenbos, Jerry Inzerillo, Solomon Kerzner, Ray Lane, JY Lee, Miky Lee, Peter Lloyd, Indra Nooyi, Denis O'Brien, Bruce Ratner, Thomas Rhee, Tomio Taki, Kaan Terzioğlu, Cathy Volker, Alain Wertheimer, Gérard Wertheimer, and Steve Wynn.

Thank you to the brilliant photographers who shared the genius of their craft with me, worked tirelessly alongside me to help realize my visions, and emboldened me to see the world—and myself—in powerful new ways: Gar Lillard, Wayne Maser, Duane Michaels, Francesco Mosto, Helmut Newton, Denis Piel, Herb Ritts, Roberto Schezen, Rune Stokmo, and Neal Slavin.

Thank you to those who work alongside me everyday and keep me curious, inspired and full of hope for the future: John Andre, Terah Coleman, Charles Doomany, John Eastridge, Gina Fujita, Hanaco Fujita, David Goldstein, Jin Seok Hwang, Jooeun Jung, MJ Jang, Kaitlin Miciunas, Dominik Ratz, McKenzie Rist, Madison Soffer, and the countless others who have had a hand in building what we have created together over these past several years.

Thank you to the great people who make up the fabric of the city of New York and who inspire me everyday, including the heroes of the FDNY, whose tireless work and ultimate sacrifices for our beloved city can never be repaid: Daniel Nigro, Elizabeth Cascio, Salvatore Cassano, Tom Von Essen, Pete Ganci, Louis Garcia, Paul Iannizzotto, Rex Morrison, Frank Rush, Steve Ruzow, Nicholas Scoppetta, and all the members of Rescue One.

From my studio, past and present, there are so many talented and passionate individuals who have worked tirelessly to help bring our work to life. There are truly too many people to name here, but just know that it has been the honor of my life to work alongside each and every one of you, and I owe you all an enormous debt of gratitude for helping to make all of this possible over the past forty years; their names are listed in the following pages.

As I sit here writing down these names, reflecting on all that we have already accomplished together, and thinking ahead to all that we have yet to do, what I feel above all is simply gratitude for all the extraordinary individuals I have known over the years. I see them in every project, in every image, in every concept, in every word, and have been fortunate enough to spend my life immersed with them in the creation of this work.

I find myself at a loss as to how to convey the enormity of what I feel in this moment. So, I will finish by repeating two simple words with all the love, respect, and gratitude that I can possibly summon: *thank you.*

ACKNOWLEDGMENTS

Abby Gardner
Abel Laza
Abraham Romano
Adam Carey
Adele Errichiello
Aimee Maliga
Akinah Rahmaan
Akira Nakasato
Albert Angel
Al -Khabir Richman
Alejandro Corpus
Alejandro Gehry
Alex Chang
Alex Hestoft
Alexandria Bergan
Alexei Borysenko-Anderson
Alexis McDermott
Alexis Valz
Alice Barzun
Alice Gao
Alicia Frank-Meltzer
Alicia Hoehlein
Allison Rice
Allison Spicer
Alvaro Bernal
Alvaro Muir
Amanda Lee
Amy Olsen
Ana Figueiredo
Ana Lindblad
Andrea Ruggier
Andrea White
Andrew Beemer
Andrew Conklin
Andrew Stole
Angela Dawson
Angela Lin
Angela Park
Angelica Brancoli
Anika Selhorst
Aniket Shahane
Anna Stoessinger
Anna Wolfe
Anneli Kers
Anthea Kelsick
Anthony Faria
Anthony Richards
Anthony Zurro
Antonio Hernandez
April Tsui
Armond Rivera
Arsenio García Monsalve
Asa Block
Ashley Harris
Askel Ceylan
Axil Celikates
Barbara Lederberg
Barry Balchun
Barry Lubov
Barry Wade
Bart Boehlert
Bartley Mulligan
Ben Snow
Benjamin Ferencz
Beth Blum
Beth Carmichael
Beth Nash
Bethany Harris
Betsy Blakemore
Bill Lee
Bill Tsoumas
Bill Visone
Bob Bishop
Bonnie Hung
Boris Kontorovich
Boyoung Lee
Bradley Kraus
Bradley Price
Bradley Schnur, Esq.
Brandon White
Brian Cunningham
Brian Heyman
Brian Lauzon (Jr.)
Brian Long
Brian LoTruglio
Brian Wright
Bridgett Lindsey
Bridgette Nilsen
Brina Gehry
Brooke Thomas
Bruno Asselin
Bryan Miller
Brynn Johnson
Byron Hoover
Camilo Guzman
Candace Clarkson
Carl Redling
Carlo Mantica
Carlos Acosta
Carlos Arana
Carlos Bernal
Carlos Fuentes
Carlos Guzman
Carlos Nanez
Carlos Reynoso
Carlye Clark
Carmen Lopez
Carol Son
Caroline Trinidad
Carrie Johnson
Carsten Fleck
Catherine Bergart
Catherine Canaga
Catherine Cantave
Catherine Derby
Catherine Tice
Catherine Tsui
Cathy Volk
CC Chiang
Chad Stoller
Charles Rose
Cherisse Thomas
Cheryl Kaplan
Chi-En Chen
Chris Benfield
Chris Bishop
Chris Pacetti
Chris Paul
Chris Shipman
Christina Campisi
Christina Pinco
Christine Austin
Christine Davis
Christopher Bostick
Christopher Chin
Christopher Davis
Christopher Dial
Christopher Lizza
Chuanda Tan
Clare Spigner
Claude Sturla
Clive Skeffrey
Coco Dorneanu
Colin Just
Colleen O'Donnel
Conrad Hanson
Cornelius Howland
Corrie McWilliams
Cory Dunham
Courtney Saunders
Craig Maus
Craig Sakuma
Cristina Rendate
Cristina Vilacoba
Crystal Perry
Curtis McClain
Cynthia Bianco
Daisy Schwartz
Dale Trigger
Damani Hawkins
Dan Piderman
Daniel Kocyba
Daniel Napolitano
Daniel Schwartz
Daniel Spadaro
Daniella Gonta
Danielle Robinson
Danielle Tuerk
Danke Wang
Daren Hom
Daria Osifchin
David Hartman
David Lamb
David Lau
David Malina
David Martinez
David Nater
David Pugh
Dawn Elardo
Dawn Hart
Dawn Viggiani
Denise Felloni
Dennis Gardon
Dennis Rodriguez
Derek Conde
Diana Morrow
Diane Benjamin
Dino Lauando
Dirk Kammerzell
Djamila Reed
Don Montgomery
Dorothy Johnson
Drew Napolitano
Edit Brahimi
Edrick Chua
Edward Blandford
Edward Littlejohn IV
Edwin Philogene
Eileen Navagh
Eileen Salzig
Einav Aviram
Elena Fernandez
Elena Levine
Elisabeth Ford
Elizabeth Giersch
Elizabeth Obee
Elizabeth Wunder
Emily Dulla
Enrico Scotto-Rosato
Eric Velasco
Erika Biddle
Erin Helmer
Erin Hudson
Erin Strater
Erin Walker
Errol Holt
Essie Dunnell
Esther Dixon
Ethan Burns
Eugene Remm
Eun-Jeon Choi
Eunice Kim
EunShil Chung
Eve Bailey
Faith Zuckerman
Fallon Scoggins
Fayette Hickox
Felicia DiSalvo
Fernando Music
Fernando Reyes
Finn Melchior
Flavio Stigliano
France Brunel
Frances Rizzo
Francesca DeStefanis
Francesca Zobele
Francesco Musto
Frank Temesvary
Fred Carney
Fred Miggins
Frederick Evans
Freya Wallace
Fritz Chesnut
Gabriel Peña
Gabriel Rotaro
Gabriela Tejada
Gaspard Giroud
Geoffrey Dunne
Georgette Robinson
Gerardo Vichich
German Bermudez
Gina Moran
Giovanni Bernal
Glen Fulk
Gloria Kwan
Gordon Vichich
Grace Lin
Greg Chann
Greg Kops
Greg Lafayette
Guillermo Bernal
Guillermo Bravo
Gustav Lee
Guy Kudo
Hannah Spinelli
Harry Chong
Heather Babb
Hector Santiago
Heejo Kim
Heidi DuPont
Henry Tam
Hilary Verni
Hillary Sunenshine
Holly Graff
Howard Patino
Hugo Eccles
Ian Colliety
Irving Pector
Isa Frias
Isabella Ossott
Isabelle Kuriata
Ivan Drucker
Ivan Brewster
Ivana Dasaro
Ivy Kazenoff
Jack Ader
Jack McCauley
Jackie DeKama
Jacqie Richardson
Jairo Barco
Jake Franek
Jake Weinstock
Jamelle Ross
James Condoluci
James Heaton
James Pojednic
James Tung
James Verrier
James Welch
Jamie Pojednic
Jason Daniels
Jason Ring
Jay Clark
Jay Ross
Jayzel Samonte
Jean-Jacques L'Henaff
Jeanne Keohane
Jeannie Quintero
Jeff Faria
Jeffrey Bartsch
Jeffrey Meier
Jeffrey Moore
Jeffrey Peters
Jeffrey Walburn
Jenean Chapman
Jenifer Colon
Jenine Twardzik
Jennie Pack
Jennie Quock
Jennifer D'Arrigo
Jennifer Dunn
Jennifer Moore
Jennifer Powers
Jennifer Rome
Jennifer Ruma
Jennifer Simmons
Jennifer Wittekind
Jenny Koo
Jeremy Dumont
Jeri Slater
Jerry Chu
Jerry Stifelman
Jessica Fiorella
Jessica Rommel
Jewel Yambao
Jill Oswskey
Jill Rosen
Jin Hwang
Joan Nocera
Jodie Goldberg
Joe Doucet
Joe Fortuna
Joe Puma
Joe Verni
Joey Crawford
Johan Marfey
John Demas
John Dodelson
John Englehart
John Feldman
John Matero
John McBride
John Stanier, Jr.
John Wall
Jon Angle
Jonah Wilcox
Jonas Damon
Jonathan Cohen
Jonathan Kim
Jonathan Morel
Jonathan Slikoff
Jonelle Procope
JooHong Joo
Jordan Howard
Jorge Martinez
Jorge Narvaez
Jose Baez
Joseph Meli
Joseph Oliver
Joseph Puma
Joseph Traylor
Josephine Sibley
Josh Whitlock
Joshua Nafman
Joy Allen
Joyce Shen
Juan Llano
Judith Asfour
Judith Kaelin
Julian Guerrero
Julianne Negroni
Julio Bravo
Jung-Ah Suh
Junia Brown
Junko Hino
Justin Kaswan
Kara Jankelowitz
Karen Alana
Karen Attyah
Karen Collazo
Karen Sapir
Karen Shuldiner
Karen Wankmuller
Karen Wehner
Karl Cluck
Karyn Forsyth
Kate Begbie
Kate Lutzner
Katelyn Hammond
Katerina McCarthy
Katherine Burke
Katherine Gliedman
Katherine Hilliard
Kathleen Dougherty
Kathrin Schipani
Kathryn Eaton
Kathryn Harvey
Kathy Angele
Katie Moorman
Keiko Kubota
Kelly Cook
Kelly Higgins
Kelly Klarich
Kelly McMahon
Kelly Na
Ken Burke

Ken DuBois
Kenan Aktulun
Kendra Pawlik
Kenjiro Yamakawa
Kerry Daly
Keshara Parker
Kevin Adato
Kevin Brainard
Kevin Katz
Kevin Leddy
Kevin O'Gilvie
Kevin Savell
Killian Strong
Kim Bates
Kim Berney
Kim Slicklein
Kimberly Albento
Kimberly Hunt
Kimberly Slicklein
Kris Tomasson
Kristen Althoff
Kristen McGrath
Kristina Radonjic
Kyle DeMilo
Laila Takesh
Lance Wyman
Larvina Wong
LaTonya Bryant-Johnson
Laura Bryant
Laura Dobbins
Laura Ganter
Laura Mich
Laura Roberts
Lauren DeRosa
Lauren Jancik
Lauren Joyce
Lauren La Valle
Lauren Levitt
Laurie Caban
Leah Greene
Lee Teng Hsu
Leena Laloo
Leilani Hernandez
Leo Comellas
LeRoy Brown
Liana Cassel
Liat Margolis
Liliana Benitez
Linda Dreeland
Linda Maglionico
Linda Wary
Lior Vaturi
Lisa Bridge
Lisa Daniels
Lisa Johnston
Lisa LeMonte
Lisa Monteleone
Lisa Musich
Lisa Tan
Liz Barnet
Liz Carlson
Liz Carstens
Liz Yam
Logan Ray
Loren Drabkin
Lorraine Iveli
Lou Banuelos
Louis Hernandez
Louise Sullivan
Lucy Jones
Luis Bohorquez
Luis Peña
Luis Peredes
Luis Rodriguez
Lynn Gambone
Magda De Mera
Makiko Miyamoto
Malc Zacherino
Malene Waldron
Manish Patel
Marc Mandel
Marc van Cauwenbergh
Marco Destefanis
Marcus Glover
Marcus Lui
Margaret Glaser
Margaret Gomez
Margaret Moore
Margaret Seeley Furniss
Margarita Ventura
Maria Daniel
Maria Gustafson
Maria McKenna
Maria Peña
Marian Fenner-Lee
Marianne Efthimiou-Cuddihy
Maribel Ortega
Marisa Wellen
Marivie Davis
Mark Blaisdell
Mark Demott
Mark Paredes
Mark Webb
Markus Olavsson
Marsha Cottrell
Marshall Guillory
Marty West
Marva Schomburg
Mary Fairbairn
Mary Faughnan
Mary Marques
Mary Somma
Mary Swenson
MaryEllen Blair
Matthew Caillouët
Matthew Higbee
Matthew Montanano
Matthew Talomie
Maureen Aja
Maya Stowe
Meghan Halpin
Meghan McBride
Melinda Key
Melissa Guttman
Melissa Mintz
Melissa Slifer
Melody Henry
Meredith Amato
Meredith Berg
Meredith Kirsch
Meredith Massey
Merianne Molina-Pethers
Michael Blanchard
Michael Cannamela
Michael D'Angelo
Michael Denzler
Michael Dunn
Michael Green
Michael Hickman
Michael Lewis
Michael Meenan
Michael Milligan
Michael Norton
Michael Sicam
Michaelle Pearson
Michele Robin
Michele Simari
Michelle Bergstrom
Michelle Lee
Michelle Montanano
Mike Daikibara
Mike Doyle
Miles O'Brien III
Milton Arnold
Min-Er Xu
Mina Tung
Miriam Franklin
Mitchell Seid
Mona Kim
Monica Acevedo
Monica Johnson
Monica Paganucci
Montana Cherney
Monte Bartlett
Murray Loecher
Nadia Schliephake
Nadine Wiggins
Nancy Raker
Nancy Rothberg
Natalia Velez
Nataliya Kisseleva
Nathaniel Boyer
Neena Khattar
Neil Bianchetti
Nelson Carillo
Nelson Morales
Nicholas Bullock
Nicholas Kaplan
Nicholas Krens
Nicola Pennington
Nicole Jackson
Nin Tam
Nina Markarian
Nisha Inalsingh
Niskaury Acosta
Nora Schiess
Olga Druzhbina
Olga Kolarzh
Olivia Mazzola
Oscar Morrison
Oscar Peña
Paige Nobles
Pamela Pilkonis
Pamela Vitale
Paolo Mereghetti
Paras Shah
Patricia Emery
Patrick Ramsey
Patrick Richardson
Patrick Stretch
Patrik Bolecek
Paul Carroll
Paul Pisoni
Paul Sapp
Paul Scheer
Paul Slater
Paul Taylor
Paul Wagner
Paula Bliss
Paula Daddario
Pauline George
Pedro Usuriaga
Pete Wood
Peter Curtner
Peter Hahn
Peter Poopat
Peter Woronkowicz
Petra Raymond
Philipp Eggert
Philipp Refior
Philipp Von Dalwig
Phillip Jung
Pierre Boursiquot
Piper Miller
Rachel Cohn
Rachel Derkits
Rachel Slater
Racquel Sinclair
Rajeev Thakker
Ramcess Jean-Louis
Randi Joseph
Randy Herbertson
Randy Pollak
Raphael Henry
Raymond Chan
Raymond Chou
Rebecca Lieberman
Rebecca Scuorzo
Rebecca Wright
Rene Ayala
Ria Shibayama
Richard Diaz-Granados
Richard Hernandez
Rie Nørregaard
Rob Dress
Robert Daniele
Robert Schroeder
Roberto D'Addona
Robyn Dutra
Rochelle Udell
Rodrigo Castanedo
Roger Bova
Rohini Sahni
Roi Elfassi
Romina Mosquera
Ronald Tsung
Rovimar Santos
Roxanne Holly
Ryan Gordon
Ryan Hwang
Ryan Padgett
Ryan Potocki
Sabine Dorisme
Sally Khim
Samantha Bonom
Samantha Franco
Samantha Roberts
Samantha Solmonson
Samba Gadio
Sandra Arndt
Sandra Berman
Sandra Villegas
Sandra Zuniga
Sandy Funatake
Sara Greene
Sarah Forsyth
Sarah Hendrick
Sarah Kloman
Sarah Miller
Sarah Mills
Sarah Vanderwerken
Sarah Wilkins
Sarra Angelou
Satoko Kawakita
Savannah Enright
Say Lim
Say-Kheong Lim
Scott Beckerman
Scott Chapps
Scott Giammarino
Scott Whitman
Sean Aziz
Sean Callen
Sei Cho
Selina Rutovitz
Shannon Swanson
Sharon Ullman
Shawn Brown
Sheena Demmon
Shelley Shaw
Sheri Womack
Sherwood Forlee
Sheryl Nelson
Shirley Hong
Silke Mennenga
Silva Premoli
Simon Grendene
Simone Burkley
Sonya George
Stacey Fenster
Stacey Margolis
Stacey Smith
Stanley Wagner
Stefan Knecht
Stefan Roschert
Stefanie Drucker
Stephanie Culley
Stephanie Taddeo
Stephen Lim
Steve Kane
Steve Lee
Steven Baillie
Steven Hankinson
Stuart David
Suchi Braga
Susan Conway
Susan Quinn
Suzanne Dykstra
Suzanne Mieczkowski
Svenja Leggewie
Sybil Cubilette
Sydney Axelrod
Sze Ooi
Tal Woidislawsky
Tami Dimond
Tamika Hollis-Davis
Tammy Martin
Tara Grote
Tara Heppekausen
Taryn Kavanaugh
Tatiana Khoury-Terzuoli
Tayler Kim
Temilola Agbede
Terry Kosierowski
Terry Thomas
Tesfa Brown
Tetsuya Hatsusegawa
Tiffany Griffin
Tim Coleman
Tim Daly
Tim Hahn
Tim Roarke
Timothy Daly
Timothy Macsuga
Tina Chiu
Tina Concha
Tiriree Kananuruk
Todd Samberg
Todd Wilcox
Tolga Morawski
Tom Whitburn
Tomasz Rygalik
Tomoko Hayakawa
Tomoko Higuchi
Toni Brody
Traci Barr
Traci Coulter
Trina Hill
Tryston Dietrich
Tucker Viemeister
Tyler Copeland
Tyler Walseth
Uday Kak
Valerie Galloway
Vanessa Ayala
Veronica Brown
Veronica Milo
Victoria Bundonis
Vikas Sharma
Viktor Jondal
Vincent Forestiere
Vincent Leong
Vincent LoPreto
Wanda Tirado
Wendell Thomas
Wendy Chan
Wendy Han
Wendy Sanders
Whitney Gallagher
William Conrad
William Kang
William Niemeier
William Prince
William Reeve
Yana Ushrenko
Yelany Acevedo
Yelena Leshchinky
Yelena Pavlenko
Yoo Yoon Seok
Yoon Young Lee
Yoon-Sun Chung
Yoshikazu Hyuga
Yuji Yoshimoto
Yukiko Krigh
Yuko Nakajima

INDEX

What we think, we become.
What we feel, we attract.
What we imagine, we create.

BLACKBERRY
parmalat
S. Carter
solo
SUNCLOUD
chesterfield
DE TOMASO
CE. ICEBERG
beau
DEVON
SoLeMia
MIAMI
simply samsung.
SAMSUNG
OneRedCube
ANNE KLEIN
camels AUTOMOTIVE
blade
GUCCIMUSEO
MUG ROOT BEER
GUGGENHEIM
GNC
WELLBEING
Trop50
one foundation
crush
360
ICEJEANS
ICEBERG
sweetpea
surge
HENRI
BENDEL
DORRIAN
breadblo
SAMSUNG DIGITall
RED AND WHITE PIZZA
LIFETIME
TELEVISION
VISARUNO
diet
pepsi
sweetstuff
Vita
READY